地方重塑

LOCAL REMODELING

国际公共艺术奖案例解读2

总策划　汪大伟　　主编　潘力

上海大学出版社

目录

序

这本册子是对提名参加第二届国际公共艺术奖的全球 125 个公共艺术案例的解读。

在解读的过程中，首先要回答的问题是：什么是公共艺术？公共艺术，简而言之就是公共空间中的艺术。但如果仅从这一点来理解，则会带来许多歧义。例如，出现在公共空间的一些毫无公共意识、没有品质的雕塑岂不都成了公共艺术？也有人认为公共艺术是由政府主导的工程，尤其是中国许多地方政府在城市建设过程中，在塑造城市标志性构筑物时就想到艺术。这样，公共艺术又被曲解为政府的艺术；还有人觉得公共艺术是民众的艺术，是自发于民众的一种积极性，带给民众自娱自乐或自我教育的方式。由此，众说纷纭，公共艺术到底是什么？

我认为，公共艺术就是用艺术的语言和方式解决公共问题。一般意义上的艺术只是提出问题，不涉及解决问题。这正是公共艺术与其他艺术的区别所在，也正因为这个区别，使公共艺术成为一种独特的艺术形式，是可以解决公共问题的艺术。所谓公共问题，可能是环境的功能性问题，也可能是社会性问题，亦有可能是民族信仰冲突的问题。如果用一种好的方式方法，以人对美的向往的本能为出发点，来解决这些问题，这就是我们所希望和需要的公共艺术。

什么是好的公共艺术？我认为有三个观察点。

第一个观察点：好的运作方式、方法。公共艺术不是以投入多少钱来评判好坏，而要看运作的方式、方法是否得体和有效。这个方式、方法实际上决定了策划人和组织者的智慧，是用智慧的方式、方法来协调政府、民众、艺术家三者之间的关系。我相信，如果关系处理好了，资金问题就会迎刃而解，民众自然会投入和参与，艺术家的作用也能充分发挥。

第二个观察点：地方重塑。重点在于公共艺术对其所在地区是否带来积极的变化。如果只凭个人意向认为这件公共艺术有多好多艺术，而当地民众看不懂，甚至嗤之以鼻，也得不到地方的支持，那就不能成为一个好的公共艺术。

第三个观察点：影响力。影响力有赖于好的运作方式和方法，能够给该地区带来积极的变化，而这种变化具有普遍的示范性和广泛的认同度。某种程度上也会影响和带动其他地区的发展和变化。

我认为，从这三个角度来观察什么是好的公共艺术，就比较全面。第一个观察点是从公共艺术的组织运作角度来看，第二个观察点是从公共艺术的作用来看，第三个观察点是从公共艺术的本身效应来看。这三大方面在两届的国际公共艺术获奖案例中都能看到，在我们身边也有很多好的案例在发生。

城市化进程使中国的城市规模迅速扩张。一方面，农民进城打工，给城市带来一系列问题（社会治安、农民工子女教育等）；另一方面，农村土地荒废，劳动力缺乏，农村被边缘化。我们的艺术家能不能找到一些好的方式、方法，来解决这些问题？

在这些案例中可以看到，艺术家以艺术的方式、方法解决了社会中的诸多公共问题。如果说这是成功的公共艺术，就应该调动更多艺术家积极投入到解决这些问题的事业中去。我觉得，这样的公共艺术，才是中国所需要的，也是当下社会发展所需要的。发乎本心，功莫大焉。

上海大学美术学院院长、教授

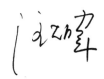

什么是好的公共艺术

WHAT'S GOOD PUBLIC ARTS

好的公共艺术要能够为社会创造积极的价值。首先有一种文化上的急迫感，社会和政治角度的迫切感；其次是对于环境敏感度以及环境当中人的敏感度的尊重；第三是创造力；第四是有变革性；最后是能够带来意义：公共作品和受众之间的联系所产生的意义。

新西兰戈维布鲁斯特艺术馆馆长
Rhana Devenport

公共艺术离开资金是不可能的，但并非资金投入得多就一定是
好的公共艺术。我们希望能够触动政府，得到政府的支持。甚
至政府也看到了公共艺术能够解决其所不能解决的事情。

上海大学美术学院院长、教授
汪大伟

必须要看到什么是公共空间，如何把公共空间跟艺术相结合。世界上不同的地域，受众的互动是多样性的。公共艺术能够让人们有意想不到的互动机会，同时能够在博物馆外、对社会造成意想不到的影响。

新加坡当代艺术中心创始董事、南洋科技大学教授
Ute Meta Bauer

谈到公共艺术经常是一个大型项目或作品。事实上有一些非常小的作品，非常小的冲动，让人感觉是最有价值的。这些非常小的火花，触发了很多世界上著名的艺术讨论和对话。

南非开普敦大学教授
Jay Pather

公共艺术就是教育、激发、组织。组织是非常重要的概念，包含着建立一个社区的组织或组织合适的力量来进行政治方面的变化等，互动并形成良好的对话。

美国独立策展人
Bill Kelley, Jr.

用一句话来总结公共艺术是一个很挑战的问题，可以说我们无法明确定义什么是公共艺术。公共艺术有很强的变革性，不存在的空间能够重新建立起来，让人们能够进行公开的讨论。

美国独立策展人
Chelsea Haines

好的公共艺术必须要考虑受众，因为受众是社区的一部分，并发挥着重要作用。公共艺术能够让不同的人发挥想象，相互交流和互动，同时让他们重新思考。对于政治、社会领域的一些话题，通过艺术的语言来思考和沟通。

印度 Khoj 国际艺术家联盟创始成员兼董事
Pooja Sood

案例解读
PUBLIC ARTS CASES

非洲
Africa

非洲

非洲的 17 个案例，让我们更多的看到，在社会矛盾尖锐的状况下，艺术家在求之于其他手段无果的前提下，公共艺术俨然已经成为传达公众心声和表达个人看法的重要手段。公共艺术是面向公众的，具有公共性的艺术形式，在艺术创作过程中需要一定的自由度，即自由的观念以及自由的话题。但是在南非这个人民生活困顿，社会问题频发，各方面问题矛盾重重，政治与经济以及文化的自由问题却一直存在争议。可以说经过各方的努力，社会尖锐矛盾有所缓解，但争论却没有停息。作为一个艺术家和公共项目负责人，参与这些公共性话题的讨论是一个非常重要的部分，对南非以及整个大陆来说，问题本身并非容易得到解决，同时也不会被轻易误解，因而，它需要解决问题的多种尝试和可能性。公共艺术的空间给艺术家和公众创造了一个互动、交流的平台，公共艺术的众多形式使融入在这个空间里的不同种族、宗教、阶层的人群可以忘记身份的差异，使他们能够明确地认识到他的想象和有限性并且了解到公共艺术所具有的力量。

但有时我们也会看到艺术家的努力与公众对作品的反映是背道而驰的，艺术家需要不断让自己的作品在公共空间得到检验，虽然这是一种非常独特的孤独体验，但是当你的一个小小的想法和火花得到公众的认可并触发了世界上不同人群的艺术讨论和艺术对话以及让一个地区有了变化的可能，你会感觉这一切都是值得的。公共艺术本身就含有各种可能性和各种待挖掘的力量。（冯正龙）

上午／下午阴影线
am/pm Shadow Lines

艺术家：凡德尔莫维·斯特赖敦
地点：南非西海岸的前德比尔斯
形式和材料：冲积岩、碎石
时间：2010 年
委托人：德比尔斯
推荐人：梅根·盖尔贝

"上午／下午阴影线"是凡德尔莫维·斯特赖敦的一个位于南非西海岸的前德比尔斯采矿营地（纳马夸兰煤矿）的一个永久地景艺术。这个艺术作品是使用采矿工作剩下的冲积岩和碎石制作而成的。这是一名艺术家对被几十年采矿活动所摧毁的毫无生气的土地的复兴。

一个多世纪以来，勘探者通过挖掘南非的土地获得高额利润，只留下不适宜居住的碎屑和荒凉的社区。通过了多项法律来确保矿业公司会在停止采矿后，负责他们营业所在土地和社区的复兴，其中最著名的是"1998 年国家环境法案"和"2002 年矿产和石油资源开发法案"。这些法律要求必须复兴被采矿摧毁的土地，并同时创造就业。德比尔斯召集了 LEAP（非洲生活边缘项目）和南非国际保护项目的协助，两个组织致力于采矿后的复兴，以形成相关复原计划。该计划由三个部分组成：

1. 在有表层土的地方，对景观进行移植；
2. 在装满海水的大洞穴中进行鲍鱼和小龙虾养殖；
3. 使用丢弃的采矿设备创建有关受损土壤的艺术作品。

凡德尔莫维·斯特赖敦被选中进行第一个土地艺术作品的创作。他以把现场现有的自然元素制作成各种几何形态而闻名。"上午／下午阴影线"的灵感来自附近城市高林奈思中小孩子所玩的游戏，在这个游戏中，使用一把沙子画出尽可能多的线。对有限材料的利用为艺术家提供了灵感，仅仅使用现场剩余的冲积岩和碎石，创造了一个直径 100 米的圆圈，由 14 米、2 米高的三角丘构成。这个作品很大，可以通过 Google Earth（谷歌地球）看到，这个作品的结构跟日出和日落相关，创造出引入注目的几何阴影，并在一天内随着时间发生变化。

在完成这个项目所需的 22 天（182 个工作小时）内，为五个等待再分配的操作员创造了新工作。

作为一个严重问题的独特解决方案，"上午／下午阴影线"脱颖而出。今天，采矿公司保存现场的表层土，以在将来恢复工厂生产，但是，一些矿区是

在这些法律执行这个实践之前建立的。在仅提供冲积岩和碎石的情况下，需要不同的方案来复兴德比尔斯的纳马夸兰煤矿。通过使用因为太贵没有运送到另外一个场地而被丢弃的机器，以及挖煤所留下的 2 000 吨冲积岩和碎石，凡德尔莫维·斯特赖敦使这些碎屑焕发新的生命。线性丘垒成为一件艺术作品，随着光线的转变发生柔和的变化。艺术给死气沉沉的大地第二次生命。

不幸的是，纳马夸兰煤矿仍然是关闭的。必须采取必要的安全措施，以防止外面的人进来挖煤，因此，只能通过预约暂时观看"上午 / 下午阴影线"，或在因特网上观看 Google Earth(谷歌地球)。即便如此，仍然制定了计划，以邀请更多的土地艺术家创造类似的作品。这些将一起形成土地艺术公园，会吸引游客来到这个已经具有漂亮海岸线的地区旅行。随着旅游业的增长，当地居民的工作和商业机遇也会得到增长。

公共地景艺术
冯正龙

一望无际的南非矿区，曾经这个地区是采矿者谋生获利的场所，在采矿者满载而归时，却忽视了地方土地的复兴问题。相关法律的出台虽然限制了采矿者，但复兴满目疮痍的土地却成了难题。德比尔斯在得到相关部门的协助下，开始了复兴土地的尝试。"上午 / 下午阴影线"就是通过公共艺术的方式，引起人们对这一地区土地现状的关注。艺术家通过就地取材创作了这一浩大具有深远意义的公共地景艺术，吸引了公众的关注。这其实仅仅是复兴土地的第一步，德比尔斯的计划是通过公共艺术的方式，在这块土地上形成供游客和公众游览的土地艺术公园，这样不仅复兴了土地，而且也为当地民众带来就业和收入。艺术家最终使这片荒芜的土地重新焕发生机与活动，让人们对土地的未来利用充满希望。

Artist: Strijdom Van der Merwe
Zone:The West Coast of South Africa
Media/Type: Land art installation
Date:2010
Commissioner: De Beers
Researcher: Megan Guerber

am/pm Shadow Lines is Strijdom Van der Merwe's permanent land art installation located in a former De Beers mining camp (Namaqualand Mines) on the West Coast of South Africa. This work is created from the alluvial rock and gravel left behind by mining operations. It is an artist's rejuvenation of the dead earth destroyed by many decades of mining.

For over a century, prospectors have dug their profits out of South African land, leaving uninhabitable detritus and desolated communities behind. Laws have been passed, most notably the National Environmental Act of 1998 and the Mineral and Petroleum Resources Development Act of 2002, to ensure that mining companies are responsible for rejuvenating the land and communities of their operations once they shut down. These laws require that land destroyed by mining must be revitalized, as well as new jobs created.

De Beers called in the assistance of LEAP (The Living Edge of Africa Project) and Conservation International South Africa, two organizations dedicated to post-mining regeneration, to form a plan of rehabilitation. The plan consists of three parts:
1. To replant the landscape where topsoil is available
2. To establish abalone and crayfish farming in large holes that were flooded with seawater
3. To use the abandoned mining equipment to create artworks with the damaged soil

Strijdom Van der Merwe was chosen for the first land art commission. He is known for his earthworks that manipulate the

existing natural elements of a site into geometric forms. am/pm Shadow Lines takes after a children's game played in the nearby city of Koingnaas where one tries to draw as many lines as possible with a handful of sand. This manipulation of a limited amount of material inspired the artist to build a 100 meter-diameter circle composed of 14 long, two-meter high triangular mounds using only the site's leftover alluvial rocks and gravel. The work, large enough to be seen via Google Earth, was configured in relation to the sunrise and sunset in order to produce dramatic, geometric shadows that change throughout the day.

During the 22 days (182 working hours) it took to complete the project, new jobs were created for five operators awaiting reassignment.

am/pm Shadow Lines stands out as a unique solution to a considerable problem. Today mining companies preserve a site's topsoil to enable the restoration of plant life at a later date, yet several mines were started long before laws enforced this practice. With only alluvial rock and gravel available, a different answer was needed for regeneration of the De Beers' Namaqualand Mines. By using the discarded machinery too expensive to bring to another site and the 2,000 tons of alluvial rock and gravel leftover from digging, Van der Merwe was able to bring new life to this debris. The linear mounds have become a piece of art that changes softly with the shifting light. Art has given dead earth a second life.

Unfortunately, Namaqualand Mines remains closed. Security is necessary to deter further digging from outsiders, making am/pm Shadow Lines temporarily available solely via appointment and the Internet (Google Earth). However, plans have been in place to invite more land artists to create similar works. Together these will form a land art park that will draw tourists to the area already blessed with a beautiful coastline. As tourism grows, so will job and business opportunities for local residents.

马可可浮动学校
Makoko Floating School

艺术家：昆仑·阿德耶米
地点：尼日利亚拉各斯
形式和材料：剩余材料和竹子
时间：2013 年
委托人：艺术家发起
推荐人：海伦·勒塞克

在 2011 年，昆仑·阿德耶米接受挑战，为位于尼日利亚拉各斯的贫困滨水区建造教室。他主要在尼日利亚和荷兰从事设计工作，昆仑·阿德耶米进行不依靠土地地基的可建造结构设计，以获得结构完整性。

"马可可浮动学校"是一个针对位于拉各斯泻湖的一系列校舍的原型设计和实现，这是一个人口稠密的都市滨水区，干燥的土地非常珍贵。浮动学校是针对可支付建筑用地挑战的巧妙解决方案，以支持低收入家庭的初等教育。该项目是一个宏大的都市设计方案，由技艺精巧的工匠在拉各斯的滨水区实现。

该项目最突出的特征是概念上的。昆仑·阿德耶米把水重新定位为他项目的施工场地。他设计结构来利用和改进滨水区工匠的手艺。根据撒哈拉以南地区利用废物的传统，该项目使用了一个当地锯木厂捐赠的剩余材料和本地生长的竹子。基于个体浮动结构的传统，把他们建造为社区中心。这个结构要比当地的渔船以及用于居住和本地业务的倾斜结构大 10 倍。

这个项目影响了设计的潜在受益目标。解决了希望为适龄儿童提供教育的无土地社区的需要，浮动学校提供了一个结构设计，来利用当地手工艺人的技能。

还不知道学校是否投入了运营，以及建筑中是否提供相关设施，如电、风扇、洗手间、内置座椅或停泊区。一个原型是一个四层的 A- 框架结构，具有太阳能动力系统，于 2013 年首次推出。

这个项目很成功，因为它提供了一种低成本、不需要地产的设施向当地年轻人提供教育。从城市规划的角度来看，它在很明显无法使用的水域表面成功地提供了可使用的空间。但是，没有提到覆盖一大片水域会对鱼类、甲壳类动物和水生植物产生什么样的影响。遮住阳光会影响水面下的群落以及依靠水产养殖或捕鱼的人类社区。

主设计师昆仑·阿德耶米是一名建筑师和城市规划师。他的公司是一个叫做 NLÉ 的建筑设计工作室，根据公司网站的介绍，NLÉ 在约鲁巴语里是"在家"的意思。昆仑·阿德耶米获得了数个设计创新方面的建筑奖项。他在尼日利亚的拉各斯大学和美国的普林斯顿大学学习建筑学，然后在雷姆·库哈斯成立的大都会建筑事务所（OMA）工作。

项目声明并未直接或间接说明项目的公共艺术方面。在这个建筑规划项目中，没有艺术家的声明或有关专业艺术家参与的证据。

项目展示了将水作为场地的创新使用，并通过雇佣当地的商贩来完成这个非传统的设计，从而使社区参与到这个项目。但是，这个项目的提交没有被认同的公共艺术家或公共艺术目标，项目简介强调了设计，但是忽视手工艺人工作的细节或学校成品和总体安装情况。

水上学校
冯正龙

尼日利亚拉各斯是一个贫困的滨水区，特殊的地理环境使土地显得特别珍贵，但也造成了当地居民要想法设法保护仅有的土地和利用水上区域。这个项目的成功之处就在于成功利用了水上区域，为当地适龄儿童提供了上学的场所。在专业设计师的带领下，建筑材料就地取材，建造过程手工艺人积极参与，最终使这座水上建筑在周围环境的映照下十分简洁漂亮。但任何项目最终都不是完美的，这个项目也是如此，虽然这个项目既富有挑战性又具有公益性质，但是人们还是会关注这个项目的安全系数和学生的反馈，但是遗憾的是，到目前为止还未知。并且，这个项目还有相关的商人参与，不得不让我们关注这个项目的最终目的，尽管项目简介中没有提及这是公共艺术项目，但我们已经看到公共艺术的运作机制在发挥作用。

Artist: Kunlé Adeyemi
Zone: Makoko Waterfront Community, Lagos State, Nigeria
Media/Type: local sawmill and locally grown bamboo
Date: 2013
Commissioner: Artist initiated
Researcher: Helen Lessick

In 2011, Kunlé Adeyemi took on the challenge of creating classrooms for impoverished waterfront communities in Lagos, Nigeria. Working in Nigeria and the Netherlands, where he maintains his design practice, Adeyemi developed designs for buildable structures not reliant on land-based foundations for structural integrity.

Makoko Floating School is the design and realization of a prototype for a series of schoolhouses located in Lagos Lagoon, a densely urban waterfront where dry land is at a premium. The Floating School an ingenious solution to the challenge of affordable building lots to support primary education for the lowest income families. The project is an ambitious urban design solution realized by skilled artisans in the waterfront community of Lagos.

The most compelling attributes of the project are conceptual. Adeyemi reimagined water as his project's construction site. He designed structures to harness and improve the skill sets of the artisans of the waterfront. The project was created from leftover materials donated by a local sawmill and locally grown bamboo, harnessing a Sub-Saharan tradition of utility from waste. It took the tradition of personal floating structures and made them into a community center. The structure is ten times larger than the local fishing vessels and stilt architecture used for residences and local businesses.

The project impacted the people for whom it was designed. Addressing the needs of a landless community interested in

educating their school-aged children, the Floating School supplied a structural design harnessing the skills of local craftspeople.

It is not known whether the school is yet in operation, nor is it known if there are any amenities in the structure, such as electricity, fans, restrooms, built-in benches or desks, or mooring. One prototype four-story A-frame structure, complete with solar-powered systems, debuted in 2013.

The project is a success in that it provided a low-cost, property-free facility to educate local youth. From the urban planning perspective, it successfully provided useable space on the apparently unusable water surface. However, there is no mention of how covering a large plot of water will affect fish, crustaceans and aquatic plant life. Taking away the sunlight affects subsurface communities in water, and the human communities who rely on aquatic agriculture or fishing.

Kunlé Adeyemi, the principal designer, is an architect and urban planner. His firm is an architectural studio named NLÉ, which according to his firm's web site, means' at home' in Yoruba. Adeyemi has won several architectural awards for design innovation. He studied architecture at the University of Lagos in Nigeria and Princeton University in the Untied States, and then worked at the Office of Metropolitan Architecture (OMA), founded by Rem Koolhaas.

The project statement does not address the public art aspect of the project directly or indirectly. There is no artist's statement or evidence of professional artist involvement in this architectural and planning achievement.

The project demonstrates an innovative use of water as site, and engaged the community by employing local tradespeople in completing the untraditional design. However, the project was submitted without an identified public artist or public art goal, and the project brief emphasized design but omitted details of the artisans' work or of the school finishes and overall installation.

桑德兰反射
Sunderland Reflections

艺术家：布朗·温雷士和马库斯·纽斯泰特
地点：南非北开普省那马夸区市
形式与材料：多种艺术形式
时间：2010 年
委托人：德比尔斯
推荐人：海伦·勒塞克

"桑德兰反射：艺术和科学在黑暗和安静的地方交会"是对原住民土地的眷恋以及使用这片土地进行科学研究的复杂和诗意的思考。从 2009 年到 2012 年，南非艺术家布朗·温雷士和马库斯·纽斯泰特开发了一系列场景，来探索南非被取代的原住民社区和天文观测台工作人员之间的距离。

艺术家去桑德兰考察隔离情况。到达时，他们发现南非种族隔离时期遗留下来的本土和多民族遗迹。这个社区仍然受传教士实践、殖民主义和 1994 年正式终止的南非种族隔离法的影响。

虽然南非种族隔离法被取消已经有一代的时间了，根据艺术家的报道，隔离对该社区仍然非常困难。失业、种族冲突和滥用药物很普遍。贫穷加上种族隔离时期遗留下来的二等和三等公民的分类，跟这个地区的科学价值形成鲜明的对比。布朗·温雷士和马库斯·纽斯泰特开发了一系列公共艺术项目，如创意新年节日、干预措施以及同都市南非人以为已经消失的社区达成合作伙伴关系。

桑德兰的祖先是卡鲁沙漠的早期"奎奎市"或丛林居民。这些猎人和采集者被从他们的传统居住地驱逐，直到南非人可以合法猎杀丛林居民的最后一个许可证在 1938 年被撤销。

在 2008 年 12 月 31 日的新年除夕，温雷士和纽斯泰特开始了他们的实验干预，标志 2009 年的国际天文年。他们的干预导致这个社区的在全年参与这个项目，产生了永久的土地艺术、放风筝的景象、博物馆展览、纪念场所和活跃分子的表演，该社区在 2012 年全年参与了这些活动。

"桑德兰反射"旨在创造一个社区驱动的体验。艺术家的意图是针对桑德兰贫困社区和邻近国际天文台之间的当前态度和关系。这个艺术干预项目缩短了两个社区之间的距离——这个距离好像地球跟星星之间一样遥远。

这个项目成功实现了艺术家的意图，即探索和影响被放弃改进的以及在市政报告中被遗漏的社区。在艺术家、居民和好奇的天文学家之间建立起人文联系。

这个项目让人想起早些时候的美国火把节，在内华达州的黑岩沙漠举行，如果仅包括两个艺术家，节日在部落保留地举办的话，该保留地由资助的科学设施围绕，并且可以忽视当地人生活在这些干净的星空下具体化的快乐。

遥远的问候
冯正龙

如果我们要问：在一个公共艺术项目中，艺术家发挥了怎样的作用？公共艺术对一个社区的改变又会发挥怎样的作用？"桑德兰反射"这个项目就给出了让我们出乎意料的答案。可以说，这是一个很神奇、又充满诗意和关怀的公共艺术项目，艺术和科学在黑暗与安静的南非桑德兰汇聚，艺术家利用艺术的力量改变了这个仍然沉溺于落后历史传统的社区，让社区在艺术的氛围里再次充满生机。我们不会想到，相隔万里的两个社区怎会通过天文观测就发生了联系；我们更不会想象到，一位艺术家怎会用艺术的莫名力量改变了一个曾被隔离许久的原住民社区；这一切在我们感叹之余，也让我们深深体会到人类在起源之始就已经与艺术产生了千丝万缕的关系了。"桑德兰反射"就是艺术家通过公共艺术干预社区重建的典型例子。

Artist: Bronwyn Lace and Marcus Neustetter
Zone: Koingnaas, Namakwa District Municipality, Northern Cape, South Africa
Media/Type: The community-driven experience
Date: 2013
Commissioner: Artist initiated
Researcher: Helen Lessick

Sutherland Reflections: Art meet's Science in a place of Darkness and Silence is a complex, poetic meditation on aboriginal land attachment and use of that same land for scientific research. From 2009 to 2012 South African artists Bronwyn Lace and Marcus Neustetter developed a series of spectacles exploring the distance between displaced communities of aboriginal South Africans and those working in astronomical observatories.

Sutherland, Namakwa, Northern Cape, South Africa is remote mountains in a southwestern stretch of that country. Regionally noted for emptiness, the area's untouched land supports crystalline skies with no urban light, smoke, or auto pollution. The night skies are among the world's clearest and darkest. These qualities drew astrophysicists to erect several multinational and international astronomical observatories in Sutherland.

The artists went to Sutherland to explore isolation. On arrival they discovered an indigenous, multiracial remnant from the Apartheid regime. This community still lives the repercussions of missionary practice, colonialism, and Apartheid law that officially ended in 1994.

Though a generation had passed since Apartheid was outlawed, the artists reported that isolation is very difficult for the community. Unemployment, racial strife and substance abuse are rife. Poverty mixed with the heritage of the Apartheid second- and third-rank citizen classification, contrasts sharply with the scientific valuation of the place. Lace and Neustetter developed a work of public art as a series of creative New Year festivals, interventions, and partnerships with a community that urban South Africans believed had disappeared.

Sutherland's ancestors were the early 'kwe kwe' or bushman people of the Karoo desert. These hunter-gatherers were constantly driven from their traditional lands until the last license for South Africans to legally hunt human bushmen was revoked in 1938.

Lace and Neustetter launched their experimental interventions during the night of New Year's Eve, December 31, 2008, marking the 2009 International Year of Astronomy. Their interventions led to annual engagements with the community resulting in permanent land-art, kite-flying spectacles, museum displays, memorial sites, and activist performances with and by the community through 2012.

Sutherland Reflections aimed to create a community-driven experience. The artists' intention was to address the current attitude and relationship between the disadvantaged communities in Sutherland and the neighboring international telescopic observatory. The artistic interventions closed the gap between these two communities—a distance seemingly as vast as between the earth and the stars.

The project succeeded in realizing the artists' intention, namely, to explore and impact a group that was disregarded in community improvements and left out of civic and municipal reporting. A human connection arose across a gulf of artists, residents, and the occasional curious astronomer.

The project invokes the early days of the America's Burning Man festival, which takes place in Nevada's Black Rock desert, if the organization was just two artists, the festival was on a tribal reservation, and that reservation was surrounded by well-funded scientific facilities who could overlook the local people reifying joy in living life under those clear, close stars.

Sutherland Reflections explores the process of how artists change a place and how a place and project changes artists. The power of art and the presence of artists are lenses for investigation and catalysts for community change and self-reflection. Returning to a place for four years and creating festivals and earthworks, the artists became part of this distant community.

聚会树
Palaver Tree

艺术家：弗雷德里克·凯费尔
地点：喀麦隆杜阿拉市的社区
形式和材料：雕塑
时间：2007 年
委托人：Doual'art 空间
推荐人：海伦·勒塞克

2007 年，艺术家弗雷德里克·凯费尔在 Espace de Douala 针对喀麦隆杜阿拉市的社区脱离现象采取了行动。Espace' Douala 艺术中心由喀麦隆法国文化研究院资助，隶属于法国文化全球推广组织。喀麦隆是一个法语区国家，曾在 20 世纪初是法国海外殖民地之一。它与尼日利亚、乍得、中非共和国和刚果接壤。

早在项目计划阶段，在这个植被稀疏且经济落后的西非国家，凯费尔注意到人类和森林社区发生的种种变化。一个关于古代硕大遮荫树的故事让他获得灵感，虽然这棵名树早已不复存在，但是这里的人们还是乐于回忆过去的美好，津津乐道于它为大家提供了一块聚会的场地。

在非洲传统文化中，这棵"聚会树"被赋予了自然的含义，它为人们提供了一处聚会场地，家长里短、故事会、解决问题和节日庆祝都成为了常态。从功能上来讲，它大致相当于市民广场、中央广场或集会地点，为公众提供了一处舒适的聚会地点。凯费尔了解到自从这颗树木消失后，社区间的联系越来越少。

"聚会树"是一棵人造树干，由回收的蓝色环保塑料材料拼凑在一起，组成一块被褥状物体，高高固定在一个钢结构支架之上。雕塑树被安装在艺术中心基础设施上，使文化中心成为街头巷尾热议的话题。

"聚会树"项目得益于艺术家细致入微的观察力，考虑到公共空间、非洲文化和人类互动，成为引人注目的焦点景观。为将这棵"新树"安装在旧址上，艺术家特意培训、雇佣游荡于杜阿拉市街头巷尾的无业青年，共同加入到这个项目。

在这次艺术创作过程中，青年们学习到了宝贵的技能，包括电焊、绘画和根据现场需求而想出的创意设计等。此外，凯费尔还颁发了官方证书以肯定他们的工作，这些证书对他们来讲显得尤其珍贵，因为只有上流社会的孩子才能上小学。他们感到弥足珍贵，这反过来促使他们做出更多的贡献。艺术创作合作伙伴为当地的小规模商业团体。为保证项目的顺利实施，杜阿拉当地的垃圾回收人从垃圾堆中淘出一些金属、玻璃和塑料材料。

一直以来，该项目影响着目标受众。2007 年，项目完成时恰逢一年一度的 SUD 节日（Salon Urbain de Douala）。官方认可树木创作和社区努力。2014 年，树木仍在使用。

凯费尔运用创新、社会和政治策略，找到了一处最合适的安装地点，既可以体现浓厚的历史背景，还能反映出城市生活。虽然这个寻址的过程比较艰辛，但是最终在城市社交网络中成功地种下了艺术创作的种子。

"聚会树"充满创意，同时满足了社区需求。做到尊重当地传统，雇佣当地青年，并循环利用回收塑料，"聚会树"兼具美好、实用和公共关注度，并转换成创作者的可利用场地。艺术家成功地跨过文化界限，尊重公共艺术精神并达到了最广泛的目标。

凯费尔是一位法国艺术家，在全球范围内从事艺术创作，其个人工作室位于法国斯特拉斯堡市附近。他的艺术创作特点鲜明，主要以人体形态和环境创作为主。

聚会之所

冯正龙

社区在当代社会是一个十分重要的概念，按照社会学的阐释，社区就是一个地域群体，它以一定地理区域为基础，这个区域的居民有着共同的意识和利益，有着较为密切的社会交往。同时，地域的概念也很重要但更具延展性，它可以是社区的，也可以是城市和乡村的，甚至是国家与民族的；社区、地域以及相关复杂的文化与地理环境构成了人们从事各种活动的背景，也构成了公共艺术赖以生存的场所。不同民族国家所特有的文化背景、地理环境呈现出这个民族国家特有的地域、社区问题，公共艺术的关注视角也从公共空间的艺术作品发展到介入社区问题的解决。公共艺术自身的意义与含义也多了更多的层面，成为一个和地域发展、社区公众沟通、互动的有效形式以及运作的过程。"聚会树"项目反映了公共艺术介入社区的重要意义。

Artist: Frédéric Keiff
Zone: B. P. 650 Douala, Cameroon
Media/Type: The sculptural tree
Date: 2007
Commissioner: Artist initiated
Researcher: Helen Lessick
Hiragino Sans GB

In 2007 artist Frédéric Keiff took on the challenge of community disengagement within Espace de Douala, in Douala, Cameroon. Espace' Douala is an art center supported by the Institut Français du Cameroun, part of the global organization for French culture. Cameroon, a French-speaking country, was in the umbrella of French colonies in the 1900s. It borders Nigeria, Chad, the Central African Republic, and Congo.

In designing the project Keiff observed changes in both the human and arboreal community in this deforested, financially strapped region of West Africa. He was inspired by the story of a historic large shade tree long gone but fondly recalled as a gathering space from better times.

In African tradition a palaver tree is a natural gathering spot for community discussions, storytelling, problem solving, and festivals. Roughly equivalent to a civic plaza, zocalo, or agora, the palaver is a place for comfortable community engagement. Keiff understood that the loss of the tree contributed to the unraveling of the fabric of the community.

Palaver Tree is an artificial canopy made of recycled green and blue plastic pieces stitched together to resemble a quilt, lifted high overhead with a structural steel armature. Installed on the infrastructure of the art space, the sculptural tree inserted the cultural center into the community conversation.

Thanks to the artist's rigorous observation of public space, African culture, and human interaction made Palaver Tree's execution most

compelling. Building a new tree based on historic roots, the artist intentionally hired and trained unemployed young men on the streets of Douala to build community engagement.

The artwork taught valuable skills to the youth including welding, painting, and creatively responding to a site's needs. Keiff granted official certificates to the men, recognizing their work, a valued credential in an area where primary schooling is for the upper classes. The men felt valued, and in turn valued their contribution.

The artwork partnered with established, though marginalized, local businesses. Doualan recyclers supplied the iron, glass, and plastics needed for the project, keeping those materials out of landfills.

The project continues to impact the audience for whom it was created. The tree was completed for the Salon Urbain de Douala (SUD festival) in 2007. Officials past and current acknowledged the creation of the tree and the community effort. Enjoyment of the work continues in 2014.

Using creative, social and political strategies, Keiff identified a suitable place where the roots of the tree would take a strong hold in the historical background and the urban life. This process has been long and painstaking, planting the seed of a work of art in the city's social network.

Palaver Tree exceptionally addresses a community need with a creative solution. Honoring local traditions, employing local youth, repurposing recycled plastics, Palaver Tree made something beautiful, useful, engaging to the public, and transformative to the participants who created it. The artist successfully crossed cultural boundaries and honored the spirit and broadest purposes of public art.

Keiff is a French artist working internationally and maintaining a studio near Strausborg, France. His practice is based on the human form and the built environment.

天上独木舟
La Pirogue Céleste

艺术家：埃尔韦·加布里埃尔·亚森
地点：喀麦隆杜阿拉
形式和材料：博览会
时间：2010 年
委托人：西蒙·恩贾米（杜阿拉城市博览会 2010 的策展人）
推荐人：里奥·谭

杜阿拉现人口已突破 300 万，是喀麦隆最大的城市。它坐落于武里河河口，延伸到几内亚湾中。杜阿拉的传统社区，通过捕鱼、乘舟在不同的河流上航行维持自己的生计。考虑到这些水道对杜阿拉人生活的重要性，水和河流在当地的传统仪式中占据着重要地位是可以理解的。自喀麦隆在 1960 年脱离法国，1961 年脱离英国独立后，这座城市迅速发展，成为这个国家的商业首都，以及非洲最富有、生活成本最昂贵的城市之一。杜阿拉的现代经济主要依赖于自然资源的出口，包括石油、木材、金属、可可粉以及咖啡。

杜阿拉城市博览会是这座城市每三年举办一次的公共艺术节日。"天上独木舟"是埃尔韦·加布里埃尔·亚森的公共作品，2010 年度杜阿拉城市博览会策展人是西蒙·恩贾米。"天上独木舟"，正如雕塑名称所表明的，雕塑由一艘令人回忆起 Bonapriso 地区的船夫所使用的传统船只的，10 米长、0.9 米宽的独木舟构成。该独木舟，由木头和金属制成，坐落于 Bonapriso 的公园中。具有象征意义的是，船只的船头代表了城市的过去和未来，而雕塑边缘上站着的耕作者使人们想起了这个区域的植被。一张长凳也为该独木舟附近的游客提供了休息和静思的场所。

该项目的一个亮点在于其唤起了与水相关的传统仪式活动。这位艺术家告诉我们天上独木舟的船头使人们想起了萨瓦人使用的传统船只。萨瓦人每年都会举办一个称为恩贡多节日的水仪式，在该仪式举办期间，会派遣信使去拜访武里河的水神。在这样的背景下，"天上独木舟"可被描述为拜访萨瓦神的当代信使，从而吸引大众关注这座城市的仪式历史和传统以及

节日的主题——杜阿拉与水的紧张关系。对于很多居民来说，很不幸地，由于给水装置的维护不当以及水的公共分配不佳，水供给形势岌岌可危。而更为糟糕的是，由于排水系统的不健全或缺乏，在季风季节，这座城市的很多地方会被淹没，从而造成严重的公共卫生问题。

以水为源

冯正龙

一件公共艺术作品的创作是以特定地域文化与地理环境为背景的，这也说明公共艺术作品具有场所特征，而且它的场所特征不是放之四海而皆准，普适性的。公共艺术的相关项目都是建立在对特定地域和社区文化挖掘上，针对这些地域或社区的敏感话题和关注焦点来创作作品，有时这些作品会因意见的不统一而引起争议，有时也会因为作品所具有的社会关怀而受到关注和喜爱。"天上的独木舟"这个作品就深刻的挖掘了当地人对水的强烈感情和关系以及水在日常生活和社会中的重要性，然后以作品之形式吸引当地居民关注。总之，公共艺术不能仅仅追寻表面的光鲜，而忘记作品本身所应该承担的内涵，否则，作品随着时间的推移，也会慢慢地被公众所忽视和遗忘，形单影只地矗立在公共空间中。

Artist: Hervé Gabriel Ngamago Youmbi
Zone: Bonapriso District, Douala, Cameroon
Media/Type: Exposition
Date: 2010
Commissioner: Simon Njami (curator for the Salon Urbain de Douala 2010)
Researcher: Leon Tan

With its population of over 3 million, Douala is Cameroon's largest city. It sits at the mouth of the Wouri River, opening out into the Gulf of Guinea. Douala's traditional communities sustained themselves through fishing, navigating the various rivers by canoe. Given the significance of these waterways in Douala life, it's understandable that water and the rivers feature strongly in the traditional rituals of the locale. Since Cameroon achieved independence from France and the UK in 1960 and 1961, the city has developed rapidly, becoming the business capital of the nation, and one of the wealthiest and most expensive cities in Africa. Douala's modern economy is based largely on exports of natural resources including oil, timber, metal, cocoa, and coffee.

Salon Urbain de Douala is the city's triennial festival of public art. La Pirogue Céleste was a public work by Hervé Gabriel Ngamago Youmbi, curated by Simon Njami for the Salon Urbain de Douala 2010 edition. Translated, La Pirogue Céleste means the celestial canoe, and as the title suggests, Youmbi's sculpture consisted of a 10-meter-long by 0.9-meter wide canoe reminiscent of the

traditional watercraft used by boatmen of the Bonapriso district. Made of wood and metal, two of the city's major export products, the canoe was situated in a park in Bonapriso. Symbolically, the prow of the boat made reference to the city's past and future, while the accompanying planter on the edge of the sculpture evoked the region's vegetation. A bench provided seating for audiences in the direct vicinity of the canoe for rest and contemplation.

One point of excellence of this project is its evocation of traditional ritual practices connected to the water. The artist tells us that the celestial canoe's prow recalls those used by the Sawa people. The Sawa held annual water celebrations known as the Ngondo Festival during which time messengers were sent to the water gods in the Wouri River, connecting human and divine realms. Against this backdrop, La Pirogue Céleste could be characterized as a contemporary messenger to the Sawa gods, drawing the attention of the public to the ritual history and tradition of the city, as well as to the festival's theme—Douala's troubled relationship with water. For many residents, water access is unfortunately precarious, due to the poor maintenance of water works and poor public distribution. To make matters worse, during the monsoon season, many parts of the city are flooded, posing serious public health problems because of the pollution resulting from poor or nonexistent mains drainage systems.

极致创意墙
Wide Open Walls

艺术家：恩约古·托雷、劳伦斯·威廉姆斯
地点：冈比亚农村民谣保护区
形式和材料：壁画
时间：2010 年
委托人：当地房屋、商业和学校持有人
推荐人：里奥·谭

"极致创意墙（WOW）"倡议自 2010 年启动，旨在举办一年一度为期 14 天的盛会，吸引来自冈比亚和海外的艺术家在冈比亚 Ballabu 保护区的各大村庄的房屋、学校和商业建筑上创作永久性壁画。2010 年至 2012 年，艺术家们总共创作了超过 100 部壁画作品，遍布 Kubuneh，Galoya，Makumbaya 以及 Bufalotou 村庄。国际艺术家包括 ROA（比利时），Know Hope（以色列），Remed（马德里），TIKA（瑞士），Freddy Sam(南非)，Selah(南非)和 Best Ever(英国)都前往冈比亚参观该项目，并与冈比亚本土艺术家 Lawrence Williams 和 Njogu Touray 一起创作壁画作品。所有的壁画创作均事先得到了当地居民和建筑产权人的准予。

在艺术工作室系列创意墙中，艺术家们专门为当地儿童创作的作品成为重要的一部分。由于冈比亚农村学校未设置艺术课程，因此这些工作室为当地儿童们提供了学习新艺术技能的机会，而且当地居民可以与这些来访艺术家们进行更深入的互动。值得一提的是，一些壁画作品由学生们和艺术家们共同完成。2011 年以来，"WOW"项目启动了一年一度的英国学校交流机制，因此来自英国的学生们与冈比亚学生建立了联系，并且可以亲自来到冈比亚与这些小伙伴们共同绘制绘画作品。

长期以来，"WOW"的宗旨是促进当地区域和村庄的经济发展，鼓励艺术和生态旅游业，通过销售壁画艺术创作的画册以及国际展会创收。艺术家们表示："计划建造一个游客中心、咖啡屋、艺术教室、社区中心和公寓楼。其中，公寓楼可以为艺术家们提供临时居住点。咖啡屋将为当地社区创收，同时教室将对外完全开放，欢迎广大青年、老年展示画技，体现创意灵感。"

虽然目标定的有点高，但是我们注意到该项目获得了广泛支持。2012 年，英国高级专员公署指派艺术家们在办公室作画，以此作为庆祝"五十年节"；2014 年，伦敦美术馆将所有展出作品销售收入捐赠给冈比亚社区。另外，2014 年 11 月之后，各大旅游公司将启动小型艺术旅游专线，将引领游客

参观该区域的作品，这些村民将因门票收入获益。大量游客的纷至沓来使得许多农村儿童获得教育专项资助。基于这些成功事例，"WOW"相信未来计划可以付诸实现。"WOW"的成功部分应归功于将艺术作为可持续发展的重要手段。当地风景迷人的村庄部落，富有视觉美感，吸引着当地居民、外国游客、艺术家以及参观者的目光。它还为当地儿童带来了一些教育和艺术上的有形成果。

此外，"WOW"的成功还应归功于组织者对 Ballabu 保护区和社区作出的长期承诺。作为国际文化联盟和交流的合作伙伴，我们为使该项目具有国际影响力，因此越来越多的个人或团体开始关注冈比亚农村区域。在当地居民看来，"WOW"帮助减少了青壮年外出打工人员的数量，并提高了人们的生活质量，尤其是当地儿童。

乡村壁画
冯正龙

"极致创意墙"这个项目的运作方式与日本新泻"大地艺术三年展"有很多相似之处，都是通过邀请国际知名艺术家在场所进行艺术创作，然后吸引公众对这一地区的关注，从而推动此地区经济的发展和居民生活品质的改善。"极致创意墙"更深层的意义在于对边缘群体的关注，并将艺术作为地区复兴和可持续发展的重要手段，这也表明公共艺术所具有的强大魅力，它不仅可以尊重和传承地区文化传统，而且也可以在有限的条件下尽可能的利用条件帮助地区焕然一新。同时，任何公共艺术项目仅靠艺术家的力量是很难完成的，它需要民间个人或国际团体组织的合作与协助，需要社区居民和地方民众的积极支持与参与。"极致创意墙"的成功离不开各方的关注与支持。

Artist : Njogu Touray Lawrence Williams
Location : Rural villages in the Ballabu Conservation Zone, the Gambia
Media/Type : Mural (painting)
Date : 2010
Commissioner : Owners of local dwellings, businesses and schools
Researcher : Leon Tan

Launched in 2010, Wide Open Walls (WOW) is an ongoing initiative that brings international and Gambian artists together annually for a fortnight to create permanent murals on homes, schools, and local businesses in rural villages in the Ballabu Conservation Project zone of the Gambia. Between 2010 and 2012, upwards of 100 murals were created in the villages of Kubuneh, Galoya, Makumbaya, and Bufalotou. International artists including ROA (Belgium), Know Hope (Israel), Remed (Madrid), TIKA (Switzerland), Freddy Sam (South Africa), Selah (South Africa), and Best Ever (United Kingdom) visited the Gambia for this project, working closely with the Bushdwellers (Gambia-based artists Lawrence Williams and Njogu Touray) on each mural. All murals were created with the permission of the dwelling and building owners.

An important part of WOW has been the series of art workshops that artists conducted for local children. Since art is not taught in schools in rural Gambia, workshops resulted in new artistic skills for the children and immersive engagement with the local community for visiting artists. Significantly, several murals were created by the school children in collaboration with the artists. From 2011, WOW began to incorporate an annual school exchange with the United Kingdom, so that children from the United Kingdom were linked with children from the Gambia, and traveled to the villages to paint murals on schools with their local counterparts.

Over the long term, the intention behind WOW is to contribute to the economic development of the region and its villages, encouraging art and ecotourism, and generating revenues through

the sale of photographs and books featuring the mural artworks, as well as through international exhibitions. According to the artists, "The plan is to build a visitor centre, with a cafe, art classroom, community centre and apartment. The apartment will be for artists to stay during residencies here. The cafe will generate money for the community and the classroom will be open for all people, young and old, to come to paint and express their creativity."

While the objective may seem ambitious, it is notable that the project has received considerable support. In 2012, the British High Commission appointed the artists to paint their offices as part of the Jubilee celebrations, and in 2014, an exhibition in a London gallery is scheduled, where proceeds from the sales of works will be donated to Gambian communities. Additionally, starting in November 2014, tour operators will actually begin small art tours, bringing visitors to the region to see the art, with the villages benefitting materially from the entry fees into the zone. The influx of tourists has already led to the sponsorship of education for many village children. In light of these successes, WOW and its future plans seem to be within the realms of the possible.

WOW's excellence consists of its success in using art as a vehicle for sustainable placemaking. It beautified local villages, making them aesthetically enjoyable for local residents as well as international visitors, artists and tourists. It also produced tangible educational and artistic outcomes for local children. Part of its success relies on the long-term commitment its organizers have demonstrated to the Ballabu conservation zone and its communities. The skillful brokering of international cultural alliances and exchanges must also be acknowledged, as these partnerships brought international exposure to the project, and consequently, to neglected rural regions of the Gambia. In the opinion of locals, WOW helped to reduce urban drift (migration of young people to city centers), and improved the quality of life, particularly for children.

UT772 DC10 纪念碑
UT772 DC10 Memorial

艺术家：吉拉姆·蒂诺克斯·圣马可
地点：撒哈拉沙漠的 Ténéré 地区
形式和材料：纪念碑
时间：2007 年
委托人：爵士 DC10 轰炸的家庭
推荐人：梅根·盖尔贝

"UT772 DC10 纪念碑"用于纪念法国联合航空公司 UTA772 次航班因遭受恐怖袭击而导致 170 人丧生的恐怖袭击事件。该航班于 1989 年 9 月 19 日当天，于荒无人烟、条件极其恶劣的沙漠地区——撒哈拉沙漠的 Ténéré 地区发生爆炸并坠毁。事发当天该航班正从巴黎飞往刚果共和国。利比亚恐怖人员事先策划了此次恐怖袭击，将炸弹藏匿于手提箱中，随后在飞行过程中引爆炸弹。机上人员全部遇难。

遇难者家属曾向利比亚政府提出索赔。在一名失去父亲的法国销售及市场营销人员吉拉姆·蒂诺克斯·圣马可的带领下，最终于 2004 年为所有遇难者家属争取到了总额为 1.7 亿美元 (1.04 亿英镑) 的赔偿金，并平均分配给了 170 名遇难者的家属。

吉拉姆·蒂诺克斯·圣马可建立起了非政府组织——UTA DC10，以便查找每名遇难者的家属并进行赔偿金的分配。他利用该笔巨额资金所得利息为基金成立了该组织。2007 年，吉拉姆·蒂诺克斯·圣马可组织了一场针对坠机现场的勘察之旅，陪同勘察的还有另外两名 UTA772 次航班的遇难者家属。在坠机现场，置身于飞机残骸之间，吉拉姆·蒂诺克斯·圣马可受到启发后决定建立一座纪念碑，让所有过往飞机都能看得到。正式执行程序之前，由 UTA DC10 组织负责相关审批工作。

此次恐怖袭击事故中遇难的埃克森公司雇员，其同事，先前已经在事故所在地树立起了一块小型的纪念物：由飞机右翼形成的一块刻有其遇难公司成员姓名的牌匾。吉拉姆·蒂诺克斯·圣马可受到启发后设计自身的纪念碑，如今他设计的纪念碑上刻满了 170 名遇难者的姓名。

"UT772 DC10 纪念碑"的外形与尺寸模仿坠毁飞机的外形及尺寸，由大型石块砌成。众多石块堆积成一个直径为 200 英尺 (约 70 米) 的圆石堆。飞机的外轮廓指向巴黎——原事故飞机的飞行目的地。170 面破碎的镜子环绕事故现场，每面镜子象征着一名遇难者。所有镜子被完美地放置在平坦的石头基座上，然后全部击碎，象征逝去的生命。环绕纪念碑一圈分别设计勾绘了四个点，形似指南针，指向飞机飞往的方向。纪念碑轮廓后面，

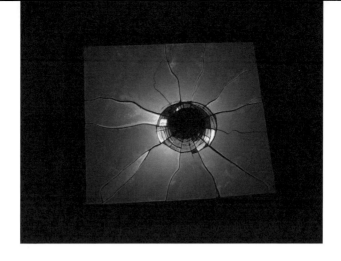

分布着三条直线，朝向空中，预示飞机前进的动作。尽管纪念碑坐落于人迹罕至、极难到达的地区，通过互联网却很容易查询到其相关信息。其尺寸足够从太空中观测到，因此通过谷歌地图也能搜寻得到。因为该纪念碑随时提醒着过往航班，此次事故可以说是一场幸福的事故。

纪念碑多由人力手工堆砌完成，石块由距离事发地 70 公里外的地方运来。由 140 名当地人耗时六周时间建造完成。完成纪念碑的建造工作之后，由赔偿金所产生的利息还剩余不少金额，足以供吉拉姆·蒂诺克斯·圣马可为恐怖主义遇难成员设立基金会，也就是其经营至今的基金会。该基金会向所有遇难者或其家属开放，不分地区及国籍。由此看来，如今 UT772 DC10 发生的悲剧事故有助于对抗恐怖主义。

"UT772 DC10 纪念碑"为一种国际性标志，可以通过互联网获取有关恐怖袭击事件之后的遇难成员家庭生活点滴。从高空中可以看见的设计理念，为公共艺术界的一种创新方法，其中含有的静止标志，如牌匾和破碎的镜子，则构成了永恒性而又富有诗意的印象，这种印象唯有靠近方能看得到。纪念碑上的标志性语言，激起了很多观众的共鸣，其外形则提醒我们铭记 1989 年 9 月 19 日消逝于 Ténéré 沙漠的生命。

无声的呐喊

冯正龙

公共艺术理念的传达可以通过多种方式，可以说纪念牌是最发人深思的一种，当人们看到纪念碑时，总会内心自然而然的产生这样或那样的情愫，也会迫切地去追问：以前在这里发生了什么？为什么会发生？为什么没有避免？但当你了解了一切，惋惜之余，你会感激并庆幸这个纪念牌的存在，是它让现在的人和以后的人记住了过去发生的事和当时的人，是它在警示人们不要忘记历史，并以史为鉴。"UT772 DC10 纪念碑"是为了纪念空难中的遇难者，但更深远的意义在于以纪念牌的形式让人们共同对抗恐怖组织和恐怖主义。它现已成为国际上的一种含有深刻内在意义的标志，其外在形式的每一部分都刺痛着看到它的人，如果说，它为什么会有如此之力量，那答案就隐藏在这座纪念牌的背后。

Artist: Guillaume Denoix de Saint Marc
Zone: Ténéré Desert, The Republic of Niger
Media/Type: Memorial
Date: 2007
Commissioner: Les Familles de l' Attentat du DC10 d' UTA
Researcher: Megan Guerber

UT772 DC10 Memorial commemorates a terrorist attack on UTA Flight 772 that resulted in 170 casualties. On September 19, 1989 the plane exploded and crashed in the Ténéré region of the Sahara, an inaccessible and harsh desert terrain. The flight was heading to Paris from the Republic of Congo. Libyan terrorists had planted a suitcase bomb that detonated during flight. There were no survivors.

Relatives of the victims fought for restitution from the Libyan government. Led by Guillaume Denoix de Saint Marc, a French salesman and marketer who lost his father in the attack, $170,000,000 (£104,000,000) was finally awarded in 2004. This entire large sum was divided equally among the relations of the 170 victims.

Denoix de Saint Marc set up the NGO Les Familles de l' Attentat du DC10 d' UTA in order to track down the victim's relatives and distribute the money. Enough money was accrued in interest to fund the organization. In 2007, Denoix de Saint Marc organized a reconnaissance trip to the crash site with the relatives of two other UTA Flight 772 victims. There, among the debris, Denoix de Saint Marc was inspired to build a memorial that would be visible

by planes passing overhead. Permission was sought from Les Familles de l' Attentat du DC10 d' UTA before proceeding.

Colleagues of Exxon employees who had died in the attack had already erected a small memorial in their honor: a plaque inscribed with their names on the starboard wing. This element was

incorporated in Denoix de Saint Marc's design and today bears the names of all 170 victims.

UT772 DC10 Memorial is a life-sized silhouette of a plane outlined by large stones. These stones fill a circle around it that is 200-foot in diameter (roughly 70 meters). The plane's outline points toward Paris, the original destination of the flight. 170 shattered mirrors encircle the site, one for each victim. These were placed intact upon flat stone pedestals, then broken to represent the lost life. Four points are also drawn in stone around the circle's circumference, creating a compass that shows the direction the plane is headed. Behind its silhouette, three straight lines project outward to indicate the plane's forward movement.

Although the memorial is in a secluded, hard to reach area, it is easily accessible via the Internet. It is prominent enough in size to be visible from space and can be viewed via Google Earth. This was a happy accident, as the memorial was originally intended for aircraft passing overhead.

The memorial was largely hand built, with rocks traveling in from 70 kilometers away. 140 locals spent six weeks building it. After its completion, enough money was left over from the accrued interest that Denoix de Saint Marc was able to set up a foundation for victims of terrorism, which he now runs. This foundation is open to all victims or their relations regardless of religion or nationality. In this way, the tragedy of UT772 DC10 helps to fight against terrorism today.

UT772 DC10 Memorial is an international sign, available via the Internet, which demonstrates the continuation of life after terrorism. Its design to be viewed from overhead is an innovative approach to public art, yet its quiet symbols, like the plaque and the broken mirrors, make a lasting, poetic impression that is intended to be seen up close. The memorial's symbolic language resonates with a large audience while its size forces us to remember the lives lost on September 19, 1989 in the Ténéré desert.

梦想时间舱
Dream's Time Capsule

艺术家：伊娃·弗拉皮奇尼
地点：埃及开罗
形式和材料：充气式梦想时间舱
时间：2010 年
委托人：开罗联排别墅画廊
推荐人：里奥·谭

心理分析学家卡尔·古斯塔夫·荣格也许和他的同行弗洛伊德一样著名。他们两人都发展了无意识论，这是一种针对人文主义者—理性主义者思想现状提出的理论。该理论认为，对于我们的所说和所做，如果我们不愿意相信，那么我们就会较少意识到它们，较少承担责任。许多知识分子认为对无意识的这种发现严重打击了人类的自恋心理。它开启了一个深入研究隐藏深处的思想、感觉和知识的年代，鼓舞了 20 世纪的超现实主义者和我们当代的艺术家。伊娃·弗拉皮奇尼的"梦想时间舱"是一个扎根于历史的雄心勃勃的艺术研究项目。它把荣格的集体无意识概念作为其概念基础。荣格认为，这种原始的无意识包括通过神话人物或原型人物来表达的宇宙能量。"梦想时间舱"于 2011 年启动，是对集体无意识的国际跨文化分析，与公众梦想的集萃相关，它被设在一个建造的充气式白色圆顶内。在不久的将来，录音会作为一份公共档案公开。

该充气式"梦想时间舱"看起来像一个胚胎或贝壳，由这位艺术家与米歇尔·塔瓦诺合作设计。然后，装有录音站的时间舱被安装到不同城市的公共场地中，包括图灵、开罗、斯德哥尔摩和里加。在此其间，这位艺术家邀请了路人来录下对梦想的讲述。2012 年，在开罗联排别墅画廊主办了这一项目。在这三天期间——上午 10：00 到下午 8：00，参与者进入"梦想时间舱"，录下过去的梦想。共用不同语言录下了 90 个梦想，包括英语和埃及阿拉伯语。据这位艺术家所说，这个项目吸引了许多人，因为"梦想在埃及社会和传统中总是非常重要的。"

很明显，"梦想时间舱"（开罗项目）发生在埃及政治历史一个特别动荡的时期。埃及的"阿拉伯之春"开始于 2011 年，2012 年举行了首届民主选举，2013 年的民众集会抗议后，这个国家在政治上仍然不稳定。从荣格的观点来看，人们可能期望通过与叛乱、暴力和解放相关的原型人物与神话人物的形式，将这些动乱反映在梦想中，因为社会动荡是与这些原型人物或神话人物相关的宇宙能量占主导的表现。当然能否找到这种关联性还有待观察，因为档案直到 2016 年才可供查看。届时，集萃的梦想会通过仪式来密封，然后由该艺术家开启，而梦想录音会作为礼物归还给参与者。

从场所营造方面来看，该项目是创新的，它使公众面临经常被忽略的日常生活的一个层面，也就是，逃离我们没有意识到的事物。它通过记录下感觉，暂时解开了当代社会和空间中日常生活和工作所施加的强烈束缚—理性功利主义思想。在艺术上，它的抱负，以及迄今为止取得的成功证明了它的卓越。尽管将"梦想时间舱"作为一个项目来充分评估是不可能的，因为它还未完工，但是我们可以说这一创意的实现会为公众提供一份档案，从而能对比分析跨越不同阶层、种族、民族、性别、宗教和政治立场的不同个人和社群的梦想生活和符号系统。这种档案会成为公开激活、分析集体无意识的宝贵模式。

梦想的珍藏
冯正龙

梦想是每一个人儿时向往憧憬的地方，随着生命历程的不断变化，梦想也在不断改变着。可以说梦想是现在与未来的互动，是意识与无意识空间的对话，是一种集体无意识的特殊体现。但没有人会用声音去记录梦想，更不用说把它当档案珍藏。当我们走进"梦想时间舱"，就感觉进入了自己梦寐以求的梦想中，轻声欣喜的述说着未来将要发生的一切，更让我们出乎意料的是，在某个时间，你会意外收到一份礼物，这个礼物承载着以前的梦想，这时你会恍然大悟，原来我与梦想失之交臂了，这次你只能直面自己，不会再去逃避了，这次你会哭泣，也会彷徨，但更会珍惜这份来之不易的梦想礼物。"梦想时间舱"的功能还不仅仅是这些，它更成为反映不同阶层、种族人群所呈现的梦想差异，从而分析出集体无意识的差异性与相似性。

Artist: Eva Frapiccini
Zone: Cairo, Egypt
Media/Type: The inflatable dream capsule
Date:2012
Commissioner: Townhouse Gallery, Cairo
Researcher: Leon Tan

The psychoanalyst Carl Jung is perhaps as renowned as his colleague Sigmund Freud. Both developed theories of the unconscious, theories that suggested, against the current of humanist-rationalist thought, that we are much less conscious of, and responsible for, the various things we say and do than we would like to believe. The discovery of the unconscious is considered by many intellectuals to have dealt a blow to human narcissism. It inaugurated an era of investigation into the hidden recesses of thought, sensation and knowledge, inspiring the surrealists of the twentieth century as well as artists of our own time. Eva Frapiccini' sDreams' Time Capsule (DTC) is an ambitious artistic research project located in this century-long trajectory. It takes Jung's concept of the collective unconscious as its founding idea. Jung believed that this primordial unconscious was populated with universal energies, taking expression through mythological figures or archetypes. Launched in 2011, DTC is an international cross-cultural analysis of the collective unconscious, involving the collection of dreams from the public, housed within a built inflatable white dome. In the near future, recordings will be made available as a public archive.

The inflatable dream capsule, which looks like an embryo or seashell, was designed by the artist in collaboration with Michele Tavano. It was then installed, with an audio recording station, in public sites in various cities, including Turin, Cairo, Stockholm, and Riga, during which time the artist invited passers-by to record a narration of a dream. In Cairo, it was hosted by the Townhouse Gallery in 2012. Over three days, between the hours of 10:00 a.m. and 8:00 p.m., participants entered the capsule and recorded a dream from the past. 90 dreams in total were recorded in various languages including English and Egyptian Arabic. According to the

artist, many were captivated by the project because "the dream has always been important in Egyptian society and traditions."

Notably, DTC Cairo took place in a particularly turbulent period in Egyptian political history. Egypt's Arab Spring began in 2011 and was followed by the country's first democratic presidential election in 2012, installing Mohamed Morsi in government. Morsi was deposed by the military after protests in 2013, and the country continues to be politically unstable. From a Jungian perspective, one might expect these upheavals to be reflected in dreams, taking the form of archetypes or myths associated with rebellion, violence, and liberation, since social unrest is a manifestation of the predominance of universal energies associated with those archetypes or myths. Whether or not such a correspondence can be found remains to be seen because the archive will not become available until 2016. At that time, the dream collection will be ritually sealed and then opened by the artist, and dream recordings will be returned as a gift to the contributors.

Considered in terms of placemaking, the project is innovative, generating public encounters with an often-neglected dimension of everyday life, namely, that which escapes consciousness, that of which we are unaware. It temporarily undoes, through the register of sensation, the intense stranglehold rational-utilitarian thought exerts on the organization of daily life and work in contemporary societies and spaces. Artistically, its ambition and the success with which it has been met to date speak to its excellence. While it is not possible to fully evaluate DTC as a project, since it is incomplete, it is possible to say that the realized idea will provide to the public an archive that can yield intriguing comparative analyses of the dream lives and sign systems of different individuals and communities across the categories of class, ethnicity, nationality, gender, religious, and political affiliation. Such an archive would be an invaluable mode of publicly activating and analyzing the collective unconscious.

"光之地"城镇艺术体验

Title: Maboneng Township Arts Experience

艺术家：西菲韦·恩圭尼亚
地点：各个镇，包括亚历山德拉，夸祖鲁—纳塔尔的纽卡斯尔和
南非的豪登省
形式和材料：多种形式
时间：2000 年
委托人："光之地"城镇艺术体验
推荐人：杰奎琳·怀特

""光之地"城镇艺术体验"的标志上有一个灯泡——灵感的典型标志。对于这个南非的年度艺术节，灯泡还有一个更加老套的意思。Alexandra——创办这个节日的人口众多的城镇——是一个靠近约翰尼斯堡的区域，在种族隔离时期，南非黑人被迫生活在这里。这仍然是全国最贫穷的地方之一，由于这个地方没有通电，长久以来被称为"黑暗城市"。

通过把这个镇转变为画廊，把街道转变为表演场地，""光之地"城镇艺术体验"旨在驱散都市地区的黑暗，其中，城镇居民可以观看到他们通常看不到的艺术作品，外部人士会给这个城镇加上危险地区的污名——而非充满活力的文化中心。Maboneng 的意思是"光之地"。

这个现在一年一度的活动于 2001 年由西菲韦·恩圭尼亚在 Alexandra 创立，但是他最知名的身份可能是饶舌歌手 Master Sip，Natives from the Sun（来自太阳的原著民）乐队的主要成员以及嘻哈乐队 Skwatta Kamp 的前成员。当他在 Alexandra 长大时，西菲韦·恩圭尼亚的妈妈在家里烘烤和出售面包，这让他有了混合家庭生活和商业的早期体验。他是""光之地"城镇艺术体验"的制作人和馆长，Zipho Dayile 是副馆长。

在 2012 年，Alexandra 的一百周年纪念，在 Alexandra 的 ""光之地"城镇艺术体验"安排和九月的"开普敦艺术周"一起举办，在此期间，美术馆协调它们的开馆日期。在 Alexandra 三小时的徒步旅行中的亮点包括有关 Alexandra 无名英雄的历史剧，以及足球舞者有关小镇生活的表演。一种切分音、快步的舞动形式进化成一种社会现象，足球舞起源于 Alexandra。

在 2012 年，这个节日拓展到另外两个地点：""光之地"城镇艺术体验"于 10 月的一个周末在 KwaZulu 的 Newcastle 举办；在一个 11 月的周末，30 个 Gugulethu 的家庭成为 ""光之地"城镇艺术体验"的展览区。居民和参观者可以品尝小镇的美味，如南非三明治和全脂带高，观看 Theo

Ndindwa 从 iKapa 舞剧院策划的舞蹈表演，观察地区最出色涂鸦艺术家的涂鸦作品以及一个非盈利照相培训项目所制作的照片。

在过去 10 年，这个项目将 70 多个家庭转换为临时美术馆，展示过 50 多位艺术家的作品。该节日展示了小镇的艺术家，鼓励当地儿童把艺术视为一个可行的职业选择，并且通过艺术品和事物的销售以及美术馆场所的租赁，向居民提供多种收入来源。另外，"'光之地'城镇艺术体验"还诱使非本地居民和甚至外国游客到访他们通常不会去的地方，帮助促进文化交流，并终止有关这些城镇的负面印象。

黑暗中的"光之地"
冯正龙

在传统意义上，人们通常把艺术的功能和作用当作是陶冶情操、培养修养，提供审美认知能力等，但当我们看到"'光之地'城镇艺术体验"这个项目时，我们看到的是艺术功能的转变，艺术开始降低姿态，摆脱高高在上的独尊地位，开始与社会和公众发生关系，这些改变是时代的需要，是社会与人的需要。同时，公众离不开艺术，社会发展也离不开艺术，这也就孕育出带有公共性质的公共艺术。公共艺术利用多种形式，以理念为前提，以创意为动力，以艺术为抓手，以地区复兴为目的，这就是"'光之地'城镇艺术体验"项目的精髓所在。这个节日在不断拓展艺术形式的同时，也在不同地点进行新的尝试，利用这种尝试让不同家庭亲身体验艺术，让他们感受到艺术就在身边，艺术可以让生活有所改变，之前的黑暗之地在艺术的照耀下已变成了现在的"光之地"。

Artist: Siphiwe Ngwenya
Zone: Various townships, including Alexandra, Newcastle
in KwaZulu-Natal, and Gauteng, South Africa
Media/Type: This annual South African art festival
Date: 2007
Commissioner: The Maboneng Township Arts Experience
Researcher: Jaqueline White

The logo for the Maboneng Township Arts Experience features a
light bulb, the classic symbol for inspiration. For this annual South
African art festival, the light bulb also has a much more a clichéd
meaning. Alexandra, the overcrowded township in which the festival
was founded, is an area near Johannesburg where black South
Africans were forced to live under apartheid. Still one of the poorest
areas in the country, it was long known as the 'Dark City' because
of its lack of electricity.

By turning township homes into art galleries and streets into
performance venues, the Maboneng Township Arts Experience
aims to dispel a metaphorical darkness, in which township residents
can view art as beyond their reach, and outsiders can stigmatize
the townships as dangerous locales—not vibrant cultural centers.
Maboneng means' place of light.'

The now annual festivals were founded in Alexandra in 2001 by
Siphiwe Ngwenya, a fine artist and sculptor, who is probably best
known as rapper Master Sip, frontman of Natives from the Sun and
former member of the hip-hop group Skwatta Kamp. When he was
growing up in Alexandra, Ngwenya's mother baked and sold bread
from their family home, which gave him an early appreciation for the
potential mixing of domestic life and commerce. He is producerand
head curator of the Maboneng Township Arts Experience; Zipho
Dayile is co-curator.

In 2012, the 100-year anniversary of Alexandra, the Maboneng Township Arts Experience in Alexandra was scheduled to coincide with Cape Town Art Week in September, during which galleries coordinate their openings. Highlights of a three-hour walkabout in Alexandra included the staging of a history play about Alexandra's unsung heroes and a performance by pantsula dancers about township life. A syncopated, quick-stepping dance form that has evolved into a form of social commentary, pantsula originated in Alexandra.

The festivals expanded in 2012 to two additional locales: the Maboneng Madadeni Arts Experience took place during an October weekend in Newcastle in KwaZulu, and over a November weekend, 30 Gugulethu homes became exhibition spaces for the Maboneng Gugulethu Arts Experience. Residents and visitors could sample delicious township cuisine such as bunny chow and fat-cakes, watch captivating dance performances curated by Theo Ndindwa from iKapa Dance Theatre, see graffiti by some of area's finest graffiti artists, as well as photos produced through a nonprofit photography training program.

Over the previous decade, the project has turned over 70 homes into temporary galleries and exhibited over 50 artists. The festivals showcase township artists, encourage local children to see art as a viable career path, and provide residents numerous revenue streams through the sale of art and food, as well as rental of their gallery homes. In addition, the Maboneng Township Arts Experience has enticed nonresidents and even foreign tourists into areas they might not typically visit, helping to facilitate cultural exchange and end negative perceptions of townships.

智能技术 / 分享文化
smARTpower / Sharing Culture

艺术家：布雷特·库克
地点：尼日利亚拉各斯及伊巴丹的不同地址
形式和材料：一系列团体艺术工作坊
时间：2012 年
委托人：美国国务院及布朗士艺术博物馆
推荐人：里奥·谭

"智能技术 / 分享文化"合作项目由布雷特·库克于 2012 年在尼日利亚接手。该项目由一系列团体艺术工作坊组成，而这些团体艺术工作坊则由包括创新性艺术部门、拉格斯大学、Wy 艺术基地以及所有圣学院在内的发起人资助。这些工作坊代表了全球范围内数量不断增加的可参与性的或相关联的艺术实践。在该智能技术案例中，不仅涉及到了群体对话，还涉及到了个体间以及合作性表达。来自各个领域的参与者，包括中学生以及成年人，他们基于各种不同的实践方式而着手创作公共艺术。其中具有超现实主义意义的室内游戏——cadavre exquis 或美艳僵尸，游戏参与者们逐个叠加组成艺术作品，共同产生完整的人体肖像。另一个实践则邀请参与者们在镜子前创造自我肖像之前首先书写其自我感觉。然后将此过程产生的作品安装在公用场所，以供观众和参与者们观摩。

由布雷特·库克主导产生的工作坊类型，使参与者们参与到与学校学生活动或治疗团体活动差别不大的各种活动中。例如：美艳僵尸游戏，则通常由发达国家的小学生在艺术课或写作课上进行，而通过书面材料以及绘画 / 喷涂材料为媒介展开的身份探索游戏，则是艺术治疗专家们工具箱中的主要工具。从这种意义上来讲，智能技术并非绝对的原汁原味或新颖。库克的项目，相反则具有显著意义，并非由于项目本身构建的各种实践确定，却由项目本身所处的地域政治环境决定。尽管尼日利亚被认为是新兴市场，单纯从 GDP 意义讲，其已超过南非总体水平，但其公用医疗及教育基础设施却仍处于低等发展水平。据当地评论家评论，其上述体系事实上处于不正常状态。更不用提其艺术及文化环节，需要不断激活并灌输营养。基于上述各种现状，尼日利亚在种族及语言方面的多样性，使得其人口始终面临着宗派暴力及冲突。

而在此种情形下，智能技术在使得来自不同种族及语言群体的成员，基于其各自的表达及对个人经验的分享而进行的合作性努力方面，的确可圈可

点。正是身份上的显著差异引发了各种宗派暴力及冲突，这种通过中介性质而使得各种不同的"自我感觉"相遇的过程，为不同成员及不同文化之间建立沟通桥梁提供了机遇。或许还能带来自我了解程度的加深以及更进一步的求同存异。通过对一组记录镜头的分析发现，对于部分参与者来说，这的确是事实。通过鼓励并邀请参与者们进行自我表达，一起在公用氛围中见证并适应不同个人及不同文化方面的差异，在民主教育方面，智能技术的确作出了一定贡献。如果回想一下过去乃至当前大规模发生在尼日利亚的贪污腐败事件，该种教育方式的重要性则显得尤为突出。通过上述方式，库克的项目成功改善了参与者们的生活质量，使得他们直属的社团变得更加和谐欢愉，最起码在整个项目进行期是这样的。

差异中的互动

冯正龙

公共艺术不是一种艺术形式和流派与思潮，它是基于社会意义上的不同地域文化地理差异的互动、交流与共享。对个体而言，它会让个体成为集体；对社区而言，它会让社区融洽和睦；对地域而言，它会搭起差异身份交流沟通的场所，在求同存异的原则下自由生活。"智能技术／分享文化"以艺术工作坊为媒介，最大限度地让不同年龄、地域、阶层的人群进行互动交流，让矛盾冲突在互动过程中烟消云散，最终让各方受益并改善关系。也让我们看到，公共艺术所营造的分享、交流与互动平台的重要性。"智能技术／分享文化"是在社会存在宗族冲突的情况下，让成员们在忽视不同种族及语言差异的身份下交流互动，这是一种既勇敢又冒险的尝试，这个项目让社区与项目参与人员都有明显的改变，这个项目的最终结果还是比较成功的。

Artist: Brett Cook
Zone: Various locations in Lagos and Ibadan, Nigeria
Media/Type: A series of community art workshops
Date: 2012
Commissioner: US State Department and the Bronx
Museum of the Arts
Researcher: Leon Tan

The smARTpower / Sharing Culture Collaborative Project was undertaken by Brett Cook in Nigeria in 2012. It consisted of a series of community art workshops supported locally by hosts including the Department of Creative Arts, University of Lagos, Wy Art Foundation, and All Saints College. Workshops are emblematic of an increasing number of participatory or relational art practices across the world. In the case of smARTpower, they involved group dialogue as well as individual and collaborative expressions. Participants from different walks of life, including middle school children and adults, worked on the creation of public art installations based on a variety of exercises. One such exercise was the Surrealist parlor game cadavre exquis, or exquisite corpse, in which participants added to compositions in sequence, collaboratively generating full body portraits. Another exercise invited participants to first write about their sense of self before creating self-portraits on mirrored surfaces. The resulting works were then installed in a public site for audiences and participants to view.

The kinds of workshops facilitated by Cook engage participants in activities that are not very different from those of school children or therapeutic groups. The exquisite corpse game, for example, is commonly played in art or writing classes in elementary schools in developed countries, while the explorations of identity through

written and drawn/painted mediums are a staple in the toolkits of art therapists. In this sense, smARTpower is not especially original or innovative. Cook's project, however, is significant not due to the kinds of exercises it deploys, but because of the geopolitical context in which it is situated. Even though Nigeria is considered an emerging market, overtaking South Africa in GDP terms, its public healthcare and education infrastructure is under-developed to say the least. Local commentators have actually described these systems as dysfunctional. Needless to say, its arts and cultural sectors remain to be activated and nurtured. On top of all of this, Nigeria's ethnically and linguistically diverse population faces continuing sectarian violence and strife.

In this context, smARTpower is commendable for engaging diverse participants in collaborative and cooperative endeavors based on the expression and sharing of individual experiences. Since it is precisely differences in identity that fuel sectarian violence and strife, this brokering of encounters between different "senses of self" provided opportunities for building bridges across personal and cultural differences. Perhaps it even led to increased self-understanding and greater acceptance of differences. A survey of documentary footage suggests that this was the case for some participants. By encouraging and inviting participants to express themselves, and to witness and accommodate personal and cultural differences in the public sphere, smARTpower made a small contribution to democracy education. The importance of such education becomes clear when one takes into account the widespread corruption in Nigeria's political past and present. In these ways, Cook's project improved the quality of living for participants, making their immediate communities more convivial, at least for the duration of the project.

解包
Uitpak

艺术家：汉内莱·库切
地点：约翰尼斯堡美术馆的国王乔治街入口
形式和材料：石材
时间：2010 年
委托人：汉内莱·库切和南非国家艺术委员会
推荐人：杰奎琳·怀特

对托臂房屋的艺术重现，"解包"由南非艺术家汉内莱·库切于 2010 年在约翰尼斯堡美术馆的国王乔治街入口创建。虽然一开始作为临时展览，它建造所用的 330 块砂岩板的重量（以及相关移动附随成本）使之成为半永久展览。

汉内莱·库切在 Naboomspruit 的一个采石场发现神秘的条板，并随后识别为纪念牌匾，进行了喷砂，带有南非种族隔离时期的"南非自由斗士"的字样，创建用来在自由公园展示，在比勒陀利亚的纪念碑用来纪念在南非战争、第一次世界大战、第二次世界大战和南非种族隔离时期被杀害的人。但是，包括牌匾复制品在内的条板并未能在自由公园展示，至于原因，库切也无法确定。

汉内莱·库切自 2008 年开始从石材行业收集废材，自费购买了八吨石材，为采石场节约了把石材运送到垃圾填埋地的费用。把这些板条堆叠起来建造房屋，库切有意地安排石材，这样仅能看见部分名字，以此模仿有关石材出处的信息抑制。

汉内莱·库切最初为摄影师，在一个历史遗址的地图上看到一个识别的托臂小屋后，对此产生兴趣。这个建筑形式在西北半荒漠卡鲁地区被视为独特的形式，托臂小屋完全是由石材建成的，形状类似蜂窝。它们很可能是由早期的南非白人（游牧牧民）建造的，他们无法获得木材建造桁架。虽然库切拍摄这些房子有 10 年之久，部分是为了帮助保存被忽视的南非白人遗产，但是最终她觉得照相不足以表达跟这些结构互动的身体感官。

汉内莱·库切本身就是南非白人后裔，于 4 月 6 日安装了"解包"，1652 年的这一天被视为 Jan van Riebeeck 登陆现在的南非领土创建第一个荷兰殖民地的日子。首次建立这种托臂小屋的游牧牧民最可能是这些早期殖民者的后裔。虽然库切以前从未尝试堆叠石材，她发现好像天生能够堆叠"解包"。小屋足够两到三人弯腰进去。

在开幕式上讲话的 Mphethi Morojele 是自由公园的一名咨询建筑师，他指出原本为旅行者提供的传统农村建筑被建立在了城市。

作为使用废弃石材建造的结构，"解包"提出了可持续性的 WNET。它还引出了跟传承相关的许多方面，如库切自己的历史、没有充分保留的独特建筑形式以及阵亡的战士被抑制的遗产。

"小小"托臂屋

冯正龙

艺术家创作一件公共艺术作品的动机有很多，为表现自己对生活的强烈感受，为传达一种艺术理想，也为生存所需。但"解包"这件作品却带有神秘的厚重感，艺术家的灵感来源于一块纪念牌匾和记忆中的托臂小屋。纪念牌匾承载着太多故事和叙事，托臂小屋却用其独特的形式再次传承这些故事，纪念牌匾和托臂小屋在艺术家的手里汇聚在一起，在公共空间，用公共艺术的形式和方法，让公众再次铭记和接受那些过去的历史和遗留下的宝贵遗产。可以说，"解包"这件作品是令人深思的，公共艺术也是一种有效的形式，它让过去的已被遗忘的历史再次被发现，被回忆，被传达，也许，这些历史是屈辱的，但通过公共艺术的形式放置在公共空间中，其实，让公众看的不是它表面的形式，而是它隐藏的内容。

Artist: Hannelie Coetzee
Zone: Joubert Park, Johannesburg 2044, South Africa
Media/Type: discarded stone
Date: 2010
Commissioner: Hannelie Coetzee and National Arts
Council of South Africa
Researcher: Jacqueline White

An artistic recreation of a corbel house, Uitpak (Unpacking) was built by South African artist Hannelie Coetzee inside the King George Street entrance of the Johannesburg Art Gallery in 2010. Although intended as a temporary exhibit, the weight of the 330 sandstone slabs from which it is constructed (and the attendant cost of moving them) has led to it becoming a semi-permanent exhibition.

Coetzee discovered the mysterious slabs at a quarry in Naboomspruit and subsequently identified them as commemorative plaques, sandblasted with the names of South African freedom fighters from the apartheid era, which had been created for display at Freedom Park, the memorial monument in Pretoria that honors those killed in the South African Wars, World War I and World War II, as well as during apartheid. However, the slabs, which include some duplicates of plaques already at Freedom Park, were never delivered, for reasons Coetzee has been unable to discern.

Coetzee, who had been collecting waste from the stone industry since 2008, offered to rescue the eight tons of stone at her own expense, which saved the quarry the cost of transporting the stone to a landfill. In stacking the slabs to create the house, Coetzee intentionally arranged the stones so that the names were only partially visible, thus mimicking the suppression of information about the provenance of the stones.

Originally a photographer, Coetzee became interested in corbel huts after seeing one identified on a map of a historical site. An architectural form considered unique to the northwestern semi-desert Karoo region, corbel houses were built entirely of stone in a shape that resembles a beehive. They were most likely built by early Afrikaner trekboers (nomadic pastoralists) who didn't have access to wood to make trusses. Though Coetzee photographed these houses for ten years, partly to help preserve an overlooked aspect of Afrikaner heritage, she eventually came to feel photography was inadequate to convey the physical sensation of interacting with the structures.

Coetzee, who is of Afrikaner heritage herself, installed Uitpak on April 6, the date (in 1652) on which Jan van Riebeeck is thought to have landed on what is now South African soil to establish the first Dutch colony. The nomadic trekboers who most likely built the corbel huts would have been direct descendants of these early colonists. Although Coetzee had never attempted dry stone stacking before, she found that packing Uitpak came naturally to her. The hut is only large enough for two or three people to stoop inside.

Speaking at the opening, Mphethi Morojele, a consulting architect at Freedom Park, noted that a traditionally rural architectural form intended for travelers had been built in a city.

As a structure constructed of discarded stone, Uitpak raises issues of sustainability. It also evokes many layers of heritage, in regard to Coetzee's own history, that of a unique architectural form that was not being adequately preserved, as well as the suppressed legacy of the fighters who died.

沉默的大多数
The Silent Majority

艺术家：阿道弗斯·奥柏拉、奥卢索拉·奥托里
地点：尼日利亚拉格斯岛附近的马科科
形式和材料：摄影
时间：2010 年
委托人：阿道弗斯·奥柏拉和、奥卢索拉·奥托里
推荐人：杰奎琳·怀特

外来摄影师已经多次对位于拉各斯泻湖的 Makoko 进行记录了，Makoko
是尼日利亚的一个贫民窟，拥有 10 万人口，房屋都用柱子支撑。这部纪录
片如此持久以至于当地居民已经习惯了当他们在独木舟生活的时候，不断
有镜头记录他们的生活。当地居民许多是渔民，他们来自其他地方或邻近
多哥和贝宁在拉各斯寻求机会。

但是当尼日利亚拉各斯的摄影记者阿道弗斯·奥柏拉和视觉艺术家及儿童
权利活动家奥卢索拉·奥托里第一次访问 Makoko 时，他们决定采取不同
的策略。他们看到当地的孩子正面临极端贫困，他们在水里玩耍，水道可
兼作开放的下水道。如果这些被阿道弗斯·奥柏拉和奥卢索拉·奥托里认
为是"沉默的大多数"的孩子被给与相机并教他们摄影技术，给他们一个
机会来讲述自己的社区会怎么样呢？

阿道弗斯·奥柏拉和奥卢索拉·奥托里首先找到 Makoko 的负责人，并给
出了详细的提案。负责人对此项目感到非常兴奋，选择了 16 名儿童参与它。
从 2010 年 2 月到 9 月每周三和周五在哇哈那小学举行两个小时的讲习班，
这个小学拥有基本的艺术和设计元素，在摄影时要把握光和色彩的关系。
这些孩子的年龄从 12 岁至 18 岁，被给与了图画和绘画材料和书写纸以及
美能达牌单透镜反摄像机。结果呢？他们拍摄出亲密的家庭和邻居的照片
以及他们的生活方式，这些被认为是"内部资料"。

四个年轻人成功从讲习班毕业，这也成为了一个目标，即教孩子们实用的
创造技法可以成为一个收益流。两个讲习班的参与者显示出特殊的希望：
Monday Asokoro (17 岁) 和 Suliaman Afose (18 岁) 入围了尼日利亚阿
联酋电信业余摄影比赛的前 25 名，并且他们都能从摄影带来收入。此外，
Mary Awajinumi (17 岁) 在社区里的一个小的产院工作并记录新生儿，
Peter Onge (13 岁)，是一个理发师，也继续着他的摄影事业。

2012 年，当文件命令 Makoko 居民在 72 小时之内腾空他们的财产时，讲

习班的毕业生运用他们的摄影技巧记录被破坏以及重建后的社区，当局认为这个社区不方便从连接尼日利亚大陆和城市富裕区的桥上被看到。

奥柏拉和奥托里在讲习班上课，这个讲习班是他们通过从事商务工作来赞助的。奥柏拉的摄影作品已经在尼日利亚国家博物馆，以及伦敦泰特现代美术馆展出；他还做过美联社和英国广播公司 (BBC) 的摄影记者。作为一个全职工作室艺术家，Otori 从 2002 年开始教尼日利亚的孩子学习艺术。

奥柏拉和奥托里是尼日利亚人，他们的项目"沉默的大多数"与其他项目不同，在这个项目中，外国人给与穷孩子相机由来记录他们的生活，这就像美国摄影师 Lana Wong 在内罗毕的"反击项目"，英国艺术家 Zana Briski 在加尔各答"带相机的孩子"和英国摄影师 Julian Germain 的在巴西作品"流浪儿童"。来自"沉默的大多数"项目的照片在 2013 年被送进尼日利亚歌德学院进行展览。

定格的瞬间感动

冯正龙

摄影是一个记录生活并发现美的技术，在这个偏僻贫穷的平民窟里却成为"大多数沉默"孩子谋生的手段和工具，他们用自己的视角来观察生活的变化，记录和保存社区里的记忆。最终，四位年轻人成功从讲习班毕业，拍摄的作品受到了人们喜爱，但他们却继续着奥柏拉和奥托里的工作，继续为平民窟里的孩子教摄影，使这一实用的摄影技法让孩子们快乐的生活着。奥柏拉和奥托里的这个项目用公共艺术的方式，关注处于社会边缘的弱势群体和各种社会问题，在边缘视线中体现了人文关怀。通过孩子们拍摄的照片，我们可以看到他们生活的环境和他们和睦乐观的生活状态，虽然生活很贫穷，但只要心中充满爱和阳光，才能够拍摄出这样让观众感动的照片，这些爱和阳光就存在于这些"大多数沉默"孩子的心中。

Artist: Adolphus Opara and Olusola Otori
Zone: Makoko, near Lagos Island, Nigeria
Media/Type: Photograph
Date: 2010
Commissioner: Artist initiated
Researcher: Jacqueline White

Photographers who are outsiders to the community have repeatedly documented Makoko, a Nigerian slum built on stilts that houses over 100,000 people in Lagos Lagoon. The documentary presences is so persistent that the residents have become accustomed to an intrusive lens being pointed at them while they navigate their lives in wooden canoes. Many residents are fishermen who have come to Lagos seeking opportunity from elsewhere in Nigeria or from neighboring Togo and Benin.

But when Lagos-based Nigerian photojournalist Adolphus Opara and visual artist and child rights activist Olusola Otori first visited Makoko, they decided to take a different tact. They looked to the children, who are contending with extreme poverty, playing in water that doubles as an open sewer. What if these children, who Opara and Otori deemed "The Silent Majority," were given cameras, photography lessons, and an opportunity to shape a narrative about their own community?

Opara and Otori first approached the Baale (or chief) of Makoko with a detailed proposal. The Baale was excited by the project and selected 16 children to participate. The two-hour workshops, which were held on Wednesdays and Fridays from February through September of 2010 at Whahina primary school, covered basic elements of art and design, as well as principles of light and color as it relates to photography. The children, who ranged in age from 12 to 18, were given drawing and painting materials and writing pads, as well as manual film single lens reflex Minolta cameras. The result? Intimate portraits of their families, their neighbors, and their way of life—deemed an "Insiders' Account."

Four young people successfully graduated from the workshop, which also had as an objective teaching the children a practical creative skill that could become an income stream. Two of the workshop participants showed exceptional promise: Monday Asokoro (17 years old) and Suliaman Afose (18 years old) were both shortlisted among the final 25 in the Nigerian Etisalat Amateur Photo Competition, and have both been able to generate income from photography. In addition, Mary Awajinumi (17 years old) works and documents newborns in a small maternity home in the community, and Peter Onge (13 years old), who works as a barber, has also continued with his photography.

In 2012, when Makoko residents were served with papers giving them 72 hours to vacate their properties, the workshop graduates were able to use their photographic skills to document the destruction and later rebuilding of their community, which authorities considered inconveniently visible from the bridge that connects the Nigerian mainland to the city's rich island districts.

Opara and Otori taught the workshops, which they personally funded through their commercial work. Opara's photography has been exhibited at the National Museum in Nigeria, as well as the Tate Modern in London; he has also worked as a photojournalist for the Associated Press and the BBC. A fulltime studio artist, Otori began teaching art to Nigerian children in 2002.

That Opara and Otori are themselves Nigerian sets The Silent Majority project apart from other projects in which poor children have been given cameras by foreigners to document their lives, as in American photographer Lana Wong' s "The Shootback Project" in Nairobi, British artist Zana Briski' s "Kids with Cameras" initiative in Calcutta, and British photographer Julian Germain's work with street children in Brazil. Photographs from The Silent Majority project have since been organized into an exhibition at the Goethe-Institute Nigeria (2013).

革命之夜的亮光

Light in the Revolution Night

艺术家：卡里姆·贾巴里
地点：突尼斯卡塞林，Zouhour 市烈士广场
形式和材料：书法和涂鸦
时间：2012 年
委托人：艺术家发起
推荐人：里奥·谭

卡里姆·贾巴里的作品 "革命之夜的亮光" 是属于古阿拉伯文书法传统的一个公共作品。书法当然是书写的学科。任何形式的象征比喻，都将有着共同伊斯兰信仰的艺术社区中书面文本的盛行解释为可能是盲目崇拜的。在基督教社团的艺术史中也可以发现相同的观点，如破坏圣像运动。贾巴里将"革命之夜的亮光"描述成一个书法涂鸦作品——是书法和涂鸦的融合。

对这名突尼斯裔加拿大艺术家来说，这个项目是对突尼斯 "阿拉伯之春" 的欲望和暴力做出妥协的一种途径，尤其是为获得公民自由和更好的生活条件而公开游行示威的抗议者（包括他的叔叔）被政府军击毙的烈士广场事件。按照贾巴里自己的说法，当他从他居住的加拿大目睹这些事件时，他感到无助和愤怒，决定在一年后（2012 年）来到该广场，"在他们被枪杀的相同地点，通过光书法重塑他们的名字，从而唤起对故去的这些勇敢人民的记忆。"

关于该项目的光书法部分，这名艺术家在夜里打开手电筒，将书法 "画" 在了现场：他将三角架式相机设置成长曝光模式，将电脑灯凝聚成文本 / 图像。这些结果从视觉上看是颇为壮观的，人们能想象通过光来绘画的公开演出同样吸引了路人 / 观众。在白天，贾巴里则从事其项目的另一个部分，主旨为 "希望在这里诞生" 的烈士广场中一个更为传统的街头艺术作品。事实证明，这一项目在这个区域的年轻人之中颇受欢迎，有很多人加入了这名艺术家，共同在墙壁上作画。

卡里姆·贾巴里力图将年轻人与阿拉伯文化和语言更加密切地联系起来，从而纪念突尼斯阿拉伯之春。总的来说，似乎 "革命之夜的亮光" 是相当成功的。尽管利用光在空间 / 时间中创造无形图片，可以在有 "烟火" 的任何地方看到，但贾巴里通过阿拉伯文书法的风格，巧妙地诠释了这种受欢迎的消遣活动，从而创作令人难忘的、美观的工艺品（照片）。该项目也拉近了这位艺术家与其祖国文化的距离，使他能表达出身在遥远的、相

对安全和稳定的加拿大的他会与仍在苦难中挣扎的故乡人民休戚与共这么一种情感。

值得一提的是，该项目使贾巴里也采取了其他几项举措，包括"走向光明"——突尼斯最长的涂鸦墙，以及"街道"——2013 年在卡塞林（艺术家故乡）举办的一个嘻哈艺术节。街道汇聚了突尼斯的说唱歌手、霹雳舞者以及视觉艺术家，他们为卡塞林年轻人表演，举办讲座。国际新闻媒体的积极新闻报道，如 BBC（英国广播公司）和半岛电视台，表明了这些活动的成功。据半岛电视台的新闻报道："出席讲座的那些年轻人说，他们会继续实践其艺术，不论是霹雳舞、说唱、D 舞还是涂鸦。"

异乡的艺术

冯正龙

艺术的表现形式不是一成不变的，在艺术实践的路上充满尝试和挑战。"革命之夜的亮光"这件作品将书法、涂鸦、摄影融合在一起，作品呈现出奇特、神秘的效果，不仅唤起了人们的记忆，同时也弘扬了阿拉伯文化，让身在异乡的游子和移民找到一条途径与故乡人们休戚与共，心在一起。可以想象到，当你深夜走在异乡的国度里，看到了空间中用书法书写的闪闪发光的本国语言，你会在惊讶之中感到不可思议，就像走进梦境回到了阔别已久的家，惊喜之余又不敢相信眼前的一切。公共艺术有时给予公众的就是这样的感动，与众不同的奇特形式、简单中又蕴含深意的理念、挑战中又富有争议的效果，这是一种奇特的交流、互动之旅，也是一种让你难以忘怀的体验之感。

Artist: Karim Jabbari
Zone: City, Kasserine, Tunisia
Media/Type: A fusion of calligraphy and graffiti.
Date: 2012
Commissioner: Artist initiated
Researcher: Leon Tan

Karim Jabbari's Light in the Revolution Night is a public project belonging to the ancient tradition of Arabic calligraphy. Calligraphy is, of course, the discipline of handwriting. The predominance of the written text in the arts of communities sharing the Islamic faith has been explained by the conception of symbolic figuration of any kind as always potentially idolatrous. The same conception can also be found in the art history of Christian societies, for example, in Byzantine Iconoclasm. Light in the Revolution Night is described by Jabbari as a work of "calligraffiti," a fusion of calligraphy and graffiti.

For the Tunisian-Canadian artist, the project was a means to come to terms with the desires and violence of the Tunisian Arab Spring, particularly the events of Martyrs Square, in which many protestors, including his uncle, demonstrating publicly for civil liberties and better living conditions, were killed by government forces. By his own account, Jabbari felt helpless and angered by the events he witnessed from his adoptive Canada, and decided to visit the square a year later in 2012, "to revive the memory of the fallen brave people by recreating their names with light calligraphy in the same spots they were shot dead."

For the light calligraphy component of the project, the artist "painted" calligraphy onto sites with a flashlight at night, condensing the moving light as a text/image through long exposure

with a tripod-mounted camera. The results were visually spectacular, and one can also imagine that the public performances of painting with light were equally arresting for passersby/audiences. During the daytime, Jabbari worked on the other component of his project, a more conventional street art piece in Martyr Square with the message "Hope was born here." This proved to be extremely popular among the area's youth, many of them joining the artist to paint the wall.

Jabbari sought to connect young people more closely with Arabic culture and language and to memorialize the Tunisian Arab Spring. Taken as a whole, it seems that Light in the Revolution Night was reasonably successful. While the use of light to create immaterial images in space/time can be seen wherever "sparklers" are available, Jabbari skillfully interpreted this popular pastime through the idiom of Arabic calligraphy to produce haunting and aesthetically pleasing artifacts (photographs). The project also brought the artist closer to his own culture, enabling him to express solidarity with what could have remained a struggle witnessed from afar in the relative safety and stability of Canada.

It is worth noting that this project gave rise to several other initiatives by Jabbari, including Towards the Light, the longest graffiti wall in Tunisia, and Streets, a hip-hop art festival held in Kasserine (the artist's hometown) in 2013. Streets brought rappers, break-dancers and visual artists together in Tunisia to perform and conduct workshops for Kasserine youth. The success of these activities is indicated by positive media coverage in international news outlets such as BBC and Al Jazeera. According to the Al Jazeera news report, "Those youth who participated [in the workshops] said they will keep practicing their art, whether breakdancing, rap, DJing, or graffiti."

梦之城
Dream City

艺术家：塞尔玛、索菲恩·桂塞
地点：突尼斯及斯法克斯
形式和材料：当代艺术展
时间：2012 年
委托人：未知
推荐人：里奥·谭

"梦之城"为以两年一次的频率在突尼斯公用场所举行的当代艺术展览。
第三版于 2012 年的 9 月及 10 月在突尼斯的麦地那以及斯法克斯两地举行。
2012 年的主题为"面对自由的艺术家们"，此次展览由来自突尼斯、阿尔
及利亚、刚果民主共和国、贝宁、埃及、伊朗、中国、西班牙、法国、巴
勒斯坦以及荷兰的艺术家们带来的以现场为主题特色的作品构成。参与其
中的艺术家们选择了众多系列的媒介，但最成功的或许是那些坐落于开放
性公共空间的作品以及那些强调观众参与度的艺术作品。

罗杰·博纳特的"公共领域"则是坐落于梦之城的参与性公共艺术的出色
代表。观众需事先注册，确保获得位置，并能按约定的时间出现在法庭街上，
在这里，观众可以选择交换个人价值选项并以此获得一副耳机。初步交易
完成后，参与者们会游览公共空间，回答通过耳机讲话的叙述者们提出的
问题。比如，回答完有关出生地的问题，观众们会以左转或右转的方式相
互分散开，分开的标准取决于他们是否为突尼斯人或外来人员。"公共领域"
通过问题作为参与度激发因素，以此为参与性社会艺术设立舞台。观众们
则通过更换不同的服装来扮演 / 体验不同的角色——抗议者、医护人员及
安保人员。他们还会受到启发而对人体及历史是如何与社会及空间分化产
生共鸣并作出反映。

Raeda Saadeh 的"愿望树"则具有类似的参与性空间。在此次表演中，
观众们面临着引人注目的视觉景象：一名女性穿着一身看起来长无止境的
白色连衣裙（直径 60 米），被别的不断分发彩色手绢及标记的妇女们环绕着。
观众们被邀请写下他们的愿望，然后将手绢系成结并扔到白色长裙上。据
艺术家所言，该表演的初衷是对处于该地区的妇女们进行角色方面的重新
想象。强大的、核心的妇女形象与性别方面的规范形成鲜明的对比。鉴于"面
对自由的艺术家们"这一主题，艺术导演们构思出以政治为主题的节目并
不足为奇。当地大的环境当然是极其显著的，在当时突尼斯环境中，穆罕
默德·布瓦吉吉在 2010 年将自身置于绝望的境地，直接点燃了所谓的"阿
拉伯之春"。来自博纳特及 Saadeh 以及很多其他地方的作品，貌似都在

突尼斯及斯法克斯的公共场所中创造了亲密无间的相遇。记录录像显示，当时参与到梦之城的观众们的热情度极高以及此次表演如何在短时间内转变了突尼斯的古麦地那及斯法克斯进。

如果我们将这种两年一度的政治讯息设定为一种有关居民对能够提供团结度而非导致布瓦吉吉自身毁灭的疏远的权利的所求，则公众亲密性的出现则可被视为制造政治诉求一致性的合理方式。在两个麦地那公共空间通过各种不同的艺术发明而提供的各种不同的亲密相遇，如果没有一种可靠的信任氛围的话，则几乎不可能发生。正是信任的品质所在，使得来自突尼斯地区的不同个人以及团体之间的团结性在未来可能实现。

梦幻之地
冯正龙

也许，对于政治诉求传达最有效的方式应该是公共艺术的方式，温和又不具有威胁性是其良好的特征。尤其在非洲，营造梦幻般的城市氛围，让不同种族、地域、阶层的人们能够摈弃差异，打破隔阂，在融洽和谐的氛围中互动、交流是如此之难，更难想象。但"梦之城"却冲破我们想象的藩篱，以当代艺术展的形式，自由的主题，汇聚来自全世界的著名艺术家，以自己擅长的创作媒介，在公共场所中进行公共艺术作品创作。这些艺术家都深知公众互动参与的重要性以及公共艺术的问题解决意识，因此在创作作品之前，会通过多种渠道了解这个地区经济、政治、文化等相关的一切，这一切都为作品的最终呈现打下良好的前提与基础。"梦之城"与其说是一个当代公共艺术展，不如说是公众参与艺术创作的盛会。

Artist: Selma &SofianeOuissi
Zone: Tunis and Sfax, Tunisia
Media/Type: Contemporary art
Date: 2012
Commissioner: Notprovided
Researcher: Leon Tan

Dream City is a Tunisian biennial of contemporary art in public space. The third edition took place in September and October 2012 in the Medinas of Tunis and Sfax. The theme for 2012 was Artists Facing Freedom, and the biennial consisted of site-specific works by artists from Tunisia, Algeria, the Democratic Republic of Congo, Benin, Egypt, Iran, China, Spain, France, Palestine, and the Netherlands. Participating artists chose a variety of mediums, but the most successful were perhaps those sited in open public spaces, and those emphasizing audience participation.

Roger Bernat's Public Domain was an outstanding example of participatory public art in Dream City. Audiences registered beforehand to secure a place and turned up at an appointed time on Tribunal Street where they were given a choice to exchange a personal item of value for a pair of headphones. After this initial transaction, participants navigated the public space in response to questions from a narrator speaking through the headphones. For example, in response to a question about place of birth, audiences divided themselves up by going right or left, depending on whether they were Tunisian or foreign. Public Domain used questions as action triggers, thereby setting the stage for a participatory social choreography. Audiences found themselves acting/experimenting with different roles—as protesters, medical staff, and security forces—through the inclusion of costume props. They were also provoked to reflect on how human bodies and histories coincided with social and spatial divisions.

Raeda Saadeh's Tree of Wishes had a similar participatory dimension. In this performance audiences were confronted with an arresting visual spectacle, a female character in a seemingly unending white dress (60-meters in diameter), attended by other women who gave out colored handkerchiefs and markers. Audiences were invited to write down their wishes and then to tie the handkerchiefs into knots to throw onto the white dress. According to the artist, the performance was intended as a re-imagination of the role of women in the region. The powerful, central, female character presumably contrasted sharply with gendered norms.

Given the theme Artists Facing Freedom, it is unsurprising that the artistic directors conceived of the programming in political terms. The regional context was, of course, highly salient, Tunisia being the place where Mohamed Bouazizi set himself alight in desperation in 2010, literally igniting the so-called Arab Spring. Works like Bernat's and Saadeh's, and many of the other pieces in the biennial, seemed to create intimate encounters in public spaces in Tunis and Sfax. Video documentary footage provides evidence of enthusiastic audiences taking part in Dream City, and of the biennial temporarily transforming the ancient Medinas of Tunis and Sfax.

If we conceive of the biennial's political message as a claim about the right of citizens and residents to publics that afford solidarity, rather than the kind of alienation that contributed to Bouazizi's self-immolation, the emergence of public intimacy may be taken as reasonable measure of the consequentiality of political claims making. The kinds of intimate encounters facilitated by various artistic interventions in the public spaces of the two medinas would not have been possible without a reliable atmosphere of trust. It is this quality of trust that similarly founds the possibility of solidarity between different individuals and constituencies in the future Tunisia.

小型革命

La Petite Révolution

艺术家：布雷特·莫里
地点：南非约翰内斯堡，妮洛丝基地雕塑公园
形式和材料：雕塑
时间：2009 年
委托人：未知
推荐人：里奥·谭

布雷特·莫里最出名的作品，或许是标题为"矛"的对南非总统雅各布·祖玛的极具争议性的画像。该作品与莫里的其他作品一同陈列于古德曼展览馆中，其中包括"被暗箱操作的非洲国民大会"宣传海报，海报中列有类似这样的标语："曼德拉，我们要求得到威士忌、宝马车及贿赂"，"告诉我的子民我爱他们，告诉他们，他们必须为芝华士、战场之狼以及回扣而作出不懈的努力"。不出所料，此次展览引发了由非洲民族会议（祖玛政党组成）立案的法律诉讼。作品同时刻画了一名艺术家的幻想随着南非向后种族隔离时代的转变而破灭的过程。

布雷特·莫里自己曾言："尽管瓦解种族隔离的政治变迁及企图仍在，我个人对大规模的社会隔离仍持批判的态度。"在一次采访中，他重新讲述了在 1980 年期间他作为文化激进者的一次经历，当时他面对的是那些被伪革命典型行为攻陷的一群同胞们，他们在枝繁叶茂的郊区喝着拿铁咖啡，以一种玩世不恭的口吻说着类似"亲爱的请把黄瓜三明治递给我，我们正在进行革命！"这样的话。这种讽刺方式，与维多利亚上层阶级（在英属印度地区将黄瓜三明治普及化的人群）对比，事实上正是第二届 IAPA 考虑的公共艺术主题。

通过在长 30 米、高 1.3 米的帆布上作画，"小型革命"实为布雷特·莫里对 2009 年展览资源的一份贡献——坐落于尼罗科斯雕像公园风景区的当代雕塑。公园前身为一家商业运作鳟鱼农场，距离约翰内斯堡一小时的车程，公园由 15 公顷的自然风景园组成，里面有通道，一个湖及各种成年树。人们不禁想，是不是公园的田园气息促发艺术家将其描述为"一种美丽田园风景区"。鉴于我们对艺术家政治学识的了解，这种强调舒适与休闲的风景区，可能会与大的社会改革格格不入。

那么很清楚的是，"小型革命"并非对和谐自然的温和赞美，而是对社会现状的尖锐讽刺。尽管所选字体传达着一种文雅礼节信息，信息本身却承载着公众诟病的氛围。讯息的不协调性，与在这种广袤而贫瘠的大地上建

立的绿洲公园的不协调性形成回应。雕刻的内容无时不刻提醒着观众，南非社会矛盾的顽固性以及后隔离时代的不一致性。

某一项目能否改变社会，仍是大家争持不下的话题。艺术在社会中的角色问题，同样难以回答。那么，"小型革命"又能否毫无疑问地传达一种可以获取的对社会进行批判的信息，其能否以一种对公园在整个更为宽广的物理环境中所处的位置的敏锐性以及对社会虚构的敏锐性进行批判？人们不禁会想，艺术家会以何种方式，将其赖以生存的鉴定和批判生涯问题取而代之呢？

异样的批判视角
冯正龙

以讽刺的口吻，时代的语境，批判的角度，公共艺术的媒介来传达对社会现状的不公，这对公众会产生什么影响是艺术家需要考虑的问题。有时候，这一切可能只代表了艺术家的一家之言；有时候，这一切代表了公众的心声；有时候，这一切只是对相关部门发发牢骚，抱怨而已；有时候，这一切还真引起了轰动的"小型革命"。南非社会充满尖锐的社会矛盾，社会改革迫在眉睫。艺术家对南非社会现状的关注，只能通过自己的方式来表现，正面的宣传太温和，讽刺批判调侃的方式可能更具效果，最终，艺术能否改变社会，艺术在社会中又承担着什么角色，这都需要时间来检验，需要时间来回答。"小型革命"的表达方式在形式上是温和的，但背后所表达的内容是深刻的，作品艺术形式的冲突正蕴含着南非社会强烈的矛盾。

Artist: Brett Murray
Zone: Nirox Foundation Sculpture Park, Johannesburg, South Africa
Media/Type: Contemporary Sculpture
Date: 2009
Commissioner: No provided
Researcher: Leon Tan

Brett Murray is perhaps best known for a controversial painting of South African President Jacob Zuma entitled The Spear. The Spear was exhibited in the Goodman Gallery alongside other works by Murray including manipulated African National Congress posters with captions like "Amandla, we demand Chivas, BMWs and bribes," and "Tell my people that I love them and that they must continue the struggle for Chivas Regal, Mercs and kick-backs." Unsurprisingly, the exhibition provoked legal action by the ANC (Zuma's party). The works also portray an artist disillusioned with South Africa's transition to a post-Apartheid era.

Murray himself has said, 'I am critical of the massive social disparities that, despite political changes and the attempts to dismantle Apartheid, remain in place.' In an interview, he recounted the experience of being a cultural activist in the 1980s, confronting compatriots who lapsed into behaviors characteristic of the 'pretend revolutionaries' and 'latte drinking Che Guevaras' of the 'leafy suburbs' with the cynical expression, 'Pass me the cucumber sandwiches darling, we are having a revolution! This jibe, with its reference to the Victorian upper classes (who popularized cucumber sandwiches as far as the British Raj) is in fact the subject of the public art under consideration for the second IAPA.

Executed with painted text on a canvas 30 meters long and 1.3

meters high, La Petite Révolution is Murray's contribution to the 2009 exhibition Sources – Contemporary Sculptures in the Landscape in the Nirox Foundation Sculpture Park. Formerly a commercial trout farm, the park is located an hour outside of Johannesburg, and features 15 hectares of landscaped nature complete with walkways, a lake and mature trees. One wonders if it was the park's idyllic atmosphere that prompted the artist to describe it as "a beautiful Arcadian landscape." Given what we know of the artist's political leanings, such a landscape, with its emphasis on comfort and leisure, would be totally opposed to the circumstances of social revolution.

Clearly then, La Petite Révolution is no facile celebration of the harmony of nature, but rather, another acerbic attack on the social establishment. While the font chosen conveys a sense of genteel propriety, the message itself has the air of a public rebuke. The incongruity of the message echoes the incongruity of the park as a manufactured oasis in an expanse of barren earth. The monumental inscription reminds audiences of the persistence of social contradictions and incongruities in post-Apartheid South Africa.

Whether or not the project actually changes society remains a matter for debate. The related question of the role of art in society is similarly difficult to answer. There can, however, be no doubt that La Petite Révolution carries an accessible message of social criticism, and that it does so with a keen sensitivity to the location of the park in the wider physical environment as well as in the social imaginary. One wonders what the artist might propose in place of the problems that he has made a career out of identifying and criticizing.

喀麦隆杜阿拉视觉艺术展 2010
Salon Urbain de Douala 2010

艺术家：多名艺术家
地点：喀麦隆杜阿拉的各种地区
形式和材料：城市公共艺术节
时间：2010 年
委托人：杜阿拉
推荐人：里奥·谭

凭借 300 万人口，杜阿拉为喀麦隆最大的城市。该城市坐落于武里河出口，河流通向几内亚湾。自喀麦隆先后于 1960 年及 1962 年摆脱法国及英国的占领而获得独立之后，城市获得了迅速发展，一举成为非洲最富有最昂贵城市之一，并成为民族的商业资本。不幸的是，喀麦隆也因其高失业率、高街道犯罪及财产犯罪率而出名。"SUD"是由艺术中心 doual' art 于 2007 年成立的三年一度的城市公共艺术节。此次为针对公共艺术的国际性奖励在 2010 年发起的第二届 SUD 盛典。

该盛典的策划目的是改变杜阿拉并从文化层面丰富杜阿拉。其首创性是来自 doual' art 艺术中心的艺术家，玛丽琳·杜阿拉 - 贝尔·迪迪埃·肖布众多努力的一部分，目的是扶持艺术家并在公共空间制造艺术品。2007 年以及 2010 年出版的作品都包含永久性及短暂性作品，以喀麦隆及国际艺术家为特点。所有作品都是围绕包罗万象的策划主题——水来创作的。

Face à l'eau 是 Salifa Lindou 向 "SUD 2010" 贡献的作品，作品由五块木质、金属材质垂直板以及螺纹塑料板组成，上面都打上了开放式正方形窗户。3.7 米的高度，从远处观看时，这些外形迥异的屏障会让人产生一种单一屏障的幻觉，近看的话，则向观众提供了一种被屏障的俯瞰河流的感觉。Ties Ten Bosch 耗时七周(10 月至 12 月)创造了供其自身居住的艺术住宅，住宅位于没有自来水的城市相对贫穷的 Ndogpassi III。艺术家的目标是使用现场的日常用品作为建筑材料。他的干预属于以社会互动和分析为基础的不断发展壮大的关联艺术潮流。就像很多类似作品一样，该作品与我们通常认为的社会作品或社会行为相差无几。居民们被留以新的桥梁，从而能够更方便地通往集水现场。

Pascale Marthine Tayou 则建造了锅具（历史上被家庭主妇用于烹饪和饮食的工具）图腾，并将其安装在了城市的闹市区。Tayou 的雕塑，尽管获取的荒谬感要胜过每天的车水马流，仍然会被视为烹饪及集水劳力的家园——传统上妇女及女孩们的领域。

有关策划主题，Simon Njami 写道："水并不是一种永恒的礼物。它是一种我们每天必须进行的战斗。它是我们世界的未来。"每名参与其中的艺术家们，都以其独特的方式，阐释着各自以水作为现场实质及奋斗目标的潜意识。在一系列在公共场合就此潜意识进行表演的过程中，"SUD 2010"以水循环利用作为公共主题的讨论中作出了一定贡献，该等讨论为围绕对水源的公共性及私人性控制和分配而进行每日政治诉求及争论提供了关键的平台。

"在水一方"
冯正龙

这一项目，呈现了公共艺术一贯的"地方重塑"的主题意识。通过城市艺术节，使公共艺术作为地区更新与文化复兴的重要推动力，以公共艺术为手段，以地区公共空间为平台，创造一个优美、舒适的生活和居住环境，让各阶层都能享受艺术带来的美好，营造良性循环的地区发展氛围。

艺术家围绕主题，从喀麦隆文化出发，用各种材料在不同的公共场所创作公共艺术作品，希望通过艺术来干预公众的日常生活和混乱的社会现状以及沉淀社会的浮躁气息，让公共艺术成为地区文化复兴和公众生活品质改善的渠道，成为社会转变的可能。从表面看，项目是极其重视水的重要性，的确，水是人类不可缺少的一部分，没有水可能也不会有一切，此项目就是以此为项目开始的主题，让水的内涵与重要性深深呈现在每一个公共艺术作品中，让它成为社会发展的潜在的动力和社会问题传达的手段。

Artist: Multiple Artists
Zone: Various locations in Duala, Cameroon
Media/Type: The city's triennial festival of public art
Date: 2010
Researcher: Leon Tan

With its population of over 3 million, Douala is Cameroon's largest city. It sits at the mouth of the Wouri River, opening out into the Gulf of Guinea. Since Cameroon achieved independence from France and the UK in 1960 and 1961, the city has developed rapidly, becoming the business capital of the nation and one of the wealthiest and most expensive cities in Africa. Unfortunately, it has also become known for its high rate of unemployment, and for street and property crimes. Salon Urbain de Douala (SUD) is the city's triennial festival of public art founded in 2007 by the arts center doual' art. It is the second edition of SUD in 2010 that is under consideration for the International Award for Public Art.

The triennial was conceived as a means of transforming and culturally enriching Douala. The initiative was a part of wider efforts by Marilyn Douala-Bell and Didier Schaub, of doual' art, to support artists and to produce artworks in public space. Both the 2007 and the 2010 editions included permanent and temporary artwork, and featured Cameroonian and international artists. Works were commissioned on the basis of an overarching curatorial theme—water.

Face à l'eau was Salifa Lindou's contribution to SUD 2010, consisting of five vertical panels made of wood, metal, and corrugated plastic sheets, punctuated with open square windows. Set at 3.7-meters

tall, these physically disparate screens created the illusion of a single screen when viewed from a distance, and up close, provided audiences with a screened view overlooking the river.

Ties Ten Bosch created his own artist residency for seven weeks (October to December) in Ndogpassi III, a block in the poorer part of the city without running water. The artist's objective was to use the daily life on site as material. His intervention belongs to a growing tide of relational art based on social interaction and analysis. Like many such art pieces, this one is indistinguishable from what we would normally consider social work or social action. The residents are left with a new bridge enabling better access to sites for water collection.

Pascale Marthine Tayou constructed a totem of pots historically used by housewives for food and drink, installing it in a busy intersection of the city. Tayou's sculpture, despite the sense of absurdity it acquired towering over the everyday flows of traffic and pedestrians, might be considered a homage to the labor of cooking and water collection, traditionally the domain of women and girls.

Concerning the curatorial theme, Simon Njami wrote: "Water is not an eternal gift. It is daily battle that should concern each of us. It is the future of our world." Each of the participating artists, in their own way, elaborated on this consciousness of water as a site of sustenance and struggle. In staging this consciousness in public sites, SUD 2010 contributed to the circulation of water as a topic in public discussions, such discussions being a crucial platform for everyday political claims-making and contention regarding the public or private control and distribution of water.

马拉喀什装置艺术

Marrakech Installations

艺术家：克莱门斯·贝尔
地点：摩洛哥马拉喀什
形式和材料：装置
时间：2011 年
委托人：Awaln 艺术节
推荐人：里奥·谭

马拉喀什是非洲国家摩洛哥西北部第四大城市。该地区自新石器时代开始就一直有本地居民居住，城市是 1062 年由阿布·伊本·奥马尔始建。独特的砂岩墙建立于 1122 年至 1123 年间，也是马拉喀什被命名为红色城市或者赭石城市的原因。今天的马拉喀什是一个重要的经济中心与一个面向游客，餐饮业发达的充满活力的工艺品产业城市。

马拉喀什装置由八个被放置在城市不同角落的公共空间的雕塑组成，包括 2011 年 6 月 Awaln 艺术节上被策展人卡里尔·塔梅尔受邀参展的作品 Medina quarter（坚固的老城）。Awaln 艺术节是一个面向公众和街头艺术的节日，在艺术家向公众展示他们作品或者展览的时候展出，现在是第八届。克莱门斯·贝尔的作品是临时为此特别制作的，作品由从当地商店里购买的纸板和颜色拼装制成。

托马斯·赫希霍恩的公共装置艺术让人想起手工制作的美好，这八件雕塑的特色是用调色板去重现本地建筑和老城红墙的颜色，然后用抽象的几何图形和曲线形式表现这些建筑。临时搭建的外观和这些作品转瞬即逝的材料特性强化了周边建筑的年代感，同样也提醒了马拉喀什部分建筑物显而易见的年久失修。装置和装置背后的年代明显老旧的建筑物和一些当代和现代建筑以及城市发展作对比，这些作品看上去就像是这个环境中的一部分，或者从环境中剥离开来。贝尔或许只是简单的想要为了展示一种视觉冲击（这个装置仅仅耗费 2000—3000 欧元），和想要为了追溯他材料的当地性的特征（让工作的生态印记减少到最少）。

由于关于这个艺术节和参加这个艺术节人员的具体信息、艺术家本人资料的缺乏，很难去全面评估这个项目的成功与否。在形式上，这些雕塑没有明显的展示原型或者有很大的创新，也没有非常绝对的表达出这个作品想表达的概念（大部分都是，比如说艺术家赫希霍恩的这个作品）。他们没有

纪念某一个事件，他们也没有带动观众超越视觉和享受的参与性。可以说他们用他们的装置带动了人们对于不同公共艺术空间的认知，或许和艺术家本身诉说的一样为路人提供了一个感官的惊喜，即"融合的视角看城市"。

装置的易读性

冯正龙

装置是公共艺术的一种独特形式之一，常因体积巨大外形奇特得到公众的关注，但也由于一些装置外形丑陋，使其褒贬不一。我们知道，每一件艺术作品背后都带有艺术家对生活与社会的认知，带有艺术家个人思想的传达，但与众不同的是，公共艺术是面向公众的，放置在公共空间中的艺术作品，它带有更多的是一种公共性质与公共意识，可能需要艺术家舍弃自己的一些主观想法，要更加客观的去创作一件公众喜闻乐见的作品。这种"舍弃"并不是限制艺术家的想象力与洞察力，而是把关注点转移到公众的视角，这也体现出如今公共艺术的方法论转向与趋势。"马拉喀什装置艺术"是通过放置在马拉喀什不同场所的装置，为人们提供另外的一种视角观察公共空间，让他们用一种包容的心态看城市的变化和发展，值得强调的是，这些装置作品都是在考虑到城市周边环境、历史以及公众的接受度的前提下创作的。

Artist: Clemens Behr
Zone: Marrakech, Morrocc
Media/Type: Installations
Date: 2011
Commissioner: Awaln`Art Festival
Researcher: Leon Tan

Marrakech is the fourth largest city in the Northwest African nation of Morocco. The region itself has been continuously inhabited since the Neolithic period, with the city founded in 1062 by Abu Bakr Ibn Umar. The distinctive sandstone walls erected in 1122-3 gave Marrakech the name Red City or Ochre City. Today it is a major economic centre with a vibrant craft industry catering primarily to tourists.

Marrakech Installations consisted of eight sculptures installed in public sites throughout Marrakech, including the Medina quarter (old fortified city), at the invitation of curator Khalil Tamer of the Awaln`Art Festival in June 2011. The Awaln`Art Festival is a festival of public and street art, during which artists present free shows or exhibits for members of the public. It is now in its eighth edition. Behr's works were temporary and site-specific, constructed out of cardboard and paint purchased from local stores.

Assembled with a do-it-yourself aesthetic reminiscent of the public installations of Thomas Hirschhorn, the eight sculptures featured a palette reflective of the colors of local buildings and the old

city's red walls, and took on abstract geometric and curvilinear forms, with the former predominating. The makeshift appearance and ephemeral nature of the works highlighted the age of the surrounding buildings as well as parts of Marrakech that were clearly in bad repair. Set against the backdrop of a much older architecture, as well as modern and contemporary constructions and real estate developments, the works seemed at once a part of, and separate from their environment. Behr may be commended for achieving a high degree of visual impact with a minimum of means (the installations only cost 2,000-3,000 Euros), and for sourcing his materials locally (minimizing the work's ecological footprint).

It is difficult to fully evaluate the success of this project given the paucity of information available on the festival and its attendance figures, and the lack of detailed information from the artist himself. Formally, the sculptures were not especially original or innovative. Nor did there appear to be a strong conceptual premise to the work (as there usually is, in the case of an artist such as Hirschhorn, for instance). They did not commemorate an event, nor did they stimulate audience participation or engagement beyond visual registration and enjoyment. What can be said is that they stimulated different perceptions of the public spaces in which they were installed, perhaps also providing passers-by with a sensation of surprise at "seeing the city in total fusion" as reported by the artist.

东亚
East Asia

东亚

以中国为主的东亚经济和文化曾经在世界上长期居于领先地位，无论是哲学、数理亦或是各种伟大的发明都对世界的发展有着举足轻重的作用。但近百年来，曾经繁荣富饶的东亚由于世界战争的破坏，历经动荡。而如今随着科技的发展，东亚正以风驰霆击之势再次试图拨动世界格局。无论哪个国家，在经济崛起的过程中总是不可避免的面临环境破坏、资源滥用等社会问题，为了警醒人们也为了避免重蹈发达国家覆辙；第二届国际公共艺术评奖中的东亚板块公共艺术案例基本围绕着能源利用、环境保护、人文情怀，乡土重塑的话题展开。

艺术的进步推动了城市的发展和人文精神的提升。无论是哪一种形式的城市艺术都是对城市化生活方式的思索。亚洲地区的公共艺术案例充满张力与矛盾，国家的发展提倡科技使人类进步，但发展科技的过程中不可避免的造成人文精神的缺失。如今的亚洲处于舆论的顶峰，矛盾的边缘。而公共艺术作为城市公共空间中的文化审美存在实体，作为民主意识的凝聚、政府宣扬话语权的工具，协调着城市空间中人与自然、民众和政府之间的微妙关系，用反讽或乌托邦式的艺术语言调着各种虚实矛盾，尝试利用语境意识创造科技与人文、发展与保护共存的理想生活环境和文化状态。（高浅）

魔方塔
Cube Tower

艺术家：阿方索·克施文德
地点：中国长春，长春国际雕塑公园
形式和材料：雕塑
时间：2012
委托人：长春国际雕塑公园
推荐人：里奥·谭

阿方索·克施文德是 2012 年长春国际雕塑大赛的获奖选手之一，意味着他受托可永久性将其代表作"魔方塔"，一个动态雕塑永久性安装在中国东北部吉林省长春市的长春雕塑公园。事实上，克施文德早在 2011 年的大赛中就曾获奖，因此其后来的"魔方塔"连同另一部被其称为大联盟的动态作品，一起安置了 92 公顷的园林公园中。公园管理部门以及市政府（受长春市政委托）在开发园林建筑的同时，负责建造"魔方塔"，塔高 8 米，外形由五块具有折射作用的方块相互堆积而成，从塔底到塔顶位置，尺寸递减。整个造型结构由不锈钢组成，在风力带动下，每个方块逆向旋转。据艺术家称，该作品为现存于中国的少量动态风力雕塑作品之一。

艺术家保罗·克里曾这样评价："艺术并不是对可见物的再创，相反，艺术创造了可见物。"既然艺术作品并不一定代表任何事物，而是致使不可见事物变得可见，那么在这种意义上讲，我们或许就能理解克施文德的动态雕塑了。作为一种抽象形式，在旁观人员看来，"魔方塔"是一种开放式的、给人们留下象征性解释的作品。观众所注册的，正是不可见的：地球不同方位因不均匀受热而产生的空气流通。将其安装在靠近湖边、经过精心修剪的花园里，不可见事物的注册有利于对自然和谐的感知，即众多不可见力量的和谐，如热和风以及可见材料比如大地、水和钢铁的和谐。"魔方塔"是一部经过精心制作的公共雕塑作品，与公园里无数其他雕塑良好地融合在了一起。其中所蕴含的别致，包括其自身通过与不可见力量的相遇而产生沉思体验的能力。

这种对自然的体验以及与自然力量的相遇，从辩证的观点看，属于一种中国古代对水墨山水画的审美传统，即使雕塑本身其实更像是对西方美学（特别是希腊及美国现代美学）的一种衍伸。两种不同美学线条（长度相同，每条线都有大约 5 000 年的历史）的怪异叠加，使得"魔方塔"构成了一部有趣的作品。该作品项目通过在雕塑公园中创造和谐的方式，促使人们

产生一种场所意识，向公众提供各种冥思性或折射性相遇。作品坐落的场所本身就值得一提——它是世界上最大的雕塑公园之一，拥有来自世界各地艺术家的作品。它向长春市民及参观人员提供了一种与高楼大厦和车水马流组成的城市风景形成鲜明对立的景象。

等风来

高浅

失之毫厘，差之千里。微小的因素总会不经意间带动出人意料的结局，而事物发展的结果，对初始条件具有极为敏感的依赖性，初始条件的极小偏差，将引起结果的极大差异。

"魔方塔"由五个表面具有折射作用的方块由大到小，自下往上依次排列。有风吹来，建筑顶部的方块最先开始旋转，并带动其下的方块依次旋转。作品坐落于长春雕塑公园，艺术家借助自然的力量让作品与金属材料配合的相得益彰。作品运用比较的设计方式，将面光与背光、刚毅与柔和等处理得当。反光的金属材料倒映周围的自然景观，随风转动。人们在欣赏雕塑作品的同时也会静心思考作者在其内部巧妙的设计思路。

看不见的未必无形，没有生命的却能召唤生命。就如看不见风，却能从微动的草尖、扬起的沙土中掌握它的行踪。这是艺术与自然生命的交融，艺术与哲理的构建。

Artist: Ralfonso Gschwend
Location : Changchun International Sculpture Park,
Changchun, China
Media/Type : Sculpture
Date: 2012
Commissioner : Changchun International Sculpture Park
Researcher: Leon Tan

Ralfonso Gschwend was one of the winners of the 2012 Changchun International Sculpture Competition, meaning that he was commissioned to permanently install Cube Tower, a kinetic sculpture, in Changchun Sculpture Park in Jilin Province, Northeast China. In fact, Gschwend was also one of the winners of the competition in 2011, so Cube Tower joined another kinetic work of his called Union in the 92-hectare landscaped garden. Developed in conjunction with a landscape architect, park officials and the municipal government (commissioned by the Government of Changchun City), Cube Tower measures eight meters in height, and takes the form of five reflective cubes stacked atop each other on their edges, with the cubes progressively reducing in size towards the top. The structure is made of stainless steel, and each cube counter-rotates, driven by the wind. According to the artist, it is one of only a handful of kinetic wind sculptures in existence in China.

The artist Paul Klee once remarked, "Art does not reproduce the visible; rather, it makes visible." We might understand Gschwend's kinetic sculptures in these terms insofar as they do not necessarily represent anything, but rather, render visible invisible forces. As an abstract form, Cube Tower is interpretively open-ended, leaving symbolic interpretation in the eyes of the beholder. What the audience registers, however, is precisely the invisible: the passage

of air caused by uneven heating of different parts of the earth's surface by the sun. Set within a meticulously manicured garden by the edge of a lake, this registration of the invisible contributes to a sensation of natural harmony, that is to say, a harmony of different invisible forces such as heat and wind, and visible material, such as earth, water, and steel. Cube Tower is a skillfully executed public sculpture, which appears well integrated with the park's numerous other sculptures. Its excellence consists of its ability to generate contemplative experiences through encounters with an invisible force.

Such experiences of nature and encounters with natural forces, arguably belong to an ancient Chinese aesthetic tradition, that of landscape ink-painting, even if the form of the sculpture itself is more derivative of Western (specifically Greek and American Modern) aesthetics. This strange conjuncture of two divergent aesthetic lines, both equally lengthy (each approximately 5.000 years), makes Cube Tower an interesting work. This project contributes to a sense of place by creating harmony in the sculpture garden, providing the public meditative or reflective encounters. The site itself is also worthy of mention—it is one of the world's largest sculpture parks and possesses works by artists from around the world. It provides to Changchun residents and visitors a striking contrast to the urban landscape of high rises, cranes, and traffic.

NeORIZON
NeORIZON

艺术家：莫里斯·贝纳永
地点：中国上海世纪大道
形式和材料：互动行为
时间：2008年10月18—22日
委托人：上海电子艺术节
推荐人：里奥·谭

"NeORIZON"是法国新媒体先锋莫里斯·贝纳永的一个交互媒体艺术装置项目，这个项目是为了上海2008年电子艺术节而准备。这是一个由多个元件组成的装置。最明显的元件可能就是红色的"IDWorms"了。这些IDWorms的几何形体具有未来蠕虫的外形。虫子一端大，一端小。它们被放置在浦东世纪大道上的多个地点。纪录片画面显示虫子的位置是为了获得都市被未来生物入侵的画面。

通过安放在上海高度开放的公共区域，这个似科幻小说中的场景立刻吸引了行人的兴趣。公共区域的人可以随意观察每只虫子的任何一端。在小端，设置一个脸的高度，没有疑心的参观者在这里拍摄脸部照片，并转换为二维（QR）码（QR码指的是大多数商业智能手机能扫描的二维码，使用户能够连接到在线网站）。这些个性化的ID码然后在虫子的大端显示。偶尔，大端的屏幕还会显示被嵌入小端摄像头所捕捉到的游客的脸，但是这些图片总是会迅速地被新QR码的黑白线侵蚀。IDWorms把人们转换为相对匿名网络ID符号或代码。在被手机和各种数码产品淹没的社会，"NeORIZON"这个特征似乎让用户思考有关他们/我们沉浸在技术中的问题，尤其是生活本身无情的商品化（QR码让人想起所有主要大众消费商品上的产品条形码）。

IDscape'是IDWorm侵蚀中间的大屏幕，其特征是稳定增长的QR码都市建筑，每个QR码来自访客遇到的IDWorm。随着都市风景的持续增长，之前的部分后退，让位给新的QR塔，给观众留下不停增长的城市景象。不断扩张的都市，就像不断增加的利润一样，是一部流行的当代小说。但是，这并没有阻止个人基于这个不稳定的基础作出实际决策。Benayoun使用"临界融合"这个术语来描述他的项目，指的是小说和现实的融合。有关IDworms的小说和科幻小说场景在公共场所展示，与本地的现实融合，至少能产生一些关键的邂逅、想法和想象，有关构成现实生活的许多小说，如之前所提到的那些有关无尽的利益以及无穷的资源和空间的神话。

形式上来讲，"NeORIZON"是一个成功的和复杂的装置，能够根据设计运作，并且使很多人参与。概念上来讲，它通过小说进行社会或文化批判，向观众巧妙地暗示对资本主义和网络文化的批评。就场所营造威严，

　　"NeORIZON" 向观众提出他们对中国社会及其空间迅速转变的反应及责任，以及国家和（私人）资本利益之间的对齐，越来越牺牲公共利益。它让观众把自己想象成大批量生产和消费的对象，在工厂的生产线上组装，并且加盖或置入独特的识别码，如条形码。它唤起看起来似乎很"科幻小说"的未来，但是危险地徘徊在现实附近，把这个赠送给大众，作为其冥想的对象。

身份二维码
高浅

科技时代，二维码借助网络侵入人们日常生活。只要联网扫码就可以轻易迅速地得到基本信息。为了强调互联网和虚拟身份对现代生活观念的冲击及影响，艺术家莫里斯·贝纳永设计制作了立体感极强的红色大型后现代主义的数码交互装置。他们被搁置在上海浦东陆家嘴的中心地段，看似摆放随意，实则经过精心的布置。强烈的色彩和现代的设计风格映衬着周围鳞次栉比的现代及后现代设计风格的办公楼，更能凸显其颜色的鲜艳和立体的设计风格。

作品的互动功能颇有新意，只要来往的游客将脸部靠近装置较小的一端，另一端的屏幕就会运用折射原理，通过内置的软件，捕捉人脸的细节，将其迅速地转换成二维码。在经济化的社会中，我们身上各式各样的信息都似乎很容易被作为商品本身而出售。艺术家的目的或许是为了让人们正视由于现代科技的过度入侵生活而引发的自身商品化和被互联网"捆绑胁迫"等现代问题。

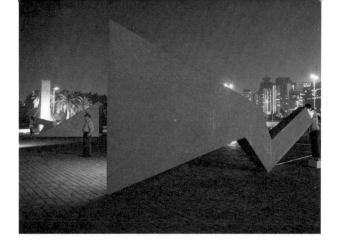

Artist: Maurice Benayoun
Location : Century Avenue, Shanghai, China
Media/Type : Interactive Behavior
Date: 18/10/2008-22/10/2008
Commissioner : eArts Festival Shanghai
Researcher: Leon Tan

NeORIZON is an interactive media-art installation by French new media pioneer Maurice Benayoun, commissioned for the 2008 eArts Festival Shanghai (October 18-22) and curated by Yan Xiaodong. As an installation, it consisted of multiple components. Perhaps the most visible of these were the red 'IDWorms.' The IDWorms took on geometric forms giving them the appearance of futuristic worms. Each worm had recognizably large and small ends. They were placed at numerous points along Century Avenue in Pudong. Documentary images reveal that the worms were sited so as to evoke the image of an urban invasion by creatures from the future.

Positioned in a highly public area in Shanghai, this 'sci-fi' scene immediately captured the interest of passers-by. Members of the public were free to look into either end of each worm. At the small end, set at face height, unsuspecting visitors had pictures of their faces taken and converted into 2D (QR) codes. (QR codes are 2 dimensional codes that can be scanned by most commercially available smart-phones in order to link the user to an online site). These personalized ID codes were then displayed at the large end of the worm. Occasionally the screen at the large end revealed the faces of visitors who had chanced upon the cameras embedded at the small end, but these images were always rapidly eroded by the black and white lines of a new QR code appearing. The IDWorms converted people into relatively anonymous networked ID signs or codes. In a society inundated with cell phones and digital codes of all kinds, this aspect of NeORIZON appeared to engage users in questions concerning their/our immersion in technology, and more especially, in the relentless commodification of life itself (the QR

codes are reminiscent of product barcodes found on all major items of mass consumption).

The 'IDscape,' a giant screen in the midst of the IDWorm invasion, featured a steadily growing urban architecture of QR codes, each QR code coming from a visitor encounter with an IDWorm. As the urban landscape continued growing, earlier parts receded to make way for new QR towers, leaving audiences with the spectacle of a never-ending and ever-growing city. The ever-expanding city, like ever-increasing profit, is a popular contemporary fiction. Yet that has not prevented individuals from making actual decisions on that shaky basis. Benayoun uses the term critical fusion to describe his work, by which he means a combination of fiction and reality. NeORIZON is a perfect example of critical fusion. The fiction of the IDworms and sci-fi scenario, sited in public space, was blended with local reality, producing at least some critical encounters, thought and imagination, regarding the many fictions structuring real life, fictions such as those previously mentioned, myths of endless profit and infinite resources and space.

Formally, NeORIZON was a successful and complex installation that worked according to design, and engaged large crowds. Conceptually, it approached social or cultural criticism through fiction, implicating audiences subtly in a critique of capitalist and networked culture. In terms of placemaking, NeORIZON confronts audiences with questions concerning their responses to, and responsibilities for, the rapid transformation of Chinese society and its spaces, and the alignment of the one-party State with (private) capitalist interests, increasingly at the expense of public interests. It asks audiences to imagine themselves as objects of mass production and consumption, assembled on a factory line, and stamped or implanted with a unique identifying code such as a barcode. It evokes a future that seems 'sci-fi' but yet hovers dangerously close to the real, and gifts this to the public as its object of contemplation.

光速
Speed of Light

艺术家：NVA
地点：日本横河，Nihon Odori 大街，港景公园，美香公园，神钢圈，
Zou-no-hana 公园以及大栈桥埠头和 Zaim
形式和材料：互动行为，临时性
时间：2012 年 10 月 31 日—11 月 4 日
委托人：日本英国文化委员会
推荐人：杰奎琳·怀特

2011 年发生在日本东部的毁灭性大地震，致使三个核反应堆的七处被摧毁，自此之后，能源效率以及接收的替代能源使得日本的利益得到强化。一年后，即 2012 年，智能照明横滨节日庆典受委托举办了"光速"的首映式。庆典集中展示持续性以及陈列的 LED 灯、太阳能板及其他产生并存储绿色能源的技术。"光速"是由苏格兰公用艺术组织 NVA 开发的一项艺术性干预，该组织的主要工作是通过集体行动寻求对城市及乡下风景的重新界定。NVA 是 nacionale vitae activa 的首字母缩略词，一个含义为"影响公共事务的权利"的拉丁词。

"光速"刻画了穿着经过专门设计的套服的跑步及行走人员，他们所穿的衣服能够照亮穿衣人的举动，该作品最早于 2012 年早期在爱丁堡国际盛宴上进行过表演。2012 年在伦敦举行的奥运会及残奥会为这种来自艺术、体育以及尖端科技的联姻提供了最初的灵感与启发，作品向非专业长跑运动员及耐力运动员提供了一个能够创造一种在视觉上具有催眠效果的景象。来自爱丁堡国际科学盛宴的一组专家，携手 NVA 的带头设计师吉姆斯·约翰逊，一起进行了无数次灯光实验，包括电激发光电线及磁带，光缆、分散及非分散的 LED。他们的目标是创造一种能够表达跑和走的动态变化的光影系统，使用最低电量但创造出尽可能显著的效果，通过跑步者及行走者的运动有效产生电能，其足够崎岖和简单，以供大量跑步者和行走者利用并挑战所有天气条件下的各种地形。

专家设计了单独通过手部运动即可点亮的闪光灯源，与包含便携式电池组及遥控无线技术的 LED 灯光组合结合到了一起。由中央系统统一控制，该灯光组合能够立刻改变颜色、闪光率以及明暗度，通过众多跑步者精心设计的动作，创造出绝妙的灯光组合效果。

为创造位于横河的"光速"，NVA 的创意总监安格斯·法夸尔与日本编舞 Makiko Izu（东京表演公司总监）展开合作，目的是在横河开发一组动态的夜景，横河拥有 367 万人口，以具有历史意义的港口建筑、一处观光海港以及风景园林为特点。作品在智能照明盛典上连续进行了两个晚上的展

出，作品的名称为"三种运动"，以100名经过精心设计的跑步者为特点，使位于横河的地标场地充满动感与活力，比如Zou-no-hana平台、美香公园以及不对称的木质通道，都延伸至位于国际轮渡码头附近的海洋中。

"光速"目前正在世界各个新地区展开旅行，从敏感的乡村环境一直到后工业城市构造的极端情形。鉴于其在位置及规模方面的适应性，"光速"与当地设计师及志愿跑步者们合作，赋予熟悉的或前所未有的空间一种新的几何结构及用途。

光，在移动
高浅

尽管从伦理道德的角度是否应该建设核电站的探讨依旧有待商榷，但出于对于国力和军事安全的考虑，核电站的建设似乎是众望所归。大型核电站的成功运转，不光可以解决居民用电问题，同时可以增强国家的军备设施。但是人类总是在获取利益的同时将自己置于险境却不自知。核电站一旦发生破坏，后果非常严重。不光是环境被破坏，同时还有大量动植物基因也将受到极大的威胁。

"光速"项目施行的初衷就源于2011年日本福岛七个核电站在大地震中的摧毁。尽管对外宣称泄漏情况并不是非常严重，但此消息依旧在日本本土及整个东亚地区引起了大面积的人心恐慌。在这种事件背景下，来自爱丁堡国际科学盛宴的一组专家，携手NVA的带头设计师吉姆斯.约翰逊策划了一场别出心裁的公共参与的唤醒行动。项目提倡大众参与、绿色生活、善待能源。参与活动的每个人身穿可以通过运动产生光的特制衣服。就像一场绿色发电运动，参与者们在黑夜中行进，通过运动发电，远远望去，好似一条长长的光带悬浮在黑夜。众志成城，不光在行为，还在理念。人类为自己作出的错误选择付出了代价，而这集合群众的绿色能源行为更像是一场黑夜中对自然的忏悔仪式。

Artists: NVA

Location : Harbor View Park, Yamashita Park, Shinko Circle, Zou-no-hana Park, and Osanbashi Pier, and Zaim, Nihon Odori Street, Yokohama, Japan

Media/Type : Interactive Behavior, Temporary

Date: October 31-November 4, 2012

Commissioner : British Council Japan

Researcher: Jacqueline White

After the devastating Great East Japan Earthquake of 2011, which caused seven meltdowns at three nuclear reactors, energy efficiency and alternative energy sources received intensified interest in Japan. One year later, the 2012 Smart Illumination Yokohama Festival commissioned the international debut of Speed of Light. The festival focused on sustainability and showcased LED lights, solar panels, and other green energy generation and storage techniques. Speed of Light, an artistic intervention, was developed by NVA, a Scottish public art organization whose work seeks to redefine urban and rural landscapes through collective action. NVA is an acronym of nacionale vitae activa, a Latin phrase describing 'the right to influence public affairs.'

Speed of Light, which features runners and walkers wearing specially designed suits that illuminate the wearers' movement, was first performed earlier in 2012 at the Edinburgh International Festival. The 2012 Olympic and Paralympic Games in London provided the initial inspiration for this marriage of art, sport, and cutting-edge technology, which gives nonprofessional distance and endurance runners an opportunity to create a mesmerizing visual spectacle.

A team of specialists from the Edinburgh International Science Festival collaborated with NVA's head designer James Johnson to conduct numerous trials with different light sources, including electroluminescent wire and tape, fiber optic fabric, and diffused

and non-diffused LEDs. Their goal was to create a lighting system that could express the movement of running and walking; use the minimum amount of power but give the greatest possible effect; efficiently generate power from movement of the runners and walkers; and be rugged and simple enough for use by a large number of runners and walkers on challenging terrain in all weather.

The specialists designed flickering light sources that could be powered by hand movement alone, which were incorporated into LED light suits containing portable battery packs and remote wireless technologies. Individually controlled from a central system, the suits can instantaneously change color, flash-rate, and luminosity, creating stunning light patterns made by the choreographed actions of massed runners.

For Yokohama's Speed of Light, NVA creative director Angus Farquhar collaborated with Japanese choreographer Makiko Izu, director of the Tokyo-based performance company Grinder Man, to develop a moving nightscape in Yokohama, a city of 3.67 million, which features historic port architecture, a scenic harbor, and landscaped parks and gardens. Presented over two nights of the Smart Illumination festival, the work, titled "3 Movements," featured 100 choreographed runners who animated landmark sites in Yokohama, such as Zou-no-hana Terrace, Yamashita Park, and the asymmetrical wooden walkways that extend out into the ocean at the International Ferry Terminal.

Speed of Light is now travelling to new locations worldwide, from sensitive rural environments to the extremes of post-industrial urban settings. Adaptable in location and scale, Speed of Light collaborates with local choreographers and volunteer runners in each country it visits, giving familiar or previously unseen spaces a new geometry and use.

再造景——三林公共艺术项目

Reconstruction Scene — Sanlin Public Art Project

艺术家：上海视觉艺术学院美术学院
地点：中国上海市浦东新区
形式和材料：临时综合材料装置
时间：2012 年 6 月 17 日—7 月 2 日
委托人：上海视觉艺术学院美术学院
推荐人：里奥·谭

　　"再造景——三林公共艺术项目"是由浦东三林镇人民政府和上海视觉艺术学院美术学院共同策划组织实施。本次展览共展示 15 件作品，50 位复旦大学上海视觉艺术学院公共艺术专业的本地学生和留学生在专业教师的带领下组成多个项目小组，对三林古镇独特而丰富的人文资源进行调研与加工提取。三林古镇历史悠久。浦东坐落在黄浦江东侧，而西侧是著名的外滩。"外滩"一词源于波斯语（意思是堤岸），19 世纪巴格达犹太人在此定居，并开展商业活动，随着历史的变迁，逐渐成为外国人的聚居地。另一方面，浦东在 20 世纪 90 年代以前经济欠发达，因此保留了一些原生态村落和自然景观，而如今浦东成为快速发展、城市化和社会变迁的代名词。

　　本次展览共展示 15 件以现场环境为思考对象的特定创作，作品以装置、雕塑、绘画、灯光、影像等各种艺术形式，试图重新解读古镇的文化传承与更新，探讨生活样式与艺术氛围之间公共性的综合命题。三林当地居民和公共艺术专业的本地学生、国际留学生共同协作，使其地域特征、特色定位得以延伸，同时充分展示了国际化大都市中传统中华古镇的变迁以及其对当地资源和居民带来的影响。

　　作品"布顶"，作者用蓝布顶烘托出传统中国建筑的样式，比如通过古典中国建筑美学和现代工业美学之间鲜明的对比，吸引了公众的目光。当游客踏入临时性"房屋"，仿佛来到一个奇妙的世界，顿时觉得身心轻松。另一件作品使用工业建筑材料和城市发展语言，充分利用交通锥标和木架展现中国人熟知的雕塑形象，一条龙在木质背景下栩栩如生。如同其他作品，该作品使大众陷入思考，当代浦东生活的自然和工业／人造元素怎样才能达到和谐统一。

　　著名哲学家、汉学家 François Jullien 认为，中国之所以不能在科学技术上紧跟欧洲人的原因在于其完全不同的审美观。古希腊或欧洲传统审美理

念注重足以展现自然世界的客观事物，而通过科技可以得到充分利用。相比之下，中国传统注重"能量"或"生命力"的和谐统一，通过整个自然转换过程可以联系个人与景观。此外，古希腊传统讲究直接介入自然，而中国传统理念则追求自然和谐。该项目充分体现了中华传统美学和欧洲传统美学的差异性，主张可以通过社会和空间融合来实现中庸之道。

艺术介入生活

高浅

公共艺术作为地方重塑的手段之一，以独特的视域串联社会职责与文化生活。历史文化的痕迹也在其中或多或少有所表现。始于北宋的三林古镇，人文资源丰富，文化历史背景深厚。然而，文化历史的传承和创新，千百年传承下来的生活方式因为科技文明的冲击和现代思想的植入受到了极大的冲击，其自有的生存体系不断的在现代生活中，被现代元素侵入受到破坏和改造，逐渐变异，濒临消失。

艺术介入生活着眼于在保证生活更美好的前提下，与空间内的人们产生互动，引起他们对地方保护理念的侧目。"再造景"选取了三林古镇自古以来，当地人熟知的传统资源和生活元素，将这些资源和元素扩大定位，重新加以利用。古今文化串联，用艺术的理念形成新的融合传统与现代的生活空间，让生活更加美好。

Artist: Shanghai Institute of Visual Art
Location : Pudong, Shanghai, China
Media/Type : Temporary Mix-Media
Date: June 17-July 2, 2012
Commissioner : Shanghai Institute of Visual Art
Researcher: Leon Tan

Reconstruction Scene – Sanlin Public Art Project was an ambitious public art exhibition mounted by the Shanghai Institute of Visual Art in the area of Shanghai known as Pudong. The project involved 50 local and international students from the Fine Art Department's undergraduate program researching the area and its resources and circumstances, and executing 15 public artworks on site, under the guidance of the Institute's faculty. The locale itself has a rich history. Pudong lies on the east of the Huangpu River, while the Bund lies on the west. The Bund was home to the International Settlement and derived its name from Persian (meaning embankment) after Baghdadi Jews settled there and established businesses during the nineteenth century. Pudong, on the other hand, remained relatively undeveloped until the 1990s. While it used to consist of village settlements and natural landscapes, Pudong today is a site of rapid construction, urbanization, and social change.

The 15 site-specific works, including murals, sculpture, textiles, light and video, were conceived as a means of fostering public interaction as well as personal contemplation or reverie. Bringing together Sanlin residents and local and international art students, the project is commendable both as a pedagogical activity and as a collective reflection on spatial change, particularly the displacement of traditional Chinese landscapes by industrial construction, and its impact on local resources and residents.

One of the 15 works, the blue tent which echoed the style of traditional Chinese architecture, for example, drew the attention of the public to the stark contrast between classical Chinese and modern industrial aesthetics. By allowing audiences to step into the makeshift 'home' the work facilitated reflection in a playful atmosphere, encouraging a convivial spirit in the random social encounters it facilitated. Another work used the materials and vocabulary of industrial construction and urban development—in this case, traffic cones and wooden workbenches—to stage a sculptural form familiar to the Chinese, a dragon, against a natural backdrop of trees. Like the other work, this too asked the public to contemplate the dissonances and possibilities for harmonization between the natural and industrial/artificial components of contemporary Pudong life.

In the view of the philosopher and Sinologist François Jullien, the Chinese were not as swift to develop the sciences as Europeans because of an entirely different aesthetic sensibility. The Greek/European aesthetic tradition valorized a close study of the objective domain in which the world of nature to be represented and subsequently manipulated through science and technology. In contrast, the Chinese tradition emphasized the unifying 'energy' or 'life force' that connects people and landscapes within an overall process of natural transformation. While the Greek tradition sought to confront and intervene in nature, the Chinese tradition focused instead on harmonization with nature. This project manages to highlight this aesthetic difference in a moment in history when the legacy of Chinese aesthetics confronts that of European aesthetics, suggesting that a middle path may be possible through a kind of social-spatial hybridization.

启德河绿色走廊
Kai Tak River Green Corridor

艺术家：郑炳鸿
地点：中国香港特别行政区启德河
形式和材料：互动行为
时间：2007 年至今
委托人：环境保护基金会和艺术发展协会
推荐人：里奥·谭

在 20 世纪中，有一大批艺术和环境行动主义的拥护者，其中最具代表性的
当属 Hans Haacke，他创立了莱茵河净水厂。1972 年，Haacke 成功开
发出一套方法净化克雷菲尔德污水厂的污水，从而改善了莱茵河的水质。
郑炳鸿牵头的"启德河绿色走廊"计划将这个古老的渠道引入当代视野，
唤起公众的环保意识和行动，关注曾经被称作"启德明渠"的香港污染最
重的一条臭水河道。曾几何时，每当国际航班降落在前启德机场时，游客
们都闻到渠沟散发出的臭味，这促使政府制定计划，沿河道修建混凝土围
墙。然而，一些民间人士，包括建筑学教授郑炳鸿，其作为该项目的牵头人，
极力反对这些建造计划，提出了"启德河绿色走廊"计划。

郑炳鸿本人在"启德渠道"附近度过了一段童年时光，像其他生活在黄大
仙区的居民一样，他目睹了河道的种种变迁。2007 年，在郑炳鸿的不懈努
力下，"启德明渠"正名为"启德河"，他相信这种概念的改变将重塑公
众对这条被遗忘河道的认知。通过与区议员的协商讨论，郑炳鸿应邀为政
府常务顾问，制定河道规划方案。2007 年末，区议会主席的更替为启德民
间联盟制定 KTRGC 框架方案争取了时间，最终郑炳鸿等人组织策划了一
系列艺术教育方案，并分别于 2009 年和 2011 年得到了艺术发展协会和环
境保护基金会的资金支持，邀请海外艺术家和 17 所当地学校共同协作，同
时开展一系列艺术节活动。

2011 年绿色艺术节期间，通过展示一系列文化活动，比如沿河道展览的艺
术作品，"启德河绿色走廊"计划使该区域重新焕发活力，成为当地居民
和环境之间的重要纽带。此外，香港长期以来受困于水道和空气污染问题，
KTRGC 促使学生们关注城市环境状况以及城市未来。该计划自 2007 年启
动以来，大获成功，业已成为地标性景观，是由民间组织推动实施的空间
规划范例。KTRGC 的成功归结于几大因素，首先，政府组织正式将渠道更
名为"启德河"；其次，民间人士、政府官员和组织机构建立起新型合作模式，
新型开放空间建立成为一个公共公园；最后，河道不再发出臭味，更加洁净，
沿河重现野生动物、鸟类和鱼类的踪迹。

当被问到下一步计划时，郑炳鸿谈到："如果我们能够为河道做点什么，城市或世界将变得更加美好。"不论是香港，还是世界上其他城市，都迫切需要解决污染和生态可持续性问题，这一点毋庸置疑。KTRGC展示了艺术家们和艺术类活动是如何利用其他学科知识，例如环境科学、教育、行政管理和空间规划，从而成功地转换公共空间。在该计划中，"绿化"城市即意味着使公众远离交通拥挤和钢筋混凝土丛林。

享受参与

高浅

公共行为泛指在公共空间中，引起人们注意并能吸引人们积极参与的行为，它不仅与公共空间发生关系，同时还与公共空间中活动的人发生关系。城市未来、人民生活幸福指数和城市环境息息相关。虽然污水治理、群众参与、重塑环境是老生常谈的话题，但治理与保护方式仁者见仁，智者见智。缺少淡水与绿化的香港比任何地区都注重保护自己的自然资源，也更迫切需要改造生活环境，而香港人的严谨认真与大胆创意更是直接促成了启德河项目的成功。

汉娜·阿伦特认为："'公共性'一词不光是指公共场合中的东西或者事物，而且还指人类共同拥有的世界。"启德河项目与其说是环境治理，不如说借治理环境之名，对文化进行追溯并借此机会增加群众的环保意识，增强凝聚力，对共同理想家园进行改造活动。在特定空间内聚集大批群众，并将政府职能内的枯燥行为转变为吸引周围群众主动参与且具有趣味性的日常行为。这样，环保并非为完成任务而成了乐趣。

Artist: Wallace Chang
Location : Kai Tak River, Hong Kong, China
Media/Type : Interactive Behavior
Date: 2007-present
Commissioner : Environmental Conservation Fund and Arts
Development Council
Researcher: Leon Tan

The intersection of art and environmental activism has a number of precursors in the twentieth-century, most notably, Hans Haacke's Rhine Water Purification Plant. Haacke developed a scalable prototype to clean the Krefeld Sewage Plant's polluted water in order to improve the quality of the Rhine River in 1972. Wallace Ping Hung Chang's Kai Tak River Green Corridor (KTRGC) extends this art historical precedent into our contemporary moment, bringing sorely needed environmental awareness and action to what used to be one of the most polluted stretches of water in Hong Kong, the so-called 'Kai Tak Nullah.' The odious smells from the river that once greeted international visitors landing at the former Kai Tak airport inspired government plans to build concrete over the waterway. However, grassroots activists, including the project lead, architecture professor Wallace Chang, rigorously opposed these plans, arguing instead for the waterway to be transformed into a green corridor.

Chang himself spent a significant part of his childhood growing up beside Kai Tak Nullah, and, like some others in the Wong Tai Sin district council, he saw potential for the river to be revitalized. In 2007, Chang was instrumental in having Kai Tak Nullah renamed as Kai Tak River. He believes this change to have been important in reshaping perceptions of the neglected waterway. Through negotiations with district councilors, Chang was invited to explore alternatives to the standing Government decision to cover the river. A change of chairperson in the district council near the end of 2007 provided time for the Community Alliance of Kai Tak to flesh out a framework for KTRGC. Eventually, Chang and others organized a

series of artistic and educational projects, receiving funding from the Arts Development Council in 2009 and the Environmental Conservation Fund in 2011, to instigate international artist residencies and participatory arts events in partnership with 17 local schools.

By staging cultural activities, for example, exhibiting the participatory art projects along the river during the 2011 Green Arts Festival, KTRGC temporarily brought new life to the area, building effective ties between residents and the environment. KTRGC also instilled in students an interest in the state of the environment, and the future of Hong Kong, a city with an extraordinarily heavy environmental footprint and extensive problems with water and air pollution. The project stands out as a highly successful creative placemaking intervention and an example of community driven spatial planning, lasting from 2007 to the present day. Various indicators attest to KTRGC's success, the incorporation of the new name, Kai Tak River, into official plans and documentation; the creation of new relationships between activists, government officials, and agencies; and the establishment of new open space zoning for the area, which allows it to be used as a public park. Most significantly, the river no longer smells offensive, the water is cleaner, and wildlife, birds and fish, are returning to reside in and alongside the waterway.

When asked about his motivations, Chang said, "If we can do something about this, the city/world can be better." It is difficult to disagree, since not only Hong Kong, but also many other cities in the world urgently need to address issues of waste and ecological sustainability. KTRGC demonstrates how artists, and artistic activities can work with other disciplines such as environmental science and education, political administration, and spatial planning to successfully transform public spaces, in this case, 'greening' a city that many perceive to be a congested and polluted concrete jungle. '

上海雕塑公园
Shanghai Sculpture Park

艺术家：多位艺术家
地点：上海松江区佘山
形式和材料：雕塑
时间：2003 年
委托人：由上海雕塑公园委托的单个项目
推荐人：里奥·谭

"上海雕塑公园"占地 86.7 公顷，位于上海西松江区佘山。公园由一名台湾企业家梁长草（音译）建立，公园规模宏大，吸收了来自中国以及世界各地雕塑大师的 80 余部委托雕塑，作品位于四个区域，分别代表了四个季节：春夏秋冬。公园于 2003 年开始开发，而后规模不断扩大，作品不再局限于公共雕塑，以便为来自当地以及世界各地的参与者们开发一个艺术家住宅计划以及雕塑工坊。

位于公园内的中央湖，被广阔的风景园区环绕着，最初可能激发对自然沉思的最安静传统的联想，水太空球（穿梭在位于水面上的一个膨胀球体里面）的活动以及在"Fuma"弹跳山上弹跳的各种活动除外。这种接收环境与现场的雕塑共同创造了一种引人入胜的紧张形势，很多可能被认为是 20 世纪现代主义的一种或多种趋势的延伸。Constantin Brâncusi 的印记，在 Maciej Fiszer 以及 Robert Pierresstiger 的作品中都有所体现，尽管叶鸿兴的作品可能被多次展出，并与后简单抽象派艺术及流行轨迹相提并论。

文件中的图片暗示，试运计划以一种现代主义国际调研、或对来自世界各地的各种不同现代主义（奇怪的是他们相似）进行的一种对比调研方式来运作。该公园因其与自身雕刻计划的衔接性，在某种正式意义上来说，可圈可点。没有通往现场的直接通道，很难评价各种雕塑是如何在旅游休闲性质的广阔环境下发挥作用的。但是，可以说，公园在创造独特性方面获得了一定成功，每年都吸引超过 50 万的游客至此参观。

公园的目标之一是"加深对雕塑艺术在当代建筑中作用的理解"。目标的实现则在一定程度上是通过让观众接触各种作品，还可以通过住宅计划以及雕塑工坊来实现，包括月湖艺术馆（位于现场）以及其对当代雕塑的采纳和展览。该公园的门票定为 120 元人民币的确略微高些，很可能将很多参观人员拒之门外。具体有关不同的参观人员是如何来体验公园的，我们

不得而知。是作为大规模消耗品来体验的？还是形式上对实验课题的审美性沉思来体验的？无论如何，它都比不上与其为邻的主题／娱乐公园——"上海欢乐谷"的知名度大。衡量该公园成功与否的其他标准还包括它先后于2007年、2008年以及2009年为中国的杰出公共雕塑项目获得的奖项。或许，公园与上海想要成为与柏林、伦敦及纽约齐名的世界艺术之都的愿望相适应吧。

公共与公益
高浅

并不是所有被搁置在所谓公共空间的艺术都可以被称作公共艺术。"上海雕塑公园"高达120元人民币一张的门票将不少人阻挡于公园外，那么这样的项目可否被纳入公共艺术范畴？

一件成功的公共艺术作品具有的形态和模式是被定性的基础。公共艺术是在公共领域中的艺术作品，根据哈贝马斯的考察和论述："公共领域指我们的社会生活的一个领域……公共领域原则上向所有公民开放。公共领域的一部分由各种对话构成，在这些对话中，作为私人的人们来到一起，形成了公众。那时，他们既不是作为商人或专业人士来处理私人行为也不是作为合法团体接受国家官僚机构的法律规章的规约。"也就是说，在公共领域内，参与的个人是可以不受到任何来自于利益、商业或者政府的约束与限制。"上海雕塑公园"既是公园是公共区域，就应该面向所有人开放。如果用艺术或者公共艺术作品作为其获取经济利益的手段，那便违背了公共艺术的"公共"与"公益"的基本属性。

Artist: Various
Location : Sheshan, Songjiang District, West Shanghai
Media/Type : Sculpture
Date: Established in 2003
Commissioner: Individual projects commissioned by Shanghai
Sculpture Park
Researcher: Leon Tan

Shanghai Sculpture Park is an 86.7-hectare park located in Sheshan, Songjiang District, West Shanghai. Founded by a Taiwanese entrepreneur, Rhy-Chang Tsao, the park is ambitious in scale and programming, incorporating over 80 commissioned sculptures from Chinese and international artists within four zones representing the four seasons: spring, summer, autumn and winter. The development of the park began in 2003, and has since expanded beyond the commissioning of public sculptures to include an artist residency program and a sculpture workshop for local and international participants.

The central lake in the park, surrounded by expanses of landscaped garden, might initially evoke associations with quietist traditions of nature contemplation, except for the fact that activities in the park include "water zorbing" (traveling in an inflatable ball across the water) and bouncing on the "Fuma Fuma" bouncy hill. The park is, in fact, a tourist attraction, and intentionally populist in its orientation. This context of reception creates an intriguing tension with the sculptures in the site, many of which may be considered extensions of one or another tendency in twentieth century modernism. The imprint of Constantin Brâncusi is detectable in the works of Maciej Fiszer and Robert Pierresstiger, for example, while Ye Hongxing's piece might be fruitfully located alongside the post-minimalist and pop trajectories.

The documentary images suggest that the commissioning program operated as an international survey of modernism, or rather, a comparative survey of different (but strangely similar) modernisms from around the world. The park is commendable at a formal level for the coherence of its sculptural program. Without direct access to the site, it is difficult to assess how the sculptures function within the broader context of touristic entertainment. It is, however, possible to say that the park succeeded in creating a distinctive place, attracting in excess of half-a-million visitors every year.

One of the park's goals consists of "fostering a deeper understanding of the art of sculpture in a contemporary setting." This is accomplished to some degree through the exposure of audiences to the commissioned works. It is also realized in the residency program and sculpture workshop, and through the Yuehu Museum of Art (on site) and its acquisition and exhibition of contemporary sculpture. Having said all that, it should be noted that the entry fee of RMB 120 is a little on the high side, possibly deterring access for many visitors. It is also unclear how different visitors experience the park. Is it experienced as a site for mass-consumption? Is it experienced as a site for aesthetic contemplation of experiments in form? In any case, it is less popular than Happy Valley Shanghai, the neighboring theme/amusement park. Other measures of the park's success are the awards it won for China's Outstanding Public Sculpture Project in 2007, 2008 and 2009. Presumably, the park sits well with Shanghai's aspirations to become a world art capital on a par with Berlin, London, or New York.

蒙古国两年一度 360 大地艺术展
Mongolia 360 Land Art Biennial

艺术家：多名艺术家
地点：蒙古国中戈壁省
形式和材料：大地艺术
时间：2012 年 8 月 5 日—17 日
委托人：安娜·奥雷利娅·布里茨克
推荐人：里奥·谭

蒙古国为夹在两个政治经济大国俄罗斯与中国之间的内陆国家。它资源丰富却广袤荒凉，其宽阔的沙漠腹地含有煤矿、铜矿、锡矿、钨矿以及金矿等丰富矿产。荒凉的大草原及荒漠地形历史上曾是游牧民族的家园。

在中戈壁省举行的第二届"蒙古国两年一度 360 大地艺术展"是一种集体性质的、以现场为特征同时围绕生态危机而进行的展览，由 Anna Aurelia Brietzke 发起、由众多组织机构如开放社团基金会、蒙古国艺术委员会以及歌德学院出资。就像名称中所暗示的，会展将自身置于大地艺术传统中（西部为主），将大地艺术家先驱们对人类有关自然力量、剧情及限制的解读方面的担忧进行延伸，从而将"全球气候变暖、污染的不可逆性以及对人类生态系统的最终毁灭"考虑在内。

来自当地以及世界各个国家的总共 27 名艺术家参加了此次"蒙古国两年一度 360 大地艺术展"，他们共同创作了多样化的艺术类别。蒙古国艺术家 Chimeddorj Shagdarjav 用透明塑料包装包装了一个 10 吨重的巨石，以此方式来模仿蒙古国免税购物袋的种种细节。我们可以将此解读为对观众的一种刺激艺术，刺激观众们时刻想着旅游及长途空运对蒙古国原本脆弱的生态环境的影响。Maro Avrabou 以及 Dimitri Xenakis 的作品则以水源和耕地的不断减少为主题。由六个蓝色纺织圆环（每个圆环直径都长达三米）组成的"水"，以对称的方式排列于一片平摊的沙漠外延两边。"厨房花园"则由 700 个皱皱巴巴的绿色垃圾袋组成，它们成几何图形状排列，看起来像一种庭园内种植的莴苣菜的菜头。两种类型的作品，都使用了单一颜色，以激起人们对不断减少的资源、蓝色水塘的消失以及不再可能种植沙拉蔬菜的潜意识及认识。其他的作品则针对隔壁沙漠的扩大、矿产开采的影响以及人类与地球之间的关系展开。

蒙古国 360 艺术展，作为一种在各种艰难的外界物理条件下进行的在大地艺术方面的长久计划，其贡献可圈可点。该作品成功地将人们的注意力吸引到对某一地区的艺术及生态环境上，而这些方面之前鲜为人知，当前正

经历着从由岩石、黄沙及草地构成的景色向充满意义及重要性的地方的转变过程。尽管各种干预并未直接改善该地区的生态平衡状况，他们却以一种幽默和充满想象力的方式不断强调着那些与人类及其他物种生命生存密切相关的一些事件。

往来皆自然

高浅

德·玛丽亚说过："土壤不仅应被看见，而且应被思考。"我国北方农牧交错带与蒙古国戈壁荒漠区相接壤，两地的环境发展与治理有密不可分的关系。由于近几年国家经济的快速发展，引发了严重的土地荒漠化问题，而接壤地带是沙质荒漠化土地分布面积最广、程度最高的地区，蒙古国经济以畜牧业和采矿业为主，地下资源丰富。因此蒙古国目前也面临严峻的环境保护以及能源合理开发利用等问题。

而公共艺术应该为社会舆论服务，宣传某种观点亦或是解决某种问题。因为环境问题而直言政治或许会激发两国之间的矛盾，而民间自发的艺术行为虽然未必能够直接改善环境平衡，但却是宣扬民主观念和政府观念的最佳手段。为了让两国正视环境问题对经济带来的后续影响，"蒙古国两年一度 360 大地艺术展"应运而生。发源于 20 世纪 60 年代，由最少派艺术发展而成的大地艺术极力的主张自然和艺术和谐共存，源自自然、融于自然的自由艺术观念。来自世界各地总共 27 名艺术家，其中包括蒙古国当地的艺术家在蒙古大地上围绕自然保护与合理开发能源等主题制作了一系列的艺术作品。

Artist: Multiple Artists
Location: Dundgovi Province, Mongolia
Media/Type: Land Art
Date: August 5-17, 2012
Commissioner: Anna Aurelia Brietzke
Researcher: Leon Tan

Mongolia is a landlocked nation sandwiched between two political-economic superpowers, Russia and China. Its ecology is at once resource rich and barren, its vast desert spaces harboring coal, copper, tin, tungsten, and gold. Its inhospitable terrain of steppes and desert was historically home to nomadic tribes.

The second edition of Mongolia 360 Land Art Biennial (August 5-17, 2012) in Dundgovi Province (Middle Gobi) was a collective and site-specific response to the theme of ecological crisis, curated by Anna Aurelia Brietzke and financed by organizations including the Open Society Foundation, the Arts Council of Mongolia and the Goethe Institut. As the name suggests, the biennial located itself within the (largely western) tradition of land art, extending the concerns of pioneering land artists with human interpretations of nature's power, drama and constraints to include considerations of "the impact of global warming, the irreversibility of pollution, and the final destruction of our ecosystem."

27 local and international artists in total participated in Mongolia 360, creating an extremely diverse range of work. For his contribution, Mongolian artist Chimeddorj Shagdarjav wrapped a 10-ton large rock in clear plastic packaging, replicating the details of Mongolian duty free shopping bags at larger-than-life scale. We might interpret this work as a provocation to audiences to consider the environmental impact of tourism and long-distance air travel on the fragile ecology of Mongolia. Maro Avrabou and Dimitri Xenakis'

works addressed water and arable land scarcity. Water consisted of 6 blue fabric circles (each of three meters in diameter) arranged systematically on a flat stretch of desert. Kitchen Garden consisted of 700 crumpled green rubbish bags set out in a geometric array to look like a domestic garden of lettuce heads. In both cases, a single color served to evoke the scarcity of a resource, the absence of pools of blue water and the impossibility of growing salad leaves respectively. Other works variously commented on the expansion of the Gobi desert, the effects of mining, and the relationship between human bodies and the earth.

Mongolia 360 is commendable as a coherent program of land art, curated and executed in often difficult physical conditions. It brought much deserved attention to the artistic and ecological circumstances of a region about which little is generally known, temporarily transforming landscapes of rock, sand, and grass into places rich in meaning and significance. While the interventions did not necessarily make direct improvements to the ecological balance of the region, they nevertheless highlight urgent issues pertaining to the continued existence of human and other life forms, with a degree of humor and imagination.

杜舞台
MORI BUTAI

艺术家：多位艺术家
地点：日本德岛市那贺镇
形式和材料：互动行为
时间：2012 年
推荐人：田甫律子

2011 年以来，ISADC （可持续发展艺术策略委员会）作为一个专业的艺术团队，为了探讨地区文化以及文化的魅力，以可持续发展为主题，创建了一批艺术项目。2013 年春季，以日本德岛县那贺镇拜宫地区的三个神社和传统造纸工作室为舞台，举行了名为"杜舞台·拜宫"的当代艺术展。2013 年夏季，这个展览在东京再次展出，添加了传统的木偶表演"人偶净琉璃"、"三番叟"与一场基于"地区艺术"的座谈会。

2013 年秋季，名为"杜舞台·那贺"当代艺术展开展。那贺镇决定赞助杜舞台艺术项目的实现，包括在德岛县那贺镇 Wajiki、Aioi 地区的八个神社的展览，并计划从调查地区文化特点的角度，以可持续发展为主题，对艺术进行策划。当地的文化教育委员会管理并策划这个项目。

"杜舞台·那贺"采用旅行的方式在每周末对艺术作品进行介绍。也正因此，这个艺术项目被认为是一个对提倡全球地区塑造和传统区域更替作出贡献的具有价值的项目。

"杜舞台·那贺"展览从一天延长到了两周，艺术家的停留也从一周延长到了三周，而展览的覆盖面更加广泛，数量上说，从都市区域增加到了 Aioi 和 Wajiki 等最多八个神社。这种扩展和改变不仅让艺术作品被展览的时间延长，同时也更加长时间的呈现了展览的环境。

小小的村庄中拥有大量的可以利用的舞台，就如同熠熠生辉的宝石散落在森林里。舞台位于神社的辖区内，他们致力于传统的人偶表演，俗称"人偶净琉璃"。在过去的四百年中，这些舞台对当地区域景观的塑造和引领地区特色文化具有重要的作用。嶙峋陡峭的山中气候温暖潮湿，郁郁葱葱，清流潺潺，这些舞台现在已经随着时间的逝去万物归一。因此这些舞台是最具有象征意义的地方，也是最适合用以表达对自然的感恩和敬畏的景观。这是与自然历史并存、经过时间的考验而被定义和形成的具有文化价值的地方。

这些地区如今被重新命名为："杜舞台"，这是对把一个神圣的神社区域称作舞台最精准的称谓，"杜舞台"艺术项目为了探求新的当代艺术的表演形式，提供了大量机会给面向地区的文化可持续发展艺术项目，将他们视作一个新的面向公共分享文化价值的交流方式。这样，杜舞台就是一个

为当代艺术表现所设立的新的场所，不仅致力于主张文化可持续发展的价值，同时也探索艺术的交叉表现形式，将"只在此处"转化为"无处不在"。

在"杜舞台"尝试建立新的艺术形式的时候，文化也一直在发生着变化。希望"杜舞台"可以带给艺术更多的想象力和经验。许多地区的有价值的传统文化已经被各种旅游业摧毁，"杜舞台"的设计思路或许可以为保护森林免受侵袭提供启示。另外，在这个项目过程中，最喜闻乐见的事情是人们包括年长的参与者都如此的活跃、具有活力，这就是文化给人们带来的动力。

在自然里活动
高浅
日本大地震之后，人类与自然的关系受到严重质疑。除了地震和海啸等破坏性极强的灾害会对生活环境造成严重损害外，核电站也在无形之中影响了人们的生活质量。

人类的文明发展，文化传承与生存活动空间息息相关。但凡对自然存一丝敬畏之心，便应从近些年愈发严重的自然环境问题中找到根源，自我拯救与约束。与其说"杜舞台"项目是一场文化保护运动，不如说是日本群众面向自然的自我赎罪仪式。艺术家和参与者试图借助艺术手段，唤醒人们对于自然和文化的尊重。中国从古代提倡天人合一的自然观，受中国古代哲学观念与文化影响至深的日本也不例外，人与自然本就应互相融入，互相推动，从而形成更符合生态坏境与生态规律的文明。

此外，作为一个成功的公共艺术项目，它完好的履行了聚集人类和充当政府话语权工具的职责。无论是在自然环境中展开的摄影比赛亦或是展览，都极力强调自然保护和合理生态开发等主题。而可持续性能源的开发与使用等当下热门话题也一再被提及。

Artist: Multiple Artists
Location: Naka Town, Tokushima , Japan
Media/Type: Interactive Behavior
Date: 2012
Researcher: Ritsuko Taho

Since 2011, ISADC (Inter Sustainable Art Driving Committee), an art arbitrary group, conducts art projects that create artworks under the theme of sustainability, while it aims to investigate the regional culture and its fascination. In the spring of 2013 the contemporary art exhibition entitled "MORIBUTAI Haigyu" was held at three shrine stages and traditional papermaking studio in Haigyu area, Naka Town, Tokushima, Japan. The exhibition traveled to Tokyo in summer of 2013, adding the traditional puppet play "Ningyo Joruri: Sanbaso" as well as the symposium on the theme of "Region and Art."

In the fall of 2013 the contemporary art exhibition entitled "MORIBUTAI Naka" was held. Naka Town decided to sponsor the MORIBUTAI Art Project to realize the exhibition which venue included eight shrine stages in Wajiki and Aioi areas, Naka town, Tokushima, Japan. MORIBUTAI Art Project was planned to execute the exhibition by under the theme of sustainability through investigating the regional fascination and its distribution. The Cultural Division of Naka Town Educational Committee managed to realize the Project.

"MORIBUTAI Naka" carried out bus tours to introduce artworks every weekend. It is believed that the Art Project advocate the important value of this region and distribute to the global world, succeeding the regional tradition and culture to the next generation.

"MORIBUTAI Naka" developed by extending the term of the exhibition from only a day to two weeks and the whole period of artist' s stay was increased from a week to three weeks. As the numbers of the shrine stages are also added up to eight in Aioi area and Wajiki , the exhibition became more accessible from urban area. Such development made it possible to enrich the experience of not only artworks but also the stages and their environment.

Tokushima is formerly known as Awa that means a place where wave begins. It sustains numerous stages even in small villages like scattered glittering jewels inlaid the forest of mountain. While the stage is located within the precincts of Shinto Shrine, they dedicated a traditional puppet ballad play called "Ningyo Joruri."

For the past four hundred years the stages were cultivated in the regional landscape and played a symbolic role of leading culture of the region. While the climate of warm dampness in the steep typography of the mountains resulted in clean streams and rich green, the stages have been transformed their appearance into the ones with a sense of time and beauty of its sedimentation. As these stages are the most symbolic place to represent their awe and gratitude to Nature, it is the cultural assets that could appeal the cultural value of coexistence with Nature in refined ultimate plain form to all times and places.

Re-naming these stages as MoriButai, precisely call as a stage in the sacred space of shrine, the MoriButai Art Project has aimed to explore a new form of artistic expression in contemporary art that could provide an opportunity to overcome the issue of cultural sustainability faced in the region, evolving into a new open community that share the same value. Thus, MoriButai is the site for new contemporary artistic expressions, dedicating not only to advocate the value of sustainable culture but also explore the crossing of artistic expression of the specific: "only there" and the universal: "nowhere."

While MoriButai attract them to create new relationships to the art, it is culture that always changes. It is hoped that MoriButai will bring them a moment of rich experience. There are many regions that were destroyed their valuable cultural tradition by the sightseeing industry. MoriButai however has been protected in the forest from such thread. The best thing is to see people including seniors so energetic and spirited. It is always great pleasure to learn from their cultural power.

大黄鸭
Rubber Duck

艺术家：弗洛伦泰因·霍夫曼
地点：日本大阪海滨
形式和材料：综合材料
时间：2009 年
委托人：水上大都市大阪 2009
推荐人：里奥·谭

荷兰艺术家弗洛伦泰因·霍夫曼的一生都在创作大型公用雕塑。通常他以制作动物造型的玩具为开端，然后将完成的玩具作为其艺术雕刻的模型。他曾经创作了比真实物体大很多倍的企鹅、猴子、熊，但是毫无疑问，最出名的当数"大黄鸭"。霍夫曼创作的充气"大黄鸭"先后出现在了世界各地的港口及水路，比如尾道市、圣保罗、大阪、圣纳泽尔、奥克兰、悉尼、匹兹堡、北京、中国香港特区、中国台湾桃园及高雄。正是因为在大阪举行的活动，促使该创作被第二届 IAPA 提名。

霍夫曼的"大黄鸭"受"水上大都市大阪 2009"的委托而创作，此次活动于 2009 年夏天由大阪市政府及大阪区政府联合主办，活动地址位于城市的滨海区。该创作由当地人员组成的一支团队来执行，大阪市的"大黄鸭"成为了水上大都市的耀眼之星，在实现主办方的创作目标即促使城市居民重新适应滨海生活方面，发挥了重要作用。该城市同时对近水建筑进行了改进，修建了木板桥，使得当地的公用空间更加适合休闲和娱乐。

如果说游客人数至关重要的话，大阪的"大黄鸭"可以说举办得相当成功。记录图片显示当地观众对"大黄鸭"的热情度极高，在 52 多天之内有大约 190 万游客到此参观。人们不禁会想，是不是因为观赏目标的无公害性。引用艺术家著名的言论即："'大黄鸭'不知国界，不会区分人，不承载任何政治意义。"使得它充满了如此广泛的吸引力。目标的熟知度以及它存在于所有能够被美国媒体及文化触摸到的地方之事实（想想迪士尼的大黄鸭或吉姆·亨森与提线木偶匹配的大黄鸭），都为"大黄鸭"的知名度作出了积极贡献。

"大黄鸭"不仅无公害，还好玩且能够激起人们对童年的回忆。在一次采访中，主创艺术家将"大黄鸭"比作一个大型玩具，而将自身比作一个将玩具扔向全世界的巨婴。他同时表示，"大黄鸭"的存在不仅给人们带来惊喜，还能打破他们的日常生活。类似于在他之前出现的著名艺术家，霍夫曼懂得如何创造奇观。但毫无疑问，"大黄鸭"的存在的确打破了城市

生活的沉闷，有关它下一步将做什么的问题从未被提起。据报道，霍夫曼通过"大黄鸭"在短时间内"绑架了公共空间"，而"大黄鸭"就像大量的图像一样，使得多数城市居民想要沉浸其中，除了提供视觉方面的享受外，似乎并未发挥任何其他方面的作用。

人们很容易联想起情景画家 Guy Debord 的言论。他告诉我们"奇观并不能表达任何比实际更美好的事物，而且那些外表看起来美好的，它原则上所需求的态度实为一种被动接受的态度。"很多观众会站在"大黄鸭"周围并拍合影。然而，这并非意味着他们在积极地创造任何事物。相反，他们实际上是在为过度的复制以及衍生物的产生发挥推动作用。

想回到过去
高浅

1992 年，一艘从中国出发的货轮遇到强风暴，船上的货柜坠入大海并破裂，里面的黄色橡皮鸭子漂浮到海面上，形成了一支庞大的"鸭子舰队"，从此随波逐流。若干年后，这群玩具历经风雨，零零散散从中国漂洋过海到了美国和英国海岸。在霍夫曼的眼中，这是橡皮鸭子们的童话大冒险，充满了趣味性，也成了他制作"大黄鸭"灵感的来源。

我认识一对 50 多岁，没有子嗣的英国夫妇．他们家浴室的梳洗台上摆放了一排形态各异的橡皮鸭。因为惊讶如此充满孩子气的东西会被保留至今，便问原因。他答小时候洗澡时，父母为防止他吵闹，总是往浴缸里放很多只这样的橡皮鸭。如今父母早已过世，但每每看到它们，他总会想起自己的父母，想起自己坐在浴缸里一边洗澡一边玩玩具的那段无忧无虑的美好时光。霍夫曼的"大黄鸭"是他送给全世界大人和孩子们的礼物，对年少无知童年的怀念，对成长的致敬。它放大的不仅是体积，还有美好时光和美好记忆。

Artist: Florentijn Hofman
Location: Osake, Japan
Media/Type: Mix-Media
Date: 2009
Commissioner: Aqua Metropolis Osaka 2009
Researcher: Leon Tan

Dutch artist Florentijn Hofman has made a career out of oversized public sculptures. He often begins with toy animals, using these as models for his work. He has created larger-than-life penguins, monkeys, and bears, but the most well known is undoubtedly the rubber duck. Hofman's inflatable duck has appeared in ports and waterways the world over, including Onomichi, São Paulo, Osaka, St. Nazaire, Auckland, Sydney, Pittsburgh, Beijing, Hong Kong, Taoyuan and Kaohsiung. It is the Osaka event that has been nominated for the second IAPA.

Hofman's Rubber Duck was commissioned for Aqua Metropolis Osaka 2009, a placemaking event organized by the Osaka Prefectural Government and the City of Osaka along the city's waterfront in the summer of 2009. Implemented by a local team, the Osaka Rubber Duck starred as one of the major attractions in the Aqua Metropolis, contributing to the organizers' objectives of (re)acquainting the city's residents with the waterfront. The city also carried out improvements to river-facing buildings and constructed boardwalks to make public spaces in the area more amenable to leisure and recreation.

If visitor numbers are anything to go by, the Rubber Duck in Osaka was a very successful event. Documentary images show a high level of interest from a large local audience—approximately 1.9 million visitors attended over 52 days. One wonders if it

is the innocuousness of the object—a fact celebrated in the artist's statement, "the Rubber Duck knows no frontiers, it doesn't discriminate against people and doesn't have a political connotation" —that gives it such wide appeal. The familiarity of the object, its existence in all societies touched by American media and culture (think of Disney's rubber Donald Duck or the rubber duck that Jim Henson paired with the Muppet Ernie), may also contribute to its popularity.

The rubber duck is not only innocuous, it is also playful, and evocative of childhood. In an interview, the artist likens the duck to a giant toy, and then likens himself to a giant baby throwing his toys around the world. He also says that the duck is there to surprise people and interrupt their daily routines. Like the pop artists preceding him, Hofman understands how to create a spectacle. While it is beyond debate that his duck interrupts the humdrum daily routines of urban life, the question of what it does next is never raised. Hofman reportedly "kidnaps public space" for brief periods with the Rubber Duck, but the duck, like the glut of images in which the majority of urban dwellers tend to be immersed, does not seem to offer anything beyond visual entertainment.

One is reminded of something the Situationist Guy Debord wrote. He tells us the spectacle "says nothing more than that which appears is good, and that which is good appears...the attitude which it demands in principle is passive acceptance." Many members of the audience took photographs of themselves ("selfies") with the duck. This does not, however, mean that they were actively creating anything. On the contrary, they were more likely contributing to the surfeit of repetitive or derivative imagery characterizing the network society.

龙溪动漫花谷
Anime Valley of the Flowers

艺术家：上海大学美术学院
地点：中国浙江省玉环县龙溪镇山里村
形式和材料：综合材料
时间：2011年至今
委托人：龙溪乡政府
推荐人：潘力

浙江省玉环县龙溪镇山里村是21个行政村中最偏远的山村。山里村是传统的农业村，村集体经济薄弱，2011年村集体经济收入仅19万余元，村民以农作及外出务工为主，2011年农民人均纯收入仅3 649元，远低于全县平均水平。为了开拓一条以文化产业推动乡风文明、村庄环境整治，并最终使农民致富的美丽乡村建设路子，由乡政府与上海大学美术学院合作，围绕"让农业有创意，让农村多美丽，让农民更富裕"的目标，将创意农业和公共艺术植入山里村。

山里村的公共艺术项目前期由政府牵头，整合部门项目资金，整治自然和人居环境。提出"慢生活、亲自然、更和谐"的理念，倡导"让城里人体验乡村生活，让村里人同享城市文明"。兴建的"龙溪动漫花谷"有恐龙知识科普、龙门鲤鱼、动漫小屋体验、动漫地球村、3D影屋、蜘蛛侠等二十多个游乐项目。"美丽乡村动漫文化节"开幕当天，只有一千多人的山里村迎来了远近八方逾万名游客。目前已开始吸引周边民众前往旅游，并开通了往返县城的公交车。事实证明，"美丽乡村"建设必须着眼"城乡一体化"，通过引入城市消费者，为乡村提供现金流和文化元素，山里村的"龙溪动漫花谷"在一定程度上就体现了这一思路。

龙溪镇立足于"三农"产业转型，挖掘原有的农业基础和特色生态资源，突出强调乡土特色和人文资源，通过调研和挖掘，归纳、总结、提升，从中找出不同的发展思路。美丽乡村建设也是一项重视环境保护、体现生态平衡、节约资源和能源的系统工程，需要考虑生态、建筑工艺和技术的要求。上海大学美术学院还在龙溪镇建立"公共艺术实践基地"，充分挖掘乡土民风民俗，为村民提供充满活力的生态文化空间。"龙溪动漫花谷"以旅游节庆活动为载体，"亲自然、慢生活、更和谐"的理念整合了乡村自然生态、创意农业、休闲旅游、动漫文化等资源，开始成为推进综合开发乡村生活方式，提升乡村休闲体验产业的知名度和美誉度的一个品牌。

同时，"美丽乡村"建设活动使艺术家与社会之间有了更深入接触的机会和空间。公共艺术在为大众和自然环境的关系创造出更好的理解途径的同

时，广大民众也日渐从正面回应并积极参与当代美术向日常生活的贴近。因此，艺术与社会发展的关系日益密切，与自然、环境共生的理念成为近年来当代艺术的新主题。不仅体现出后现代艺术普遍反思、回归的心态，也促使当代艺术向公共艺术转化，成为社会经济发展的推进力量之一。由于公共艺术关注的是当代乡村文化建设的可能性，对乡村文化的传统进行了启动和再生设计，因此有可能促成乡村及其文化的复兴。

另一个世界

高浅

约瑟夫·波伊斯认为 "人，必须向下与动物、植物、大自然，向上和天使、神互相维系、联接"。"慢生活、亲自然、更和谐"，城市和乡村的生活方式一体化已成众望所需，但都市中的青山绿水早已被封存在冰凉的水泥结构中，好在驱车可达距离仍有一方净土，面朝大海，春暖花开。"龙溪动漫花谷"是上海大学美术学院和龙溪镇共同合作的美丽乡村建设项目，旨在让都市中生活的人们回归自然，体验人文。此项目将现代文明和乡村生活完美的融合，艺术作品的进驻和自然结合，乡村角落所见之处更添几分文化气息。我们固有思想中对于乡村定义被完全打破并重新定义，这里一年百花斗艳，农田不仅是农田，也是艺术创作的基地，时令季节的蔬果、花卉经过设计变成了大地艺术作品。鱼塘成了孩子们玩乐的场地，还有恐龙谷、QQ 农场亲子乐园等实体还原了人们在网络生活中寻求的生态园，将虚拟搬移到现实，远离城市喧嚣，远离网络虚幻，体验淳朴民风，感受自然惬意。乡村虽远离城市，但也美好。

《山里春色》，詹周平 摄

Artist: College of Fine Arts Shanghai University
Location: Shanli village in Longxi Town, Yuhuan County, Zhejiang Province, China
Media/Type: Mix-Media
Date: 2011-present
Commissioner: Longxi Town Government
Researcher: Pan Li

Shanli is a remote mountain village in Longxi Town of Yuhuan County in Zhejiang Province, China. It is a rural region with a weak economy. Farming and working in the cities are the major sources of revenue. The combined village income was only 190,000 yuan in 2011. The net average income of farmers for the whole year was 3,649 yuan, far below the county's average income. In order to promote rural development and give villagers better living opportunities, the county government collaborated with the College of Fine Arts at Shanghai University to work towards reviving the cultural industry and driving the development of village construction. The ensuing introduction of creative culture and public art into the area enriched the villagers' lives.

The Shanli village government funded the public art project, which advocated an ideal human living environment with an improved ecological environment and a more harmonious relationship between man and nature. The project also sought to allow urban inhabitants to experience rural life and to let the villagers share in the city's more modern amenities.

In the Anime Valley of the Flowers there are more than 20 recreation projects. On the day of the Anime Valley opening ceremony, countless people visited Shanli from different provinces of China; the atmosphere was joyful and boisterous.

Experience has proved that the only way to construct a "beautiful new countryside" is by incorporating city and countryside. The beautiful new countryside construction in Shanli attaches great importance to protecting the ecological environment, keeping a balance of ecology and economy, conserving and protecting resources in order to achieve sustainable development.

The College of Fine Arts at Shanghai University established a practical teaching center in Longxi Town, letting the students explore the folk customs and the rural life, while providing a public space of cultural life for the villagers. Anime Valley of the Flowers sparked a public approach to a healthy lifestyle, integrated urban and rural development, and improved the popularity and reputation of the local district.

In the meantime, the creation of this "Beauty County" generated a fantastic opportunity and an expansive space for both artists and society. Public art is responsible for creating a good relationship between the public and the natural environment by allowing citizens to attend and receive public art activities in a positive way, day by day. Therefore, the relationship between arts and the development of society has become increasingly closer in recent years. The concept of art working with nature has become a new topic recently in the contemporary arts. This relationship not only expresses the reflection and dialogue methods of post-contemporary art, but it is also stimulating a huge change from contemporary art to public art, turning it into the main power of social and economic development. As public art plays an increasing role in rural culture construction and the redesigning of county culture and traditions, it will contribute to a renewal of the county and its culture.

静安雕塑公园
Jing'an Sculpture Park

艺术家：多名艺术家
地点：中国上海市静安区北京西路 500 号
形式和材料：雕塑
时间：2007 年至今
委托人：上海市政府
推荐人：汪单

"静安雕塑公园"是上海市中心唯一的雕塑主题公园，位于上海市静安区北京西路 500 号，始建于 2007 年 10 月，是一座现代园林风格的雕塑公园，具有大众休闲、雕塑展示、艺术交流三大功能，被评为全国优秀城市雕塑建设项目、上海市文明公园，是上海市雕塑艺术展示和交流的重要平台。国内外雕塑错落有致地点缀在绿地上树荫中，为上海这座繁华都市增添了更多趣味与亲民的艺术气息。

园区占地面积 6 万平方米，划分为入口广场、流动展示长廊、中心广场景观区、白玉兰花瓣景观区、梅园景观区、小型景观区等六大功能区，并以流动展示长廊为主线，将各个主题景观空间和不同创意的雕塑串联起来，创造不同的视觉效果。

公园已承办了 2010 和 2012 两届国际雕塑展，并致力于将"JISP"静安国际雕塑双年展打造成国际性艺术展览品牌。借助于国际雕塑展，公园陆续引进了一批具有国际影响力的雕塑作品，成为公园的"镇园之宝"，吸引众多国内外艺术爱好人士前来参观。经国际权威机构评选，园内由比利时艺术家阿纳·奎兹创作的"慈航"和法国雕塑大师菲利普·伊其理创作的"巨型风向车"荣获"上海五星雕塑"称号。

"静安雕塑公园"作为专业完整的雕塑公园在全国范围内具有独特的唯一性。围绕现代风格的主题，结合雕塑布局，在景观营造上，以现代手法建设的廊架、百米跑泉、台地园风光等特色鲜明，营造了规整、清新又丰富的公园景观环境；在植物种植上，突出了春景和秋色，春季繁花连绵、秋季红黄色叶，七彩花带四季鲜花烂漫，实现了时间和空间上的景观变化，在曲折廊架围合之中形成了现代梅园。

公园里还有塑胶篮球场、白玉兰步道等运动设施，为喜爱锻炼的市民提供了好去处。平时踏青的游客也可以到园内麓庭咖啡馆里坐下来小憩，尤其是几十座欧美艺术家创作的现代雕塑，别有一番情致。

　　"2013 上海青年蚂蚁设计节"在公园举办，着重演绎"追逐青春梦想，开启创意之门"的精神。公园还成立了青少年公共艺术教育基地，举办一系列活动，如艺术论坛、摄影大赛作品展等，逐步发展成为上海市公共文化生活中的一个品牌。

诗意的栖居
高浅

　　克劳斯·奈特说过："我们必须考虑到公共艺术与内容的关系，充分考虑到艺术必须言说的东西是什么，艺术究竟对谁说话以及它向公众要传递的是什么样的信息，为公众创造什么样的公共空间和公共领域"。海德格尔曾经对生活完美诠释"人应诗意的栖居"。但城市人口日益增多，高楼鳞次栉比，社会日渐光怪陆离。独独缺少了人与自然的零距离接触。对于生活在大城市的人们来说，一抹清新弥足珍贵。

　　一向追求生活品质的上海早就开始着手解决这个问题，始建于 2007 年 10月的"静安雕塑公园"是一座现代园林风格的雕塑公园，也是上海市中心唯一的雕塑主题公园。作为上海市雕塑艺术展示和交流的重要平台，这个项目本身就是一件绝妙的公共艺术作品。聚众分众，公园与生俱来的公共属性让人们聚集，国内外雕塑错落有致地点缀在绿地上树荫中，使人们在进入公园后又拥有自主选择欣赏艺术品的权利。对于艺术的喜好和偏爱让人们行为自由，思想自由，这也是公共艺术的功能之一。

Artist: Multiple artists
Location: West Beijing Road Jing'an District Shanghai City No. 500, Shanghai, China
Media/Type: Sculpture
Date: 2007-present
Commissioner: Shanghai municipal government
Researcher: Dan Wang

Located near the center of Shanghai, Jing'an Sculpture Park was founded in October 2007 and is one of the largest parks of its type in the region. As a modern sculpture park it serves three functions: popular leisure, exhibition, and art exchange. It has been named outstanding city sculpture construction project in the country and was also named a "Shanghai Civilized Park."

The park covers an area of 60,000 square meters and is divided into six major areas: entrance plaza, flow display gallery, and four landscape areas. The landscape space of each area incorporates creative sculpture series to create different visual effects.

The park hosted the 2010 and 2012 Jing' an International Sculpture Biennials (JISP). Through these exhibitions, the park has introduced a number of influential international sculpture works, attracting many domestic and foreign art lovers to visit. Red Beacon, by Belgian artist Arne Quinze, and Les Girouettes Monumentales, by the French sculptor Phillippe Hiquily, won the Shanghai Five-Star Sculpture title.

Jing' an Sculpture Park is unique among China's parks. Its combination of sculpture layout, landscape construction, a modern-style gallery frame, terrace garden scenery, and other distinctive features creates a clean and abundant park environment. The landscaping, which includes hundreds of trees and thousands of flowers, changes to offer the highlights of each season, from spring flowers to autumn's red and yellow leaves.

For public recreation, the park also has a unique basketball court, walking trails, and other sports facilities. Visitors can also sit and rest, and appreciate international artworks from dozens of modern sculpture artists.

The 2013 Shanghai Youth Ants Design Festival was held in the park, focusing on the interpretation of "chasing the dream of youth and opening the door to creative spirit." The park also established a youth public art education base and a series of activities, such as an art forum and a photography contest exhibition. All these factors help the park gradually develop into a brand of the Shanghai public cultural life.

新工人艺术团
The New Workers Art Troupe

艺术家：孙恒（主创人）、屈远方、许多、王德志、姜国良、宋德忠等
地点：中国北京朝阳区金山镇皮村同心学校
形式和材料：互动行为
时间：2002 年
推荐人：汪单

皮村位于北京城的东北五环外，多年前它是一座村庄，如今已被纳入北京城区的范围。由于这里紧邻首都国际机场，发展比较缓慢，房屋的租金相对较低廉，因此吸引了不少外来务工人员居住。皮村成为了外来打工者聚集的新社区，本地人口大约 1 400 人，外地人口大约有 1 万多到 2 万人，相应的配套设施如出租房和商铺也日益增多，皮村大概有大大小小的工厂企业 205 家。

北京工友之家是一家为打工者服务的草根公益机构，成立于 2002 年，长期致力于打工者聚居区的小区工作。从 2005 年在皮村开始开办同心实验学校，为打工子女入学提供方便；建立免费借阅的图书室；开展法律、计算机等培训；建立各种文艺小组，编辑皮村报等。他们成立的"新工人艺术团"经常在皮村演出。自成立以来，出版了七张原创歌曲专辑和几部电影、纪录片、戏剧作品，策划了打工春晚。"新工人艺术团"最突出的特点，一是旗帜鲜明地"为劳动者歌唱"，他们的歌曲都是与打工群体的生活状态、劳动状态密切结合在一起。二是艺术创作和社会实践密切结合在一起，他们的创作，是和建立同心实验学校、工友之家、打工博物馆、新工人剧场、同心互惠商店、工友影院等社会实践紧密联系在一起的。

2008 年"新工人艺术团"发起人自发筹建了"打工文化艺术博物馆"。"打工文化艺术博物馆"实则是这间仅一百多平方的小院落，由放映室、展厅、图书馆和在展场外面的帐篷剧社组成。"打工文化艺术博物馆"成为在北京五、六环之间广大的城中村区域的文化中心，其一方面的功能是以实物、文字、档案等记录打工人群自己的历史和文化；另一方面，则扎根于皮村社区，服务于本地居民文化生活的直接诉求，并以凝聚社区认同和促进成员沟通作为一项工作重点。

融入与归属

高浅

创建公共艺术，也许并不单指创建实体作品，也可能将某种公众行为艺术化。设计师和规划者们在开展项目之前，应对空间的主体、空间建构及周围人文环境有相当程度的了解。公共艺术建构的空间，是一定公共区域范围内人们生活世界的有机构成部分，公共环境作为艺术化的审美空间充满了对生活和人文精神的感知与感悟。

对外地打工者来说，融入一个城市的困难也许并不来自经济，更多的是城市文化对他们的拒绝，而正由于城市主流文化难以对他们产生认同，务工者则更难在精神上融入都市。"新工人艺术团"关注外来打工者的利益，尝试在他们的生活区域及生活范围内创造完美的精神生活环境，通过建立免费借阅的图书室、创立文艺小组、开展文艺活动、组织文化培训机构等项目，增强外地在京务工人员对所生活城市的文化认同感及自我归属感。

Artists: Zhang Huan (Initiator),Qu Yuanfang, Xu Duo, Wang Dezhi, Jiang Guoliang, Song Dezong
Location: Tongxin School, Picun Village, Jinshan Township, Chaoyang District, Beijing 100018, China
Media/Type: Interactive Behavior
Date: 2002
Researcher: Dan Wang

Picun used to be a village situated on the outskirts of northeastern Beijing, near Capital International Airport, but now it is considered part of Beijing city. Due to slow development and relatively low-rent housing, Picun is an ideal place for migrant residency. Only 1,400 local people live in Picun, while the migrant population numbers about 10,000 to 20,000. The related facilities such as rental housing and shops are also increasing, and there are currently about 205 factories and other enterprises.

The New Workers Art Troupe is a grassroots nonprofit organization initiated by a group of migrant workers. It is committed to the local community, serving public interests through art. The team members are migrant workers who have certain talents and who are willing to devote their spare time to other workers. Since its establishment on May 1, 2002, the Troupe has left its footprints in a wide range of places, including construction sites, factories, universities, various enterprises, schools for migrant children, and communities where migrant workers reside. It founded the Tongxin Experimental School for migrant children's education in 2005. It also established a library that offers free book lending as well as training in computer skills.

The Troupe has performed more than 500 times, reaching out directly to audiences totaling more than 100,000 people, and has gained a lot of attention and praise from both migrant workers and the general public. Since the establishment of the art troupe, it has published seven albums of original songs, several documentary

films and drama works, and planned the Spring Festival Gala especially for migrant workers. The aim of the Troupe is to provide a voice for migrant workers—to call out with their songs and defend their rights through art.

The New Workers Art Troupe has created several entities and enterprises on the basis of social practice, including Tongxin Experimental School, Migrant Worker Family, Museum of Workers, New Workers Theatre, Tongxin Reciprocity Shop, Cinema of Migrant Worker, and others. It is dedicated to building a cultural brand for migrant workers, to depict and promote the workers' healthy, active, and enterprising spirit. Various programs and worker-initiated activities help workers to establish their new self-identities and enrich their leisure lives.

In 2008, the initiators of the New Workers Art Troupe spontaneously established the Migrant Worker Culture and Art Museum. The museum is a small courtyard, little more than 100 square meters, and consists of a projection room, an exhibition room, a library, and a theater. This museum has become the cultural center of villages in the city between the 5th Ring Road and the 6th Ring Road of Beijing. On the one hand, the aim of the museum is to record the history and culture of migrant workers via material objects, written materials, files, and so forth. On the other hand, it entrenches itself in the communities of Picun village, providing services aimed at the direct requirements of local inhabitants in their cultural lives. Fostering the cohesion of recognition and promoting communication among community members are considered the key emphases of its work.

许村国际艺术公社
Xucun International Art Commune

艺术家：渠岩
地点：中国山西省和顺县松烟镇许村
形式和材料：综合材料
时间：2011 年 7 月至今
推荐人：汪单

"许村国际艺术公社"地属中国北部山西省和顺县，坐落在太行山北部最高峰的一个高山盆地里，这个具有两千多年历史的村落呈现出了它优美和丰富的东方情调，并且一直延续了自中世纪以来中国传统的民居特色与民俗风貌。和顺地区风景秀美，气候宜人，处处是桃花源般诗情画意的乡村。许村民风淳朴，历史遗存丰富鲜活，保留了从明清时期到现代完整的历史线索的建筑和民俗生态。

根据查阅资料和对村里老人的访问，许村的历史最早可追溯到春秋战国时期，是晋国与鲁国相接之处，村南边的夫子岭至今还流传着孔子"倒翻坡"的故事。许村旧村的遗址是唐朝晋王李克用驻军的营寨，村民现在都还称呼旧村遗址为"大寨"。许村的祖先是于、杨、范、王四个大姓，原来散落在山上居住，明朝开国年间迁到山下建村。

国际艺术节是"许村国际艺术公社"定期举办的每两年一届的活动，邀请世界各地的艺术家参加。2011 年 7 月已经成功举办了首届"和顺乡村国际艺术节"，开创了在中国传统文化腹地举办国际当代艺术创作活动的先例。首届"和顺乡村国际艺术节"活动邀请了 14 名国际艺术家和 6 名中国艺术家一起生活和创作 15 天。活动包含：1. 国际艺术家驻地创作；2. 民俗文化参观；3. 中西艺术家、文化学者的交流与研讨；4. 公共艺术教育辅导；5. 驻地创作交流展等内容。

许村计划由渠岩以艺术家身份的率先介入使得该项目得以启动，是一个建筑师、各自然科学和社会科学专家、村民并以政府政策和行政为中介的互动过程。渠岩说："不破坏原有的来建设一个新农村，不是表面的新，而是有内容注入，是实际生活的态。让年轻人能回来，这才是乡村复苏的根本问题。"

许村现在有 1/4 的村民在经营农家乐，许多在外打工的村民回到家乡做生意，村里的人气兴旺许多。渠岩还在思考着许村的持续发展。许村的艺术

节要能不依靠政府拨款自我运转，他们计划把 20 世纪 70 年代仿造遵义会议建筑的村中学改造成艺术家公寓，让更多的艺术家能在许村设立工作室。随着艺术节的开办，村粮仓已经放不下那么多的艺术品了。

"许村国际艺术公社"不是一项艺术家美化乡村的活动，更不是做乡村旅游，其目的是保持村民鲜活的形态，抢救该地区的文明。艺术家的行为比其他学科的专家更领先，而且艺术家的情感式介入能够召唤更多的公众热情。从细节上来说，渠岩像创作作品一样地在许村做着每一件公共事务，比如带头在许村拣垃圾，发展村民一起拣垃圾，使一个脏、乱、差的乡村变得干净。许村计划整整花了五年的时间，而和村民的关系、当地政府的关系也是错综复杂的，这种关系处于坚持和妥协之间。

思念有因
高浅

民间传说中女娲用五色石补了天，用五色土造了人，所以天地之间，每一位炎黄子孙都对乡土具有割舍不断的情怀。尽管城镇化建设曾经吸引着数不胜数的人们从村庄走向城镇工作定居，但无论是少小离家老大回，还是举头望明月，低头思故乡，诉说的都是对家乡的无尽思念。而让已经习惯城市生活的人们毫无负担和压力的回到故乡，正是未来的趋势。许村计划中试图达成的理想境界也即是此。旧屋重塑，乡村改造，塑造的不仅仅是形态，而是情感的重塑。在飞速发展的社会进程中，在金钱和利益的驱使下，很多旧时的记忆随着建筑的拆迁和重建灰飞烟灭，而叶落归根的故土情节却一直根存于脑海，难以磨灭。许村计划作为中国新农村建设的样板，必将引领城市和乡村共建共同发展的潮流，根植于这个社会并影响人们的思维和生活方式。

Artist: Qu Yan

Location: XuCun, SongYan Town, HeShun County, Shanxi Province, 032707, China

Media/Type: Mix-Media

Date: July 2011-present

Researcher: Dan Wang

Xucun International Art Commune is situated in a mountain basin of Taihang Mountain, which is located in Heshun County in Shanxi Province of northern China. The village has been here for 2,000 years and still keeps its graceful and abundant oriental tone and the characteristic Chinese traditional common-style dwellings and ethnic look. Heshun County has a beautiful landscape and paradise-like villages. The climate here is agreeable and the people are warm-hearted. The abundant historical buildings, completely preserved, trace back to the Ming or Qing Dynasties.

According to reliable documents and interviews with the elderly people in the village, the history of Xucun goes back even farther to the Spring and Autumn Period (approximately 771–476 B.C.), and the village is located at the border between the ancient states of Jin and Lu. Fuzi Ling, the south end of the village, is still spreading stories of Confucius. The site of Xucun village is the site of the army camp of King Li Ke (the king of the old Tang dynasty), and villagers are still calling the old village sites dazhai (outposts). Xu's ancestors were mainly composed of four families—Yu, Yang, Fan, and Wang—who lived scattered in the mountains and then moved to the mountain village during the founding of the Ming dynasty.

The International Art Festival is the regular activity held every two years at the Xucun International Art Commune. Artists from all over the world are invited to participate in this festival. The first Heshun County Art Festival, held in July 2011, was successful and set a precedent of conducting contemporary art creation activities

in the traditional Chinese culture hinterland. Fourteen international artists and six Chinese artists joined the first festival. Activities mainly included (1) international artists' residency, (2) visiting folk art culture sites, (3) exchanges and discussions among Chinese and Western artists and cultural scholars, (4) public art education and counseling, and (5) residents' creative exchange and exhibition.

The Xucun project is the first international contemporary art commune for artistic creation that has been established in the heart of a traditional culture area—in this case, Shanxi ancient villages. The project began with Qu Yan's artistic intervention and became an interactive process between artists, architects, scientists from different fields, sociologists, and farmers by government intervention.

Qu Yan said: "We are trying to protect, trying to keep the original conditions of the village, on the basis of no destroying to build a new village. Besides, we put the new contents into the 'new village.' " Here, the so-called "new" refers not to the outside but to the inside. Attracting younger people to come back is the essential issue of county recovery.

The village is not simply a project that artists constructed, and it is not a center for rural tourism. As an emerging nonprofit art institution in China, Xucun International Art Commune's aim is to keep the original village and its local culture. Artists' intervention left the other scientific experts far behind, and artists' emotional intervention was more inspiring than that of the others. The specific things that Qu Yan did in Xucun looked like public affairs. For example, Qu led farmers to pick up rubbish, which made the village clean. The process took five years, and the complicated relationship between farmers and local government was aided through persistence and compromise.

兼容的盒子计划
Bazaar Compatible Program

艺术家：夏意兰、保罗·德沃图
地点：中国上海安顺路 98 号
形式和材料：互动行为
时间：2011 年
推荐人：汪单

"白立方"（画廊／艺术空间）已不是唯一展示艺术的空间，越来越多的艺术创作则借由各种手段，无孔不入，渗透浸润于日常生活。在探讨全球化、网络化的今天，兼容性已经成为下载形式美学的第一个标准。"兼容的盒子计划"通过邀请来自世界各国的艺术家来考察和挖掘新颖的，与社会环境相融的"兼容格式"，同时，各种形式的艺术在这个实验杂货铺"小盒子"持续登场，潜移默化地开拓当地民众对艺术的认识。

这个迷你空间位于安顺路新华小商品市场内，是小街诸多杂货铺中的一间。没有门，没有标识，没有解释。项目负责人邀请艺术家就周边环境，提出改变这个空间、并保障不改变与周围环境和谐兼容气氛、达到融入日常生活之即时效果的个人创作想法。

"兼容的盒子计划"由习艺堂——上海国际艺术研究生院发起，已纳入法国南锡高等艺术学院"创造力与全球化"研究项目，由上海得译文化艺术交流咨询有限公司运营支持。从 2011 年 9 月至今完成的 64 个项目（2014 年 3 月 9 日），空间负责人给予艺术家两周空间的使用权，这个空间不仅是一个较知名艺术家小型、实践创意的场所，而且也是一个展示年轻艺术家习作的场地。

公共艺术是基于社会精神活动下的行为，其最终目的不仅是产生固化的物质形态，而是通过"地方重塑"的观念来催生公共空间的文化生长性，投射出当代生命意义与多元文化生态之间的密切关联。"地方重塑"不仅从物质层面将视线引向公民问题，更从精神层面引入对社区的关怀与感召。近年来，国际上的"社区艺术"（Community Art）继续得以多方拓展，通过研究和优化地域生活方式，让新的社会互动成为提升社区及居民生活品质的重要推力。

从这个意义上说，公共艺术对于地区的建构就不仅是一件艺术品或一种艺

术现象、艺术方式进入当地，内在的生长潜力会让其自然地融入到地区生态的每一个元素当中，在不断交互的过程里，催生出鲜活的良性生长机制。

在这个新机制里，地区的艺术潜力被逐渐激活，地区的发展面貌将呈现出更加灵活、开放的趋势。因此，公共艺术以创造性的方式改善当地居民的生活与生态环境，重塑地域的内在气质和风貌。因此，"兼容的盒子计划"留给地区的不是单纯的艺术品，而是一个活的成长基因。

闹市中的艺术

高浅

虽然有时候精神需求被认为比物质需求高尚，但是对精神需求和物质需求的培养往往需要同时开展。"兼容的盒子计划"是一个关注艺术如何更好融入特定阶层生活空间的艺术项目。为了提升群众在艺术行为中的参与度，为了培养特定空间内群众对多种艺术形态的认知并重塑该地区对文化的观念与态度，艺术家以居民生活环境为划分指标，将项目区域化。项目最终被落实在喧闹的小商品市场内。项目负责人邀请艺术家根据项目所处环境，提出改变这个空间、并保障不改变与周围环境和谐兼容气氛、将艺术思想与周围环境融合的个人创作理念。进驻的艺术项目按照时间定期更新，艺术形式不拘一格。此种行为和方式证明了精神文化的传播无法摆脱已固化的物质形态，也完全无法脱离群众实际生活。当前艺术所面临的很大一部分问题的根源在于精英文化与草根文化的碰撞，而此种行为将破除这种偏见，让艺术更加平民化，文化生命力更加多彩和顽强。

Artist: Xia Yilan and Paul Devautour
Location: 98 Anshun Road, Shanghai, 200051, China
Media/Type: Interactive Behavior
Date: 2011
Researcher: Dan Wang

Beyond the white cube and the quest for specificity that have characterized the history of modern and contemporary art, the latest practices of art are implementing strategies of infiltration, immersion, and encryption. Compatibility has become the first criterion of downloadable aesthetic forms.

In this experimental booth, located among the retail stores in a typical Shanghai market, various art programs continuously take place and broaden local people's knowledge of art. This mini-sized space is less than 10 square meters. There is no door or logo, nor even any labels on the space. This program is designed to explore new compatible formats developed by the artists in a global networked culture, but in a way that is harmonious and compatible with the surrounding environment.

Bazaar Compatible Program was designed by Xiyitang and Shanghai International Graduate School of Arts as part of a research program called Creation and Globalization at École Nationale des Beaux Arts de Nancy. It implemented by Shanghai Deyi Studio. The space hosted 64 projects between March 9, 2011, and September 2014. Invited artists could use this space for two weeks. It is open to different kinds of arts and is also a playground for experimental projects. The space is designed not only for experienced artists but is also a place for young artists to present their creations.

Public art is based on the behavior of the social activity of the spirit, whose ultimate goal is not only the solidified physical form, but also to give birth to the concept of public space and cultural growth through placemaking. At the same time, it also projects a contemporary meaning of life closely associated with cultural pluralism. Placemaking views civic issues not only through physical vision but pays great attention to community concerns at the level of spiritual inspiration.

In recent years, international "community arts" continue to be expanded through research and optimization of regional lifestyles. As a result, the new social interaction becomes an important thrust in improving communities and residents' quality of life.

In this sense, public art for the building of an area is not only a work of art but an artistic phenomenon or art which merges into the local environment. In addition, intrinsic growth potential will let it take its course into each element of the area's ecology. Ultimately, it will create fresh healthy growth in a process of constant interaction. Under the new mechanism, the artistic potential of an area is gradually activated and regional development will be more flexible and open. Therefore, public art improves local people's lives and the environment and restores the inherent temperament and style of a location in a creative way. Thus, Bazaar Compatible Program is not simply a work of art, but a generator of living growth.

3331 千代田艺术区
3331 Arts Chiyoda

艺术家：中村政人
地点：日本东京千代田区
形式和材料：互动行为
时间：2010 年至今
推荐人：潘力

坐落在东京千代田区一所翻修过的初中里，"3331 千代田艺术区"提供了一种新形式的艺术空间。名字"3331"来源于日本江户时代的一种民间祈福仪式，传达的思想是"艺术让我们的生活更加美好"。

作为一个公共空间，这栋建筑允许市民自由出入。人们，尤其是孩子和老年人可以在此学习或娱乐。这座四层的建筑包含了美术馆、办公室、咖啡馆和一些饭店。许多赛事和展览都在这栋建筑物内举行，包括讲习班和有关艺术的对话。建筑物隔壁是一个有特色的公园。这座兴建于关东大地震之后的建筑物，被认为是一个基本的城市设施。因此，它也是回避自然灾难的一个避难所。

中村政人是来自东京艺术大学美术学院的一位教授，在 2009 年发起了这个项目。他是一位当代的艺术家，注意发掘日常生活的现代元素，也擅长把流行文化中的符号融入到他自己的艺术作品中。近年来，他开始关注艺术教育问题，并且对传统的教育方法和手段提出了建议。他关于这个问题的想法导致了创立这个设施的过程，这是一个把他的理念变成现实的地方。

在"3331 千代田艺术区"开始运行的第一年，中村政人收到了 2010 年科学大臣奖奖项，以此奖励他作为艺术家所发起的独特的公众艺术活动并刺激了市民的生命活动。2012 年，这栋建筑荣获了日本建筑学会奖的奖项。

当今，公众艺术和日本社会之间的关系与人们的日常生活变得越来越密切，尤其是在 2011 年 3 月 11 日地震之后，围绕着自然环境和公众艺术关系的观念和社会模式已经成为日本的一个新话题。艺术的力量日益用来刺激地方改造和地方发展的方案。

"3331 千代田艺术区"对东京和其他国家都有巨大的影响。自从它于 2010 年开放以来，已经接受了超过 56 万人次的参观，其中有 1/3 的游客来自外地。截止到 2012 年，在这个建筑里成功举办了 12 次大型活动，包

括文化和体育活动，家庭活动以及不同主题的讲习班，比如插花、艺术剪纸。同时，这个建筑也为年轻的艺术家提供了一个举办展览的场所。还值得一提的是，这里还举办过特殊的活动，就是残疾人同艺术家一起参与创作并创造艺术品。这个特殊活动在2011年和2012年两年内已经举办了三次。此外，这个艺术中心也成立了"千代田艺术节"，为参与者提供了一个自由创造的空间。

小空间里大文章

高浅

随着城市范围日益扩大和商业中心的逐渐区域化，越来越多在居住区生活的人们出门行走需要依靠繁复的交通系统。在远离商业中心，交通不发达地区，外出购物娱乐的不便利性就随之增加。目的地的分散往往需要人们往返穿梭于不同地点，这样的行程费时耗力，一圈下来，通常事情还没有做完，却已经精疲力竭。但人类毕竟是血肉支撑的实体，哪怕物质欲望可以借助科技完成，身体和精神上的需求也很难通过电子化获得满足。

为了应对都市人群在快节奏生活下对于便捷性的需求，日本艺术家中村政人利用"3331千代田艺术空间"提出了有效的解决方式。日本的设计精神是在小空间里做文章也可以至善至美，他将东京千代田区一所废弃的中学校舍，改造成为一种新形式的公共艺术空间。这栋建筑允许市民自由出入。人们，尤其是孩子和老年人可以在此学习或娱乐。建筑一共有四层，不光包含商业办公区，同时还有美术馆、咖啡馆和餐厅。许多赛事、展览和其他活动都可以在这栋建筑物内举行。考虑到日本经常会发生地震的地区特征，这个建筑物的选址也是极有讲究，其隔壁是一座公园，因此它也是躲避自然灾难的避难所。

Artist: Nakamura Masati
Location: Tokyo, Japan
Media/Type: Interactive Behavior
Date: 2010-Present
Researcher: Pan Li

Located in a renovated junior high school in the Chiyoda district of Tokyo, 3331 Arts Chiyoda offers a new kind of artistic space. The name 3331 is derived from a folk blessing ceremony in the Edo period of Japan, conveying the idea that "arts make our lives better."

As a public space, this building allows citizens to enter freely. People, especially children and elderly people, can study and find entertainment here. The building's four floors contain galleries, offices, a café, and restaurants. Many events and exhibitions are held in this building, including workshops and dialogues about art. A special park is located next door. The building is considered a basic city facility, built after the Great Kanto earthquake; therefore, it is also seen as a refuge for avoiding nature hazards.

Nakamura Masato, a professor at the Fine Arts College of the Tokyo University of the Arts, initiated this project in 2009. He is a modern artist who pays attention to the contemporary elements in daily life. He is also good at incorporating the symbols from pop culture into his artworks. In recent years, he has become concerned about arts education issues, and gives suggestions for traditional education methods and approaches. His thoughts on this topic led to the process of creating this facility, a place that can bring his ideas into form.

In the first year of 3331's operation, Nakamura Masato received the Awards of Science Minister Prize 2010 for launching a unique public arts activity as an artist and stimulating citizens' life activities. In 2012 this building was honored with the Awards of Japanese Architecture Academy Prize.

Nowadays, the relationship between public art and society in Japan is becoming more and more close to people's daily life. Especially since the 3/11 earthquake, the concept and the society mode which surround the relationship between the natural environment and public art has become a new topic in Japan. Increasingly the power of art is used to stimulate the plans of place remaking and place development.

The 3331 art center has a huge impact in Japan and in other countries. More than 560,000 people have visited this place since it opened in 2010. A third of the visitors have come from outside of Tokyo. By 2012, this building had successfully held 12 mega-events, including culture and sports events, family events, and different theme-workshops such as flower arrangement and artistic paper cutting. Meanwhile, this building also offers younger artists a location for holding exhibitions. Worth mentioning also is a specific event in which disabled people work together with artists and create artworks. This event has already been held three times within the years 2011 and 2012. This arts center also founded the Chiyoda Arts Festival, which offers a space of creative freedom for participants.

GTS 观光艺术项目
GTS Sightseeing Art Project

艺术家：保科豊巳
地点：日本东京
形式和材料：综合材料
时间：2010—2012 年
推荐人：潘力

"GTS 观光艺术项目"是被当作一个新型的当地社区合作工程来创建的，它的三个组织实体是东京艺术大学，台东区和墨田区的城市。这个项目的活动包括环境艺术作品的安装、艺术活动、工作坊和其他活动。从 2010 到 2012 的三个学年里，项目一直有三重目标，把此地区变成连接晴空塔和浅草区的室外博物馆；把它改造成一个有益于本社区的艺术和环境中心；让它成为孵化新艺术的创造区域。

通过探索现在位于一个不断变化的状态，东京艺术大学的师生根据世道和文化的转变上演了不计其数的活动，并学习到隅田川两岸许多地区的历史积淀。他们通过两个项目去完成上述的活动：艺术环境项目和隅田川艺术桥，一个是永久持续的，而另一个则依存于它。

在东京晴空塔的观察要点里，艺术环境项目确立了永久的环境艺术作品。2011 年隅田川艺术桥是一组 10 个由工作坊组成的临时项目，当地居民可以参与其中，包括"东京晴空塔绘画"图片展以及相关的活动，比如在"GTS 音乐会"下的六场演唱会。

胜美达市幸运地被选为建造东京晴空塔的地点。隅田川是这个项目立项的地点，因为它刚好流经东西之间的核心位置，就是台东区的浅草寺和胜美达的东京晴空塔之间。在 2010 年，艺术家们开始进行一个三年的当地社区合作项目，并从文部科学省以及台东区和墨田区寻求融资支持。我们的计划是第一年播下种子，第二年成长，第三年收获。在项目结束之日，将会有共计 12 个艺术作品矗立在那里。

此外，还举行了两个子项目：一个工作室，当地居民可以在里面参与描摹东京晴空塔的图像；另一个国际音乐演唱会，包含四个独唱会。同时，东京艺术大学把发展 GTS 作为其教育活动计划的一部分。

对于台东区和墨田区来说，"GTS 观光艺术项目"是极其重要的一个任务。这个项目有助于提升此地区的景点。通过再一次调查隅田川两岸已经发现的艺术文化的隐藏记忆，并通过各种艺术方式将这些记忆在表达中自由运用，各个遗迹相互串联起来就创造出一条艺术之路，在对日本文化认同的过程中升华为艺术。

修补文化裂痕的方法
高浅

詹妮特·卡丹认为："公共艺术提供一种修复当代生活与我们已然失去的事物之间裂痕的方法。"而记忆重塑，历史重塑就隶属其研究范畴之中。在此之上再进一步构建文化性的空间，让人能够主动进入，参与和享受文化的世界，最终使公共艺术的参与过程成为具有民主性的审美观点的公共行为。换句话说：公共艺术应竭力促成文化认同，服务于人的精神和物质需求，例如让生活更便捷，让生活更美好。保科豊巳的"GTS 观光艺术项目"完成了以上对于公共艺术所有职责和属性的界定，旨在提倡发展公众参与并重塑记忆与历史的概念。项目的三个合作方分别是日本东京艺术大学，台东区和墨田区。作为子项目进行的工作室和音乐会整合群众积极参与过程和成果中，这不仅有助于提升观光者的人数，且可以灌输文化理念，将各个具有独特历史记忆和文化特点的经典整合、开发和利用。

Artist: Hoshina Toyomi
Location: Tokyo, Japan
Media/Type: Mix-Media
Date: 2010-present
Researcher: Li Pan

The GTS Sightseeing Art Project was created as a new local community partnership project organized by three entities: Tokyo University of the Arts and the cities of Taito and Sumida. Activities of this project include installation of environmental art pieces, art events, workshops, and others. For the three academic years from 2010 through 2012, the project's threefold aim has been to use art to turn the area connecting the Tokyo Skytree with Asakusa into an outdoor museum, to transform it into an artistic and environmental center that contributes to the local community, and to make it a creative zone that spawns new art .

The university's faculty and students have put on numerous events with chronological and cultural crossovers by discovering how the present is in a state of continuous flux, and have learned about the accumulated history of the many areas on each side of the Sumida River. They have done this through two projects: the Art Environment Project and the Sumida River Art Bridge, one permanent and the other event based.

The Art Environment Project establishes permanent environmental art pieces in key viewing points of the Tokyo Skytree. The Sumida River Art Bridge 2011 was a group of ten temporary projects consisting of workshops in which local residents participated,

including a "painting Tokyo Skytree" picture exhibition, as well as art-related activities such as six concerts under the title "GTS Music Concert."

Sumida City was fortunate to be selected as the site for the construction of the Tokyo Skytree. The Sumida River is the ideal location for the project because it flows right through the middle of the east-west "nucleus" between Senso-ji Temple in Taito and the Tokyo Skytree in Sumida. In 2010, the artists began a three-year local community partnership project with funding from the Ministry of Education, Culture, Sports, Science, and Technology (MEXT) as well as support from Taito and Sumida cities. Our plan was to sow the seeds in the first year, to grow in the second, and to blossom in the third. A total of 12 art pieces were there at the end of the project .

In addition, two sub-projects were held: a workshop where local residents participated in making portraits of the Tokyo Skytree, and a GTS International Music Concert consisting of four recitals. Meanwhile, Tokyo University of the Arts is developing GTS as a part of its educational activities program.

The GTS Sightseeing art project is an extremely important undertaking for the cities of Sumida and Taito. The event serves to highlight the attractions of this area. By researching hidden memories of art culture found on both sides of the Sumida River one more time and utilizing them freely in expressions done by various artistic methods, the vestiges of one dot are connected to another, creating an art line in the process of affirming Japanese cultural identity which is sublimated into art.

红毛港
Hongmao Harbor

艺术家：李亿勋
地点：中国台湾高雄市
形式和材料：综合材料
时间：1993 年至今
委托人：红毛港文化协会
推荐人：潘力

红毛港原为高雄港内传统渔村聚落，内临高雄湾舄湖，外濒台湾海峡，居民多数从事捕鱼及养殖相关渔业，曾经拥有富裕辉煌的岁月，1967 年高雄港第二港口开辟，红毛港被划定为第六货柜中心预定地，从此遭受禁建与限建的法令限制，最终面临迁村的命运。由于政策的延宕，40 年内红毛港处于时空冻结的状态。

1990 年代台湾政治解严后，各地小区意识逐渐抬头，红毛港地区民众经历台电储煤场污染事件、高雄第二港口封港事件等重大抗争行动，在这期间，小区民众自主成立"红毛港迁村自救会"，以争取切身权益；"红毛港文化协会"则以保存历史文献及影像文物为主要工作。在两个协会与民众的努力争取之下，让红毛港在迁村的过程中，保障了一些应有的权益，并保留了部分文化资产与共同记忆。

红毛港文化协会积极保存地方文史并参与小区营造，藉此凝聚民众小区意识，其中"小区游廊"的建造结合小区民众以马赛克镶嵌表现融入在游廊的铺面，深获好评。同年迁村计划确定执行，民众平日赖以维生的环境与网络，亦面临摧毁，基础产业是小区总体营造的原动力，因此兴起将马赛克镶嵌技术应用于小区产业，利用马赛克镶嵌技术的劳力密集特性，将小区民众因迁村而改变的失业人口，培养成具备镶嵌技术之从业人员。

高雄师范大学美术系、视觉设计系师生协助设计与技术指导，包含美感与职工训练等课程，将技术与观念转移至小区，协助解决红毛港居民因迁村而产生之失业问题，培养当地民众学习新技艺，创造红毛港小区活力，带动传统工艺技术之复兴。

2007 年多元就业开发方案成果评选，红毛港马赛克镶嵌工作室获选为特优单位。考虑到工作室未来发展的可能性，与摆脱受政府经费补助的限制性，于 2008 年工作室正式转型成立"红毛港文创公司"，成为台湾多元就业

项目转型的首例，未来的发展虽无法预料，但对愿景也充满着期待。通过本项目的分析，可了解小区营造与文化产业的概念，日本宫崎清教授提到内发性的城乡建设，而"内发性的"发展策略，是由居民内部产生出来构想和方案，这些方案也经过"外界的诱导"而产生。

在小区营造的过程中，通过民众参与，达成了小区民众的共识凝聚。并在项目实施的过程中，发展出马赛克镶嵌艺术，在经历了实验摸索阶段之后，技术逐渐成熟，然后再将专业技术导入小区，并藉由政府资源与专业艺术工作者的协助，成功推广马赛克镶嵌艺术，帮助红毛港妇女就业与技能学习。2007年，红毛港迁村完成，居民散置于各处，但在红毛港文化协会的努力下，对地方文史的保存，留下了见证与薪火，期使高雄市红毛港未来能发展成为具地方特色的文化创意产业。

拼贴的艺术
高浅

在台湾历史中但凡关于荷兰与西班牙之地方名称或器物等均冠以"红毛"，而红毛港也由此而来。时过境迁，历史翻页，红毛港当今的经济发展已经与当地文化紧密结合。任何一个凝结本土居民意识且具有文化历史的地区都应被保护。为让地区文化和文明免于规划破坏，艺术家运用马赛克作为艺术元素，在公共空间中摆放由民众自己动手制作的艺术品，这是民众参与机制的建立，有效的推动红毛港地区文化历史保护工作，深化增强小区内群众的文化凝聚力。从而从经济，社会和文化三个维度开展以历史文化结合艺术向文化创意产业转型的项目。创意产业与文化发展和社会发展的作用是相互的，不光可强调公民社区生活及身份的认同，还可以强调文化价值和创意社区的共融。而以此为基点，深度开发本地资源及民族特色，民族庆典，文化遗产，公共空间和传统等。

Artist: Yi Hsun Lee
Location: Taiwan KaoHsiung, China
Media/Type: Mix-Media
Date: 1993-present
Commissioner: Hongmao Harbor Culture Association
Researcher: Pan Li

Hongmao Harbor, located between Kaohsiung Harbor and Taiwan Strait, used to be a traditional fishing town within Kaohsiung Harbor. The local residents worked mostly in fishing and aquaculture-related business. Following the completion of the second harbor in Kaohsiung, the small town's glory years came to an end in 1967 with the imposition of a building moratorium that lasted 40 years.

In the 1990s, following a coal mine pollution accident and facing increasing unemployment and the threat of town relocation, the townspeople pulled together to help themselves and resist the relocation. The Hongmao Culture Association was formed to protect local history and promote residential construction in ways that make the population more cohesive. The association primarily focuses on the preservation of historical literacy and images.

When the plan to relocate the village was confirmed, it was clear that this would cause destruction of both the natural environment and the social network of the village. Many villagers lost their jobs because of the relocation. Thanks to efforts on the part of both the association and the public, part of the town's cultural capital and common memory was preserved during the relocation process.

The Fine Arts and Visual Design departments of National Kaohsiung Normal University took the opportunity to create a mosaic "village corridor," drawing the manpower for the project from the unemployed villagers themselves and teaching them the techniques of inlay. This not only helped ease the village's unemployment, but it also stimulated the renaissance of traditional art techniques.

In this case study, the author used public engagement to achieve the cohesion of awareness that was needed. The process evolved from the design of the mosaic inlay art to an early experimental stage. Eventually, as the project participants developed greater skill, they were then able to share their expertise with others in the region.

Support came from government sponsorship and from the professional craftsmen who successfully promoted the mosaic inlay art. The project helped reduce unemployment and improve the skills of the women of Hongmao Village. When the relocation of the village was completed in 2007, the former residents moved to different areas. However, thanks to the efforts of the Hongmao Culture Association, the local literature and history were preserved.

红土地乐园——头湖红堡垒
Touhu Red Fortress

艺术家：黄浩德
地点：中国台湾台北市
形式和材料：互动行为
时间：未知
推荐人：潘力

在台湾公共艺术发展过程中，实际执行部门包括主管机关、主办机关、策划及创作者。其中主管机关的任务在于行政立法以及实际执行面关于法令程序的审议；主办机关为实际公共艺术设置之主体单位，负责从预算编列至控管公共艺术计划执行完成；策划及创作者则实际负责创作、制作及整体公共艺术计划之执行。在依法执行的公共艺术计划中，非营利组织并不一定能直接参与，但台湾推动公共艺术二十余年来，非营利组织却始终扮演着非常重要的角色。

自公共艺术相关法令推动开始，非营利组织便肩负着倡导与推广的工作，透过案例研究、出版、研讨座谈、教育研习、进而协助兴办机关行政作业代办等。同时，随着法令不断修正，公共艺术执行方式越来越多元，非营利组织的功能也从早期单纯制式教育推广，逐渐发展出更积极更活泼的推动型态，投入推动公共艺术（艺术教育）之非营利组织越来越多，各自关注与发展的方向也越来越广。方式运作，有的以协助公共部门执行相关教育推广计划为主，有的则以自发性策划展览活动为主，彼此之间有许多共通点，却又有各自发展的方向；此外，另一些由民间组成之协会、学会，则是串联相关专业者，提供必要信息与教育的服务。

本件作品位于新北市林口区头湖国小，为一件参与式公共艺术作品。林口台地多为贫瘠的红土，先民来台拓垦时因地制宜以种茶为主，而后因红土适合烧制建筑用的红砖，开创了林口砖窑场的发展。本作品特别使用红砖，并完全以清水砖砌筑工法，展现砌砖工艺之美，在校园里塑造一个"新土地乐园"，引领大家认识这片土地的历史与文化。

作品中之"头湖红堡垒"，以等高线概念，呈现出林口红土台地、砖窑厂以及茶园意象；红砖所砌成的墙面，象征了在这片红土地上建立的家园；墙上所镶嵌的砖雕作品——"我的宝贝"，为创作营中学员们创作的成果，而一颗年逾50岁的老茶树，串联起过去与未来，娓娓诉说着岁月的故事。

"头湖红堡垒"是一个可提供活动的空间，其本身也是一个可以用以寓教于乐的教材，而以自由曲线构成的等高线层次、弧形渐变的墙面，更呈现出高难度的砌砖工法，让观赏者得以重新认识红砖。

"头湖红堡垒"位于校门口，家长接送区旁的草地上，三座以清水砖砌筑而成的地景，既是校园的精神堡垒，更重要的是提供学生、老师一个有安全感、有归属感的休憩及交流空间。

"头湖红堡垒"广场，以弧形的砖墙，围塑出不规则内聚性的空间，除了长条座椅，高低错落的等高线亦留设出许多可坐的平台，提供师生活泼有趣的上课说故事园地、小朋友下课时的游戏乐园，或是家长等候休憩的空间。而三座堡垒之间，同样刻意营造各种可对话的平台角落，目的即在于让小朋友们或其他使用者，随处都可以坐下来聊天、游戏。

红堡垒周边的数棵大树为原校地所保留下的老树，日后绿树成荫，将更有古早林口台地的氛围，更可以让这座隐身于树林里的红堡垒增加故事性，述说林口旧地名——"树林口"的故事。

因地制宜的堡垒
高浅

城市中个人空间越来越狭小，人们对于空间的要求却越来越高。起先只要有集会的场所，现在还同时要求兼具功能性、便利性、并集合历史文化和艺术感。因此利用本地特色因地制宜的创建多功能的活动空间并融合艺术、教育等文化活动对城市空间项目来说是大势所趋。每个孩子心底都有一个小小的秘密基地，"头湖红堡垒"项目选取了新北市林口区头湖国小，此地多产红土，相对贫瘠，但当地利用泥土的特点建造砖窑厂，发展红砖烧制产业。项目就地取材，利用当地特色红砖以清水砖砌筑工法建造了这个项目，即展现了工艺的历史，又展现了造物之美。一砖一瓦无不凝结了当地的文化、历史和人民智慧，任何一个细节的背后都拥有一段故事。而孩子们也可以在此新空间中观摩学习，藉此了解和认识当地的历史与文化，这也是此项目作为公共艺术考虑到了社会实用功能以及被大众接受方面的可能性。

Artist: Hao Te Huang
Location: Taipei City, Taiwan, China
Media/Type: Interactive Behavior
Date: Unknown
Researcher: Pan Li

There are two primary facets to the process of development of public art in Taiwan. First, a competent authority has a mandate to carry out a particular public art policy or task, including determining the hosting institution and establishing the budget for the public art program. Second, an execution team is responsible for the creation, production, and execution of the public art program as a whole. While it's not necessary that nonprofit organizations participate directly, they have always played a very important role in promoting public art.

The nonprofit organizations undertake advocacy and promotion of works through case studies, research, publications, seminars and discussion to help set up administrative agents, and so on. Meanwhile, thanks to an amended decree, implementation modalities of public art are becoming more and more diverse, while the function of nonprofit organizations has also extended from simple formal education to more active and lively developments. Finally, not only do nonprofit organizations attach great importance to art education, but the scope of their development directions and their concerns is also expanding.

The Touhu Red Fortress is a public art project created at Lake Elementary School in the Linkou District of New Taipei City.

The Linkou Tableland is composed of barren clay, unsuitable for farming. Pioneers to this area adjusted to the local conditions by planting tea trees and using the red soil to form bricks used in the construction of buildings.

Echoing the history of the land, the Touhu Red Fortress incorporates red bricks in a sculpted terrain, creating a "new land paradise" on the campus, a space where students and teachers can interact, rest, and play. Carved bricks, called "my treasures," were created by the students and their parents and worked into the wall. Each of these bricks tells a story of families who have lived in the area for several generations. A 50-year-old tea tree grows beside this curved and hilly plaza, providing a connection between past and future, a reminder of the stories of past years.

To promote the development of public art in Taiwan, in addition to the help of several important nonprofit organizations, some civil associations assist in the public sector; some of these focus on promoting related education plans, while others mostly hold some exhibition activities. The two groups—nonprofits and civil associations—will have many common points as well as their own respective development directions. Working together, they provide necessary information and education of service.

高雄捷运
Kaohsiung Public Transportation System

艺术家：多位艺术家
地点：中国台湾高雄市
形式和材料：综合材料
时间：2008 年至今
推荐人：潘力

"高雄捷运"自 2008 年通车以来，不仅为高雄市民日常生活带来交通的便利，也为高雄的城市形象带来重大改观。"高雄捷运"的公共艺术不仅展现了高雄的人文特性，行成特有的"高雄捷运"文化，更为台湾公共艺术实践激荡出一股火花。

高雄都会区大众捷运系统整体路网共设 23 座车站，其中 15 座地下车站，8 座高架车站，橘线全长 14.4 公里，设有 14 座车站，除大寮站外均为地下隧道。

"高雄捷运"红橘线路网采用奖励民间参与投资（一般称之为 BOT）方式，由高雄市政府捷运局制订相关技术文件，作为特许公司（以下称高雄捷运公司）的规范，对捷运系统工程的"安全、功能、质量"作必要且最低的要求。在车站构筑方面，除了考虑安全、便捷、耐久、易于维护之功能外，并强调应重视舒适度、高质感但不奢华、不浪费的原则。基于上述融合美学的工程技术与设计指导方针，将意象设计的概念和公共艺术的制度置于文件中，先行确立意象风貌主题，后续建筑细部设计、景观工程、公共艺术等乃能配合整体工程计划主要时程循序推动完成。

为提升高雄为"海洋首都"的国际化形象，全线设计采用现代科技绿色环保建筑产品，如金属、铁材、钢构、玻璃等轻量化、透明化的材料，以降低对周围环境的压迫感，且适应台湾南部阳光充足、耐长夏湿热的气候特性。不拘泥于历史图腾的建筑手法，是年轻、健康、符合时代意义的现代车站。车站建筑成为改善市容的第一步，也成为激发城市公共资源发展潜力的触媒。

"高雄捷运"公共艺术方案依据高雄捷运公司与高雄市政府订定的兴建营运合约，"文化艺术奖助条例"及施行细则、公共艺术设置办法及相关规定在车站规划制定。整个公共艺术设置案分为 O5/R10 站"光之穹顶"、R4 站"凝聚的绿宝石"、R17 站"半屏山之魂"、R23 站"天工开物"等。

"光之穹顶"直径 30 米，设置在全球最大的圆型地下车站建筑内，构图运用天体、人物、大自然、高雄土地上的特有生物等，每一块玻璃都是一小

段故事，演绎人生在世的各种情感画面；"半屏山之魂"结合水彩、摄影、数字影像手法，在采光天窗、车站主体墙以及电扶梯的艺术玻璃上处处可见花草、树木、森林，让人彷佛置身半屏山生态绿林的步道中。

高雄捷运车站为配合建筑及景观风貌，反映捷运站区环境特色，邀请国际知名艺术家对特殊车站进行规划设计，以期能有效与建筑、景观相融合，营造出独特的车站艺术与风貌，不仅提升环境美，并为高雄艺文注入活力。高雄捷运车站的公共艺术各有独创的艺术风貌与特色，结合捷运、艺术、生活与创意，以常民化方式推动（如共同创作、体验营、参访活动等），推广至趣味学习的生活体验中，造就一个新的捷运文化，高雄捷运以艺术作品串联大高雄，成为城市文化的主轴线和高雄艺文的新象征。

归途

高浅

城市空间的限制无法让自然景观时刻入眼，可倘若回家的路途充满着新奇与期待，再陌生再疲惫的归途都是美好的。自2008年通车以来，"高雄捷运"不仅为高雄市民日常生活带来交通的便利，也为高雄的城市形象带来重大改观，更为台湾公共艺术实践激荡出一股火花。为了把自然元素和低碳生活理念融入到公共艺术中，整条线路在设计中均采用现代科技的环保绿色建筑材料，如金属、铁材、钢构、玻璃等轻量化、透明化的材料，以降低对周围环境的压迫感，提高光亮度，减少能源的使用。并使其能适应台湾南部阳光充足、耐长夏湿热的气候特性。而设计风格也不拘泥于历史图腾的绘制，而是用多种现代风格诠释年轻、健康等生活理念。无论是"光之穹顶"、"凝聚的绿宝石"还是"半屏山之魂"、"天工开物"，无一不演绎人类的情感，呈现自然景观，也表现出艺术家在设计过程中对自然作为公共艺术要素，融入城市空间规划的思考。

Artist: Multiple Artists
Location: Kaohsiung, Taiwan, China
Media/Type: Mix-Media
Date: 2008-present
Researcher: Pan Li

The Kaohsiung Mass Rapid Transit (KMRT) system has 36 stations, many of them underground. When construction of the system began in 2001, with the intent to promote the image of Kaohsiung as an "ocean capital," modern high-tech green materials were used, such as metals, irons, steels, glasses, and other lightweight and transparent materials, to reduce the impact on the environment and also to accommodate the humid weather in southern Taiwan. The distinctive design of the buildings gives the stations a young, healthy, and modern image. These stations are the first step toward improving the city's image and fostering the development potential of public resources.

The Kaohsiung subway public art project was drafted under the operation agreement signed by Kaohsiung Rapid Transit Corporation (KRTC) and local government. The agreement incorporated culture art sponsorship regulation and its detailed implementation rules, as well as public art setup instructions and related rules. The public artworks include The Dome of Light, the largest glass work in the world, designed by Italian artist Narcissus Quagliata, at the Formosa Boulevard station; Emerald Laminata, by Gabriel Lomelin, at the International Airport station; and Floating Rainforest at the World Games Station, designed by the WJI team, Ron Wood and Christian Karl Janssen.

During the construction of the Kaohsiung subway public art project, there were disputes and different voices speaking against the public art institution and the enforcement process. But it all disappeared after the actual result was unveiled, as it's so stunning and impressive. Via this case study, we can see the new vitality and positive impact on city art brought by thinking out-of-the-box.

The KMRT is more than just traffic transport. It not only brings convenient transportation to people, but is also the source of shaping a modernized urban pattern and is the driving force for a better lifestyle. In addition, its incorporation of public arts helps cultivate an active and good-quality rapid transit culture. On this basis, it is hoped that a fresh style may be shaped for this harbor city in providing an environment for citizens to access art and bringing a brand-new look for the southern area of Taiwan.

The construction of the Kaohsiung rapid transit lines was announced by Kaohsiung City Government, and it encouraged private investment on a build-operate-transfer (BOT) basis. The KMRT red and orange lines network and related technical documentation was developed by the Kaohsiung City Government with "safety, functionality, quality," as necessary and minimum requirements. In addition to being safe, convenient, durable, and easy to maintain, it also reflects other principles, such as being comfortable, being highly textured but not luxurious, not wasting, and so forth. The fusion of aesthetics, engineering and design guidelines, subsequent architectural detail design, public art, and landscaping create the unique KMRT experience.

大洋洲和东南亚
Oceania & Southeast Asia

大洋洲和东南亚

大洋洲位于太平洋西南部和南部的赤道南北广大海域中。在亚洲和南极洲之间，西邻印度洋，东临太平洋，并与南北美洲遥遥相对。独特的地理位置和气候环境，为大洋洲的文化艺术造就了更为特殊和多样的文艺色彩。

文化多样，艺术丰富。大洋洲的土著民族创造了许多独具特色的艺术。不论是诉说着生活和故事的土著壁画、神秘的祖先雕像、面具、图腾，还是怪诞离奇的毛利雕像和面具，都对今天大洋洲的艺术产生了深远的影响。早期被殖民的经历，为其艺术带来了多元化的西方格调；来自世界各地的大量移民为包容多元的大洋洲文化注入了更多新鲜的血液。艺术的多样性为艺术家的创作带来更多的灵感和更为充分的可能性。

因此，大洋洲的公共艺术，在彰显现代文明的同时，却又明显渗透出土著民族的文化和西方文明的气息。通过多元文化的兼容并蓄，形成了自身的特色，铸就别具一格的艺术风格。与东南亚纯粹的、以科学理念铸就的公共艺术形式相比，大洋洲的公共艺术更多了一层悠远的文化情怀和厚重的艺术积淀。像一位曼妙而富有内涵的姑娘，惹人心动，又深深令人折服、钦佩……（李柯臻）

超级树
Supertrees

艺术家：格兰特协会景观事务所、Atelier 10 结构工程公司
地点：新加坡滨海湾花园
形式和材料：建筑
时间：2012 年
委托人：国家公园局
推荐人：凯利·卡迈克尔

滨海湾花园毗邻新加坡的金融区，是一个新兴的大型公共花园。滨海南花园是三个园林中规划最早、规模最大的一个，项目中最令人瞩目的是被称为"超级树"的公共艺术项目。超级树有着类似于未来主义森林的巨型结构。这 18 棵热带风格的人造"树"，每棵树都有着巨型钢和混凝土制成的树干和数千根盘条制成的树枝，高度都在 25—50 米左右，与城市建筑的高度比肩。其中两棵树由一条 128 米长的淡黄色空中步道相连接，步道悬浮在空中，约有七层楼高。

虽然是工业原料制成的人工形式艺术品，但超级树还是重新引入了自然元素作为其重要组成部分。超级树打造出一个种植有 200 多种植物的垂直花园，植物总数达到 162 900 棵。滨海南花园占地 54 公顷，其总体规划的设计灵感来源于新加坡的国花——兰花的生理结构。在新花园的布局和基本理念上都在强调和试图扑捉兰花的基本特性和特点，该景点由英国景观建筑和城市设计公司 Grant Associates 领衔设计，意图打造"一个融合自然、科技、环境管理和想象力于一体，以热带园艺为重点的 21 世纪独特景点体验。"

在 21 世纪，绿地和高楼林立的城区相结合的这种模式，已经成为城市可持续发展的一个日益重要的组成部分，而这个项目恰恰体现了这种想法。无论是城市和还是自然都在考虑向这方面发展。新加坡是一个土地资源有限的小岛屿，人们过着高密度的城市生活。该岛以繁华的都市旅游景点和购物而闻名，尽管城市化程度很高，但新加坡的绿化却做得出奇的好。"超级树"的建造则使这一反差更为明显。

"超级树"的高度提供了花园规模和面积，赞美着滨海湾地区的高速发展。超级树最高高度为 50 米，其规模与邻近的摩天大楼交相呼应，不仅看起来毫不矛盾，反而与城市环境相和谐。"超级树"为（人们的）生活空间带来了自然元素，为兰花，其他花卉以及各种攀援在钢架上的蕨类植物提供了一片乐土。巨大的树冠起着温度调节的作用，能够吸收和分散热量，并为新加坡热带阳光下面的游客提供庇荫。位于树冠下的空中走道则可以通向树顶酒吧和小酒馆，为人们提供了一个独特的社会空间。

紫红色的桁架，淡黄色走道和热带花卉营造出一个充满异国情调的视觉效果，使其已成为当地人和游客的休闲胜地，"超级树"的安装工作很艰难。每棵"树"由四个部分组成：钢筋混凝土浇灌的树心、树干、节能板做的树皮和树冠。"超级树"的每个结构都配有太阳能电池板和防雨台，同时还是附近温室的通气管道。这些"超级树"中将有 11 棵安装太阳能光伏电池和一系列与水有关的技术，用于发电满足照明需要和冷却温室。作为整个网站系统的一个组成部分，"超级树"采用了融合自然与环保的技术，这个项目体现了一种前瞻性的思考方式，如何在老式的公园漫步中管理我们的资源。滨海南花园和耀眼的新产物——"超级树"是一个集园艺、工程和建筑于一体的复杂的三维网络。

震撼人心的现代艺术新体验

李柯臻

"超级树"是一项于艺术、科技、自然合为一体的公共艺术建筑项目。该项目将人们置身于艺术作品之中，浑然的现代科技感给观众带来震撼的艺术新体验。走入其中，仿佛进入了植物丛生的热带雨林；登上空中通道，尽情体验云中漫步、鸟瞰全城的美妙感觉。夜幕下，超级树将白天吸收的太阳转变为七色的光亮，迷离、梦幻、美轮美奂，仿佛走进了童话王国。在休闲、娱乐中体验艺术和科技带给生活品质的新变化。环保材料和科技结合、植物与钢筋水泥和谐共处，为高速发展的城市和人们带来了未来主义的新生活。艺术家们用公共艺术的形式，营造了最新的城市生态空间，向人们传达可持续发展和节能环保的新一代生活理念。

Artist: Grant Associates, Atelier One, Atelier Ten
Location: Gardens by the Bay, Marina Gardens Drive, Singapore
Media/Type: Build
Date: 2012
Commissioner: National Parks Board
Researcher: Kelly Carmichael

Gardens by the Bay is a new, large, public garden close to Singapore's financial district. Bay South is the first and largest of three planned gardens, and it's at the heart of the project is a public art project called Supertrees. Resembling a futuristic woodland of giant structures, each of the 18 tropically-inspired and manmade "trees" range in height from 25-50 meters. As tall as city buildings, the installation has giant steel and concrete trunks and thousands of thick wire rods for branches. Two of the trees are connected by a canary yellow 128-meter-long aerial walkway, suspended approximately seven stories high.

Although manmade and industrial materials create the form, nature is reintroduced as an essential part of this artwork. Supertrees creates vertical gardens, home to 162,900 plants and over 200 species. Spread across 54 hectares, the master plan for Bay South takes its inspiration from the organization and physiology of Singapore's national flower, the orchid. An attempt to capture the essential qualities and characteristics of orchids in the layout and underlying philosophy for these new gardens underscores the approach. Led by British landscape architecture and urban design practice Grant Associates, the desire for the site was to "blend nature, technology, environmental management, and imagination to create a twenty-first century focus for tropical horticulture and a unique destination experience."

An increasingly important component of sustainable urban development in the twenty-first century is the incorporation of green spaces into densely built urban areas positions, and this

project exemplifies current thinking. Both the urban and the natural have been carefully considered in this development. Singapore is a small island with land restraints and high-density urban living. The island state is famous for its bustling city attractions and shopping but is also surprisingly green despite urbanization. The Supertrees installation extends this paradox further.

Providing scale and dimension to the Gardens, the height of Supertrees compliments the tall developments in the Marina Bay area. Towering up to 50 meters in height, the scale of Supertrees responds to neighboring skyscrapers, operating not in contradiction but in sync with the urban environment. Supertrees brings a natural element to the space, providing a home for orchids, flowers, and various ferns that climb across the steel framework. The trees' large canopies operate as temperature moderators, absorbing and dispersing heat, and providing shelter from Singapore's tropical sun to visitors below. Positioned just beneath the canopy, the aerial walkway leads to a treetop bar and bistro, offering a unique social space.

While the fuchsia-colored supports, canary yellow walkway, and tropical flowers create an exotic visual and have become a destination for locals and tourists alike, the Supertrees installation is hard at work. Each "tree" is made of four parts: a reinforced concrete core, trunk, planting panels of the living skin, and the canopy. Each structure in Supertrees is fitted with solar panels and rainwater catches, while simultaneously functioning as air venting ducts for nearby conservatories. Eleven of the Supertrees are fitted with solar photovoltaic systems that convert sunlight into energy, providing lighting for the Gardens at night and aiding water technology within the conservatories below. The trees act as an integral part of the overall site system. A fusion of nature and environmental technologies, this project embodies cutting-edge thinking about how we manage our resources with the charm of an old fashioned stroll in the park. Bay South and its star attraction, Supertrees, is a highly sophisticated and integrated three-dimensional network of horticulture, engineering, and architecture.

72DP

72DP

艺术家：克雷格·雷德曼、卡尔·梅尔
地点：澳大利亚悉尼
形式和材料：壁画
时间：2012 年
委托人：Marsh Cashman Koolloos 的建筑师
推荐人：凯利·卡迈克尔

停车场是典型的沉闷场所，尤其是地下停车场。通常它们都是褐色、灰色的，并且大小、形状和功能都是统一的。有的时候，由于灯光暗淡和缺乏视线，停车场让人觉得有点儿毛骨悚然。然而，设计师克雷格·雷德曼和插画家卡尔·梅尔设计的"72DP"地下停车场把这些缺陷都弹走了。此项工作是为澳大利亚悉尼的一个住宅区创作的沉浸式壁画，一进到车库，就能看到粗大的几何图样和令人瞠目的颜色，如同小波浪一样。一旦进入到内部，就会看到明亮的爆炸式的热烈色调。

这种感觉更像是一个时尚的酒吧或老式的溜冰场，"72DP"让日常的事物变得不平常，为你创造了一个不仅仅只是用于停车的环境。这个公共艺术品将车库的地板、墙壁和斜坡作为画布，创造出了一个具备装置功能的沉浸式壁画。不要地板和墙壁之间的界限，色彩在两者上面涂满。不是把环境作为一个空间，而是通过色调和狂野的构图方法来重新配置，与地面上方的建筑物呈一定角度。这个项目由驻悉尼的公司 Marsh Cashman Koolloos 所设计，当人们旋转穿过空间，壁画与强烈的模式相结合，并塑造出那个镜子来回应建筑物。极度华美的几何形图案诱惑地蜿蜒，伸向街道，把艺术作品和地下停车场融入到住宅环境中。

在地下车库里面，一连串的色彩从金属管道表面放射开来，即使只有一点自然光线进入，绿色、红色和黄色的拼接也会给予空间活力。整块被一个弯曲的绸带状的装置绑在一起充当中心轴，从车道导入并穿过空间到达花园外部。设计结果是一个动态的和重叠的几何图形离奇地混合，更像是热带俄尔普斯立体主义的探索，而不是标准的停车设施。有意创造一个开放式艺术作品，雷德曼和梅尔让一些墙面留白，尽管充斥着颜色，但他们打算把这些空间留给未来追加的、个人委任的作品。

复杂的理念，简单的执行，停车场的典型建筑物和临近的住宅上都是大规

模的壁画，拒绝停车场结构上传统的渲染，来创造一个能够唱歌的艺术品。这个空间是每天的幻想曲，动感音乐的享受地，在那里你可以体验色彩。"72DP"不仅仅是一个停车场，它为空间使用者提供了一个连接点。壁画变成了一个话题，为通常坐在挡风玻璃后面分开的人们创造了一种联系。这个作品迫使与公共空间和迥异的条件下必要的基础设施之间的一种互动。它提供了一种不期而遇，凸显之前忽略的景观并创造出沉浸式的独特环境。"72DP"确实为深夜外出找不到自己车的人提出了一个新的难题。

你好，色彩

李柯臻

公共场所指的是供人们从事社会生活的各种场所；公共艺术是公共场所中的艺术，是以大众需求为前提的艺术创作活动。停车场，就是生活中最为常用的公共场所。

"72DP"这个项目用艺术的形式，将停车场赋予了新的生命。艺术家以大型壁画的形式创作之后，其爆炸式的热烈色调，呼之欲出的鲜活色彩给人们欢乐兴奋的身心感受。运用活泼的色块与不扰视觉的色彩搭配，妆点了了无生机的灰色墙面，使暗淡的空间重新获得新生。为生活和心情带来更为缤纷的色彩与诗意。生活上的诗意，生活中的艺术品，使人生更有意义。公共艺术的意义并不是仅仅达到功能、装饰、商业等目的，而是在空间中，追求一个更接近于艺术与文明的新目标，为公共生活增添别样的色彩。

Artist: Craig Redman and Karl Maier
Location: Sydney, Australia
Media/Type: Mural (painting)
Date: 2012
Commissioner: Architects of Marsh Cashman Koolloos (Sydney)
Researcher: Kelly Carmichael

Parking garages are typically uninspiring, especially the underground type. Most often they are drab, grey, uniform in scale, shape and functionality, and sometimes a little creepy thanks to the low light and lack of sightlines. The underground car park work known as 72DP by designers and illustrators Craig Redman and Karl Maier, however, flips all this on its head. This work is an immersive mural created for the garage of a residence in Sydney, Australia. Bold geometric patterns and eye-popping colors begin in small waves upon entering the garage and once inside the structure, it's an explosion of bright, tropical hues.

Feeling more like a trendy bar or old-fashioned roller skating rink, 72DP makes an everyday thing extraordinary and creates an environment for more than just parking your car. This public artwork uses the floor, wall and ramp of a garage as its canvas to create an immersive mural that functions almost as an installation. Negating the boundary between wall and floor, color fields bleed over both and instead treat the whole environment as a space to reconfigure through tone and a bold graphic approach. Playing off the angularity of the residential building above ground, designed by Sydney-based firm Marsh Cashman Koolloos, the mural incorporates intense patterns and shapes that mirror and respond to the architecture as they gyrate through the space. Blisteringly colorful geometric forms seductively snake their way out to street level, connecting artwork and underground car park to its residential surroundings.

Inside the subterranean car park a spectrum of colors radiate from a metal drainage cover while splices of green, red, and yellow breathe new life into a space, which, with little inlet of natural light, previously felt dark and heavy. The whole piece is tied together by a winding, ribbon-style device which, acting as a central axis, leads in from the driveway, through the space and out to the garden beyond. The resulting design is a dynamic and surreal mix of overlapping geometric forms more like an exploration of tropical Orphic Cubism than standard car parking facility. Intentionally creating an open-ended artwork, Redman and Maier have left some walls blank—albeit awash with color—with the intention that these spaces could potentially be used for additional, individually commissioned works in the future.

A sophisticated concept with simple execution, the large-scale mural draws on the defining architecture of the car park and adjacent residence and a negation of the traditional rending of car parking structures to create a public artwork that sings. The space is an everyday fantasia, a synesthetic feast where you can almost taste the color. 72DP offers more than parking, it offers a point of connection for users of the space. The mural becomes a talking point, creating connection between those normally separated behind a windscreen. The work forces an interaction with public space and its necessary infrastructure under very different terms, it offers an unexpected encounter, highlights the previously overlooked and creates an immersive and distinctive environment. 72DP certainly offers a new twist on not being able to find your car after a night out.

遗忘之歌
Forgotten Songs

艺术家：迈克尔·希尔
地点：澳大利亚悉尼天使大道
形式和材料：装置
时间：2009—2011 年
委托人：悉尼市政府
推荐人：凯利·卡迈克尔

艺术盛行的年代，大规模丰富多彩和互动性强的公共艺术作品越来越流行，我们常常流连忘返，而能够发出轻声细语的作品实属难得。澳大利亚艺术家、建筑师迈克尔·希尔的代表作"遗忘之歌"充分体现了他细致入微和考虑周到的一面。该作品是一处永久性公共艺术装置，坐落于澳大利亚悉尼市。它离中心商业区域仅隔一条小道，由180个鸟笼组成，发出阵阵轻柔的鸟声。这是一部精美绝伦的作品，大大小小、各种形状的鸟笼高高悬挂在窄巷之上。由铁丝组成的鸟笼内部充分可见，极其有趣，而最迷人之处在于它发出的声音。

"遗忘之歌"由 50 只鸟共同"演唱"，它们的祖先曾经世世代代生活在悉尼市内部城区，可追溯到 17 世纪末期，当时一大批欧洲人来到澳大利亚开始殖民，并启动了城市化建设。日夜交替，斗转星移，鸟笼树丛中发出声声啼叫；日落之后，白鸟啼罢，夜鸟声又冲破夜晚的寂静。置身于窄巷当中，聆听着阵阵鸟叫，越过围墙，仿佛听到过去的回响，几米之外就是全悉尼最繁华的商业购物街区。

"遗忘之歌"是一部公共艺术作品，介于政治和可持续艺术品类别之间，巧妙地展现两种类别的优势特点。不同于许多同类艺术品，希尔采取的方法略有差别，他运用充满矛盾的动态表现力，时刻提醒人们去铭记殖民时期的屈辱历史。艺术家借助这部作品表达一个观点，即"通过改变环境来改变生活中隐藏的方方面面。随着原始林区面积的减少，我们以城市之声发现了声音环境的变化"。随着城市化进程不断发展，自然环境不断恶化以及野生动物栖息地减少带来更多的威胁，"遗忘之歌"表达出公众的普遍担忧。它引领着人们不断审视、思考，缺乏可持续的环境会带来的一系列重要问题。

该作品经过学者们认真调研。艺术家迈克尔·希尔与来自澳大利亚博物馆的科学家 Richard Major 共同协作，通过本地土壤鉴定植被和鸟类品种。野地勘察学者 Fred van Geseell，特别向本项目提供他曾经记录的鸟类叫

声。截止目前，新南威尔士州总共存在 129 个鸟类品种，曾被列入濒危动物名册。"遗忘之歌"重现天使大道，而该区域是这些鸟类已知的历史栖息地，后来它们被欧洲移民驱赶到其他区域。

"遗忘之歌"项目是 2009 年暂时性巷道艺术项目的组成部分。该作品受到悉尼人和外来游客的追捧，使天使大道和小巷成为一处名胜。当暂时性装置被拆除后，公众无不愤慨。悉尼当地居民强烈要求他们非常喜欢的艺术作品重新出现在大道上，于是当局不得不考虑他们的诉求。2010 年 3 月，考虑到作品大获成功，悉尼市公共艺术顾问小组向市议会提议，城市应将作品永久化，这样可以提升天使大道的整体形象。如今，作品重新掀起一股追捧热潮。最重要的是，曾经栖息在该区域的鸟类名称被印刻在鸟笼下的铺石道上。

永恒的歌唱

李柯臻

装置艺术，是"场地＋材料＋情感"的综合展示艺术，指艺术家在特定的时空环境里，将人类日常生活中的的物质文化实体，进行艺术性地有效选择、利用、改造、组合，令其演绎出新的展示丰富精神文化意蕴的艺术形态。"遗忘之歌"是一处永久性公共艺术装置项目，用充满矛盾的动态表现力，提醒人们铭记殖民时期的屈辱，引领人们不断审视、思考缺乏可持续的环境所带来的一系列重要问题。在鸟儿永恒的歌唱中，使观众全方位、立体的置身作品之中，介入和参与。侧耳聆听，阵阵鸟鸣越过围墙，带你走进曾经的时光；脚下，是块块曾经栖息在此的鸟类名称印刻的铺石小路；抬头看去，大大小小、各式各样的鸟笼高高悬挂在窄巷之上，宁静、安心、岁月静好。

Artist: Michael Hill
Location: Angel Place, Sydney, Australia
Media/Type: Device
Date: 2009-2011
Commissioner: The City of Sydney
Researcher: Kelly Carmichael

In a time of extreme art, when large scale, colorful and interactive public art works are popular and often loudly demand our attention, work that speaks in a soft voice is not common. Forgotten Songs by Australian artist and architect Michael Hill represents a subtle and thoughtful approach to placemaking. This work is a permanent public art installation in Sydney, Australia. A small lane off the central business district houses the installation, comprised of 180 birdcages emitting a soft soundscape of birdsong. This is a delicate, beautiful work where a curious variety of shapes and sizes of birdcages are hung high in the narrow alleyway. The transparency of the wire cages is ghostly and oddly intriguing, but what is remarkable is the sound element.

Forgotten Songs plays the calls of fifty birds that once lived in what is now inner city Sydney before the arrival of Europeans to Australia in the late 1700s brought colonization and the beginnings of urbanization to the area. The calls filter down from the canopy of birdcages, changing from day to night with daytime birds' songs disappearing with the setting sun and nocturnal birds which inhabited the area sounding late into the evening. It is truly magical to stand in the lane and listen to birdsongs echoing off the walls, songs from a previous time, just meters away from one of Sydney's busiest commercial and shopping districts.

Forgotten Songs is a public artwork that exists on the border of political and sustainable art, skillfully demonstrating some of the best characteristics of both. Contrary to many artworks in this vein, Hill takes a nuanced approach, exploring a contradictory dynamic

which entangles the desire for preservation and progress with the often painful history of colonization. The artist desired the work to show "that changes to the environment alter many unseen aspects of life... We have highlighted the changes to the acoustic environment—the sound of the city—through the loss of native bushland." Forgotten Songs is a response to widespread concerns regarding urbanization, diminishing natural environments and the threat of habitat loss for native wildlife. It prompts the examination of important issues without being obviously environmentally sustainable in its execution.

The work was carefully researched for accuracy. Artist Michael Hill worked with Dr. Richard Major, a scientist from the Australia Museum, who identified vegetation and therefore probable bird species via local soils. One of Major's colleagues, wildlife field recorder Fred van Geseell, had recorded the birdsongs of all these species and provided the sound files for this project. At present there are 129 species of birds native to the state of New South Wales, formally listed as extinct or threatened with extinction. The birdcalls Forgotten Songs returns to Angel Lane are amongst these species known to be from the area before European settlement gradually forced them away.

Forgotten Songs was commissioned as part of a 2009 temporary laneway art program. The work was hugely popular with Sydneysiders and visitors alike, transforming Angel Place and making the narrow lane a destination. When the temporary work was uninstalled, public outcry ensued. The people of Sydney spoke—they wanted their much-loved artwork back in the lane and the important issues it evokes back on the table. Due to the work's success, the City of Sydney's Public Art Advisory Panel recommended to the Council in March 2010 that the City should make the work permanent and incorporate it into the upgrade of Angel Place. The work is now back by popular demand. As a final touch, the names of the birds that once populated the area are now engraved on the paved stones beneath the cages.

Pou Tu Te Rangi
Pou Tu Te Rangi

艺术家：克里斯·贝里
地点：新西兰奥克兰中心戈尔大街
形式和材料：雕塑
时间：2011 年
委托人：布里托马特艺术基金会经 FHE 美术馆
推荐人：凯利·卡迈克尔

"Pou Tu Te Rangi"由新西兰雕刻师克里斯·贝里制作，该雕刻由七个特色分明的雕刻形式（被称为"pou"）构成。Pouwhenua 为毛利文化的一种艺术表现形式，是一种雕刻好的木质杆子，作为地域边界和标示具有重大意义的地域。这些雕刻都极其精美且极具艺术性，可作为某一区域的地理及符号标识，标明地名、身份、历史。Pouwhenua 这种艺术形式对于毛利人极为重要，标明人们与土地的联系。

"Pou Tu Te Rangi"拥有丰厚历史积淀。早在欧洲人征服新西兰之前，它就被认为是毛利人军事防御性居住地。当地部落对该区域具有强烈的祖先情结。布里托马特，一座港口城市，前身为英国殖民军队基地，20 世纪初发展成为一个融合商业、铁路、仓储及邮政总局为一体的中心城市。然而由于缺乏投资及日益猖獗的腐败使得该片区域衰落。许多建筑也逐渐被人们遗忘。2000 年中叶，该城市重新发展起来，现在已经演变成为富有生机的高端购物、娱乐、商业中心及交通枢纽。"Pou Tu Te Rangi"被划定为这片重燃生机区域内的一块室外庭院。

"Pou Tu Te Rangi"的功能是为空间的使用者在各种感觉下导航。同时作为港口旁边布里托马特的一个集合点，它勾起对古代毛利人导航的回忆。作品的名称翻译为"可以到达天堂的耸立标杆"，也是毛利人对河鼓的叫法，河鼓就是指引着早期波利尼西亚人穿过太平洋的恒星。传统上这标杆耸立在村子外面并且要足够的高大，以便使行人从远处能看着它，才能知道行人回家的道路。贝里打算让"Pou Tu Te Rangi"成为布里托马特港口内外的行人定位和指引的一个标杆。利用 pou 作为地方、认同历史的物理和象征性印记的作用，这个作品把人们和地点定位在一个更大的社会、历史和地理的环境中。

"Pou Tu Te Rangi"的 whakairo 被创作成直线型，以指代附近建筑物高低不同的竖向形状。木材的黑色燃料是直接参考 Te Aupōuri 部落的传统，雕刻家和理事通过它们来追踪祖先的关系。克里斯·贝里说："这个部落使用云状的黑色烟雾来逃命并保证在攻击下存活"，"今天黑色是新西兰的国色，安定被认为是新西兰力量和独特性的象征。"

雕塑把布里托马特当作融合文化和种族的一个地方，它在早期的土著的、殖民的和海事历史中扮演着举足轻重的作用。布里托马特对港口的关系有着个人的意义，因为艺术家的祖父就是码头上许多劳动者中间的一位。他说："和港口是奥特雷新西兰的公路"，"太平洋上不同的王国在此汇合。" "Pou Tu Te Rangi" 的象征是富裕和多层次，但是不同的高度和身体喜好安置家庭作为一种强大的视觉概念。Pou 字面上也代表着家族，每根杆子的底部都雕刻着彼得·库伯的名字，此人是私人布里托马特发展公司的董事长和此作品的特派员和他的家庭。"这些名字是私人的贡物，我想将他看作是一个地位高的人，一个领导，库伯的首领和公司家族。"

来自民族灵魂的声音

李柯臻

雕塑是公共艺术中一种重要的创作手段，公共艺术中的雕塑，不仅仅从视觉上与大众对话，它还可以是民族、城市精神的直接载体，凝聚着民族、文化和大众的记忆。"Pou Tu Te Rangi" 就是这样的公共艺术项目。它所代表的，是毛利民族与毛利文化；它不仅仅是一组精美且极具艺术性的作品，也是某一区域的地理及符号标识，带领人们找寻回家的路。在作品中，我们看到了毛利民族的起落兴衰，感悟到了源远流长的民族文化，更体会到了毛利人坚强勇敢的个性以及渴望找寻民族更高的发展方向的心声。这就是公共雕塑的魅力，它以独有的空间语言诠释着从人类心灵深处发出的声音，展示了灵魂交流的美妙。

Artist: Chris Bailey
Location: Gore Street, Auckland Central, New Zealand
Media/Type: Sculpture
Date: 2011
Commissioner: Britomart Arts Foundation via FHE Galleries
Researcher : Kelly Carmichael

Pou Tu Te Rangi (2011) by New Zealand sculptor Chris Bailey rises
in seven distinct carved forms known as "pou." Pouwhenua are an
art form within Māori culture, taking the form of carved wooden
posts that mark territorial boundaries or places of significance.
Often artistically and elaborately decorated, the whakairo (carvings)
act as physical and symbolic markers of place, identity, and history.
Pouwhenua are highly significant to Māori, acknowledging the
association between the people and the land.

Pou Tu Te Rangi is situated on a site of rich history. Before the arrival
of Europeans to New Zealand it is thought to have been the site of
a Māori fortified village and the local tribe claims strong ancestral
ties to this area. Positioned on the waterfront of a city dominated
by its harbor, Britomart was a base for British colonial troops and
then home to a thriving mix of businesses, railways, warehouses and
a Chief Post Office at the dawn of the twentieth century. However,
lack of investment and increasing decay saw the area decline and
much of the buildings lay derelict and forgotten. Redeveloped for
mixed-use in the mid-2000s, Britomart is a now a vibrant high-end
shopping, entertainment and business precinct, and transport hub.
Pou Tu Te Rangi was commissioned for an outdoor courtyard site
within this redevelopment.

Pou Tu Te Rangi functions for users of the space by referencing
way-finding in multiple senses. As well as a meeting point for
visitors to harbor-side Britomart, it evokes ancient Māori navigation.
The work's title translates as "standing posts that reach for the

heavens" . It is also the Māori name for Altair, the star that guided early Polynesian navigators across the Pacific. Traditionally standing outside the village to guard those within and lofty enough that travelers were able to see them and navigate their way back home, Bailey intends Pou Tu Te Rangi to be a locating and guiding devise for travelers in and out of Britomart. Drawing on the role of pou as physical and symbolic markers of place, identity, and history, this work locates people and site within a larger social, historical and geographic context.

The whakairo of Pou Tu Te Rangi were created in a linear pattern to reference the vertical form of low and high-rise buildings nearby. The black staining of the timber is a direct reference to the tradition of the Te Aupōuri tribe to which the sculptor and commissioner trace ancestral ties. "The tribe used a cloud of black smoke to mask its escape and to ensure survival when under attack," says Bailey. "Today black is New Zealand's national color and is seen as a symbol of New Zealand's strength and uniqueness."

The sculpture references Britomart as a place of converging cultures and ethnicities, and its significant role in early indigenous, colonial, and maritime history. Britomart's relationship to the port has personal significance, as the artist's grandfather was one of many Māori laborers on the wharves. "The sea and the ports are the highways of Aotearoa New Zealand," he says. "All the different kingdoms of the Pacific meet here." The symbolism of Pou Tu Te Rangi is rich and many-layered, but the differing heights and the physical likeness allude to family as a strong visual concept. The pou also represent family quite literally: carved at the base of each post are the names of Peter Cooper, Chairman of the private Britomart development company and commissioner of the work, and his family. "The names were a personal tribute. I wanted to recognize him as a rangatira, a leader, and the head of the Cooper and Company family."

Classification Pending
Classification Pending

艺术家：克雷格·沃尔什
地点：澳大利亚昆士兰伊普斯维奇绿地不来梅河
形式和材料：数字媒体
时间：2007—2012 年
委托人：伊普斯维奇市议会
推荐人：凯利·卡迈克尔

"Classification Pending" 是一个四通道的同步数字投影。该作品作为不来梅河的常设公共设施，于 2012 年在一场洪水中被毁坏。"Classification Pending" 是一件不断演化的作品，现在作为每四个月循环的三个序列而存在（每循环为时 30 分钟）。在该作品中，人造生命形式似乎作为永久居民存在于这条河流中；这些生命形式通过连接到河畔木板路上的一系列数码放映机投射在河流表面上而形成。叙事在天黑后可见，以每四个月为循环，在 12 个月期间演变，从简单动物形式开始，然后随着时间的流逝找到伴侣，繁殖下一代。这位艺术家创作的 3D 动画似乎游向了河流表面；3D 动画建立在灭绝的伍伦加龙、鳝鱼、叉尾鲶鱼、布里斯本短颈龟和乌鱼的杂交后代的基础上。在后续演变的版本中，全球的其他地点也创建了"Classification Pending"。

克雷格·沃尔什的实践通常采取将视频投影到自然或建筑上的形式。"Classification Pending"将尖端的数码技术，与精心构思的场地相关反应，和对当代艺术的其他环境的探索相结合。克雷格·沃尔什通过与社群团体讨论和协商加入了新的叙事和主题，对其公共艺术作品做出了合理的改进；随着时间的推移，其作品发生演变。"Classification Pending" 扩展了克雷格·沃尔什对肖像画的兴趣，使他通过捕捉社区的身份和感觉这一场地的过去和现在对这一方面进行探索。"Classification Pending" 作为大型重建工程的一部分，体现了重建的其中一个关键目标——将城市与河流连接起来。不来梅河，由于深受煤矿开采和其他工业的污染，不仅在空间上因陆坡路堤而隔绝，从社区思想这一心理层面上来讲也与外隔绝。该设施，在该区大型重建工程的协助下，对人们回归河流，使这一地方变得更有活力起着核心作用。

"Classification Pending" 在一个具有史前外貌的身体中融合了先进的计算机生成的人造生命形式，它让观众思考神话、进化和遗传工程之间的联系或边界。在生态问题处于公众争论的风尖浪头，可持续艺术的崛起使全球的画廊业和公共艺术工程面临这种问题之时，"Classification

Pending"鼓励观众反思物种的灭绝和环境退化。但是，有趣的是，该作品通过投射出先前看不到的河流杂交生物，使人想起了一点——尽管物种灭绝速度越来越快，但同时也发现了新的动植物物种。

"Classification Pending"与许多公共艺术项目不同，尽管具有极强的视觉冲击力，涉及到一些重要问题，随着时间的推移会失去其新颖性，被人们忽视，但它通过不断演化吸引了人们的眼球。该设施为常设和演化的公共艺术设施融入到环境中，带来一种长期关系提供了新的可能性。该作品还提示人们审视技术如何能刺激观众了解自然，与自然交互以及技术是否取代真实的体验。"Classification Pending"并非作为一个孤立元素存在它的存在通过与不来梅河区当前以及历史上的动植物的创新对话技术而实现，从而使该设施能表达、反映地方发生的大事件以及不断演变的故事——这是一种真正的场地相关反应。

数字媒体时代
李柯臻

无论何种时代，艺术的创作都依赖于当下的艺术语言所赋予的前提和特征。对于公共艺术而言，以新的技术为支撑的各类新数字媒体艺术的发生为其在数字媒体时代的新发展提供了肥沃的土壤。"Classification Pending"是一项同步数字投影常设公共设施项目，以数码技术和极强的视觉冲击力，诠释着生命的繁衍生息，鼓励观众反思物种的灭绝和环境的退化。随着数字媒体时代的迈进，数字媒体艺术为艺术创作提供了更为广阔的发展空间和发展形势，越来越多的不可能在技术的支持下，改变影响着人们的生活，不断带给人们更多的艺术感受和生活启发。该项目以全新的形式阐述了艺术家的心声，唤起人们对所忽视的物种的关注和环境意识，成功地展现了数字媒体所发挥的重要作用和深远意义。

Artist: Craig Walsh
Location: Bremer River, Ipswich Parklands, Queensland, Australia
Media/Type: Igital media
Date: 2007-2012
Commissioner: Ipswich City Council
Researcher: Kelly Carmichael

Classification Pending (2007) is a four-channel, synchronized digital projection. Intended as a permanent public installation for the Bremer River, it was destroyed by a flood in 2012. Classification Pending is an evolving piece, existing as three, four-month sequences in 30-minute loops. In this work, artificial life forms appear to exist in the river as permanent residents, generated by a series of data projectors attached to a riverside boardwalk and trained onto the river's surface. Visible after dark, the narrative evolves over twelve months in four monthly cycles, commencing as a single animal form that over time finds a mate and breeds young. The 3D animation the artist has created appears to be swimming towards the surface and is based on a hybrid of the extinct Woolungasaurus, the common eel, fork-tailed catfish, Brisbane short-necked turtle, and the mullet. Classification Pending has been created in subsequent editions for other locations internationally.

Craig Walsh's practice often takes the form of video projection onto nature or architecture. Classification Pending combines cutting edge digital technologies with a carefully crafted site-specific response and an exploration of alternative contexts for contemporary art. In what he sees as a logical progression for public art, Walsh's artworks evolve over time with the incorporation of new narratives and themes the artist uncovers through discussion and consultation with community groups. Classification Pending extends Walsh's interest in portraiture, exploring this through capturing the identity and feel of a community or site's past and present. Commissioned as part of a larger redevelopment, Classification Pending embodies one of the

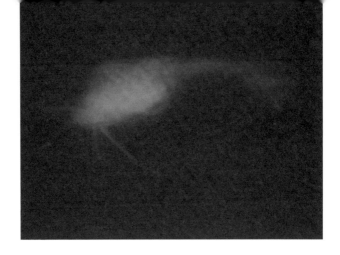

redevelopment's key objectives—to connect the city with its river. Heavily poisoned by coal mining and other industries, the Bremer River was separated not only physically by a steep embankment, but also psychologically in the minds of its community. Assisted by a major redevelopment of the site, the installation plays a central role in bringing people down to the riverbank, enlivening the space and suggesting life returning to the river.

A curious mix of advanced computer-generated artificial life form in a pre-historic looking body, Classification Pending asks audiences to think about the links or boundaries between myth, evolution, and genetic engineering. At a time when ecological issues are at the forefront of public debate and the rise of sustainable art brings such issues into galleries and public art projects internationally, Classification Pending encourages the audience to reflect on the extinction of species and environmental degradation. Intriguingly, however, the work brings to mind an interesting point through its projection of the previously unseen hybrid river creature—although species are becoming more and more rapidly extinct, at the same time new examples of flora and fauna are discovered.

Unlike many public art projects that, although visually stunning and engaged with important issues, lose their novelty over time and become overlooked, Classification Pending holds interest through continuous evolution. The installation presents new possibilities for how permanent and evolving public art installations can be integrated into environments and provide long-term engagement. The work also prompts an examination of how technology can stimulate audiences to learn about and interact with nature, and if technology can replace or substitute an authentic experience. Classification Pending exists not as an isolated element, but through technology in creative dialogue with the existing and historic flora and fauna of the Bremer River site. This allows the installation to express and reflect a bigger, evolving story of place, a truly site-specific response.

雨之舞
Kinetic Rain

艺术家：德国 Art+Com 设计公司
地点：新加坡樟宜机场第一航站
形式和材料：多媒体互动装置
时间：2012 年
委托人：新加坡樟宜机场集团
推荐人：凯利·卡迈克尔

"雨之舞"由德国 Art+Com 设计公司创作，是通过两个元素，由 1000 多颗起起落落的镀铜轻质铝液滴形成的设施。闪闪发光的雨滴悬挂在钢丝上，隐藏的发动机利用计算机实现控制，从而使其以一种精心设计的模式上下移动。这两大元素，通过精心设计的 15 分钟的运动，从抽象形态演变成立体丰富的形态。人们可以从上方、下方以及各方面来观察"雨之舞"。根据观看者的位置，其对计算机设计的复杂运动的视觉体验也不同。关于位于新加坡樟宜机场（亚洲主要机场枢纽之一）第一航站候机楼中的"雨之舞"，还增加了一个沉思元素。该设施跨越了 75 平方米的区域，其高度超过 7.3 米。其创作者声称它是世界上最大的动态雕塑。

该作品多次参照了自然现象，最为明显的是通过 1216 颗雨滴的形状反映了新加坡的热带气候模式。产生的波浪式运动模式描绘出海洋的运动，而"雨之舞"流动、变换成各种形状的方式使人想起一群鱼的翩翩游动，或一群鸟儿如芭蕾舞般优雅的飞舞。"雨之舞"悬吊在两个相对的自动扶梯上，反映出忙碌的机场候机楼（如樟宜机场）内不断变化的状态。其创作者 Art+Com 评论说，该作品是关于"飞翔的梦想"的，飞翔是该设施的一个重要的视觉主题，这一主题将该作品与机场背景密切联系在一起。一些较易被识别的形状包括飞机、热气球和风筝。这 16 个精心设计的不同分段中的其他形状还包括龙和群鸟。动作和形态的变化具有催眠的功效，且它们是流动的、微妙的。

"雨之舞"作为大型翻新工程的一部分，是一个委派的场地定制项目。在樟宜机场翻新期间，机场一个称为 Mylar Cord 喷泉的颇受喜爱的艺术品正式停用。"雨之舞"与 Mylar Cord 喷泉位于同一场所，加入了使人联想到 Mylar Cord 作品的水特征，实现了翻新工程早期的愿望。但是，该新作品不仅采纳了水的创意，通过水滴加入了水的概念从而将这一创意发扬光大，而且将其与先进的计算机技术和工程学相结合。

和一些无固定场所的地方一样，它们从来不是人们的最终归宿，机场可能也不是一个理想的场所，但它却是一个快速崛起、发展的设计行业。机场，遍布着各种精品店、酒吧、咖啡馆、酒店和其他休闲场所，正迅速成为后现代第三空间的一个经典典范，渐渐成为新兴地球村日常生活的一部分。强调设计的机场，如樟宜机场的改造，创建了小城市，而其目标在于创建一个更加舒适的、更有利于社交互动的环境，而这正是场所营造的主要元素。尽管出境站会充满各种情绪，有离开亲人的悲伤，有沮丧和无聊，但雨之舞提供了另一种选择：15 分钟优雅的、迷人的缓缓流淌，从而使旅客在其旅途中能感受到内心的宁静。

科技与艺术的完美诠释

李柯臻

"雨之舞"是一组利用数字媒体艺术诠释的动态雕塑艺术项目。艺术家运用水滴的概念，设计出了赋予流动舞蹈美感的公共艺术作品。在特定的时空环境里，将人类日常生活中的的物质文化实体，进行艺术性地有效选择、利用、改造、组合，令其演绎出新的展示丰富精神文化意蕴的艺术形态。在最新的计算机技术和工程学的支持下"雨珠们"优雅的、迷人的缓缓流淌，变换着多样优美的步子，为旅途中的人们带来内心的宁静。艺术，在数字媒体提供的平台之上不断成长、创新，运用新的科学技术为公共艺术的发展与转型创造了新的机遇。也许在未来的某一天，结合了数字媒体的公共艺术将会成为我们生活中不可缺少的一部分，成为我们传达城市气质最为鲜活生动的符号。

Artist: German design collective Art+Com
Location: Terminal One, Changi Airport Singapore, Singapore
Media/Type: Interactive multimedia device
Date: 2012
Commissioner: Changi Airport Group Singapore
Researcher: Kelly Carmichael

Created by German design collective Art+Com, Kinetic Rain is an installation formed in two elements from over one thousand rising and falling copper covered lightweight aluminium droplets. Suspended by steel wires, the gleaming raindrop-like shapes are computer-controlled by a hidden motor to move up and down in choreographed patterns. The two elements move in dialogue through a fifteen-minute choreographed piece, evolving from abstract to figurative, three-dimensional forms. Kinetic Rain can be viewed from above, below, and all sides. Visual experiences of the complex computer-designed movement are different depending on the viewer's position. Kinetic Rain was commissioned to add a contemplative element to the lively transit space of the Terminal One departure hall in Singapore's Changi Airport, one of the major airport hubs in Asia. The installation spans an area of more than 75 square meters and is over 7.3 meters in height. Its creators claim it is the world's largest kinetic sculpture.

The work makes many references to natural phenomena, most obviously Singapore's tropical weather patterns in the shape of its 1,216 droplets , and title. The wave-like motion patterns generated evoke the ocean, while the manner in which Kinetic Rain flows and morphs into various shapes brings to mind the motion of a school of fish or the elegant ballet of a flock of birds in flight. Suspended above two opposing escalators the kinetic artwork both reinforces and is reflective of the state of flux within which a busy airport terminal like Changi exists. Its creators, Art+Com,

have commented that the work is about "the dream of flying" and flight is an important visual motif in the installation, connecting the work strongly to its airport setting. Some of the more recognizable shapes include an airplane, hot air balloon, and kite. Other shapes in the sixteen differently choreographed segments include a dragon and a flock of birds. The motion and changes in form are hypnotic, fluid, and refined.

Installed as part of a major refurbishment, Kinetic Rain is a commissioned, site-specific project. A much loved artwork at the airport known as the Mylar Cord fountain was decommissioned during Changi's upgrade. This work is commemorated in Kinetic Rain, located in the same place and fulfilling an expressed desire early in the refurbishment plans to incorporate a water feature reminiscent of the Mylar Cord work. The new work, however, takes the idea of water and carries it forward, incorporating the concept of water via its droplets but combining this with advanced computer technology and engineering.

As somewhat placeless places—they are never one's final desitnation—airports may be an undesirable location, but one with a rapidly emerging and evolving design specialty. Populated with boutiques, bars, cafes, hotels, and other attractions, airports are rapidly becoming a classic example of postmodern third-space and an increasingly common part of life for an emerging global community. Design led airports such as the Changi redevelopment are creating small cities, targeted to make your time there more comfortable and more conducive to social interaction—major components of placemaking. While the departure terminal can ring with emotion ranging from sadness at leaving loved ones to candid frustration and boredom, Kinetic Rain offers an alternative— an elegant sixteen minutes of mesmerising fluidity to transport and soothe travellers along their journey.

数字奥德赛
Digital Odyssey

艺术家：克雷格·沃尔什以及当地的社区团体
地点：11 个旅行展示 / 居住地点：澳大利亚的穆雷桥、昆士兰州温顿、凯恩斯、麦基、伍德福德、格莱斯顿、新南威尔士杰林冈、维多利亚州巴拉腊特、塔斯马尼亚霍巴特、北领地埃利斯斯普林斯、新南威尔士阿米代尔
形式和材料：数字媒体
时间：2010—2011 年
委托人：澳大利亚当代艺术博物馆
推荐人：凯莉·卡迈克尔

　　"数字奥德赛"讲述的是澳大利亚艺术家克雷格·沃尔什的为期 18 个月的旅程经历。在旅途中他参观访问了 11 个澳大利亚的偏远地区。该艺术家改编和重新执行了该项目的 4 个主要作品 "Humanature"、"交叉引用"、"入侵"和 "待定分类"——并将在途中创建新的合作作品。2010 年 1 月，克雷格·沃尔什和他的家人开始在可变换的移动房屋 (同时也兼做工作室) 里开始了旅行。克雷格·沃尔什按照预先的计划，每次花几个星期的时间在全国的某一个地区居住一阵子。艺术家与社区团体进行咨询和协商，创建并展示数字作品，作品可源自改编他的一个现有的主要作品，或是根据当地环境进行新的创作。澳大利亚当代艺术博物馆 (MCA) 旅游项目 "数字奥德赛"与每一个地方的当地场馆合作，进行包括公共会议、会谈、研讨会和宣讲课在内的活动。

　　克雷格·沃尔什被大众所知晓，是因为他把大规模计算机合成图像投影到现成的城镇和乡村的环境和公共空间上，包括建筑、树木、河道和商场。沃尔什有一个混合的艺术实践，可以理解为在非正统的场景里的奇思异想。他的实践需要安装临时或永久的投影装置，使用尖端的数码技术，创造出超大尺寸的作品。克雷格·沃尔什的实践以体验、实地而具象的景象和瞬间来展示艺术。

　　通过 "数字奥德赛"项目，克雷格·沃尔什在路上进行工作，因此，乡村和偏远地区不仅能够在当代艺术中展示出来，还可以通过高新、先进的技术积极参与到艺术品的生成和展示中去。在对艺术家先前主要作品进行重新创作的同时，"数字奥德赛"使新一代反映特定团体顾虑或者利益的艺术品得到了展示。通过和沿途的团体进行咨询和合作，该项目培养了传统形式的文化交流，但采用的是先进的数字技术。该项目描绘了当地历史、当前问题或当地的故事，创造的艺术反映了社区本身的东西，创造了一个特定地点和局部的反映。

　　住在房车里，"家"的概念，成为一个新的艺术品的沃土。在从不丹、中国、欧洲、韩国和土耳其的新移民邂逅以后，一个名为 "家"的作品被创造出来。

在该作品中，来自不同文化背景的人表达他们对于"家"的想法。该项目是克雷格·沃尔什、艺术家 Hiromi Tango（沃尔什的妻子）和当地社区团体进行合作的。一个在"数字奥德赛"之旅中创作的名为"谁的平均数？"的合作作品展示了一个社区的肖像。克雷格·沃尔什邀请人们站在两根白色柱子后面拍照，这两根白色柱子分别代表了澳大利亚人的平均身高：男性为 178.4 厘米，女性为 163.9 厘米。在某一个地方，一共有 120 个男人、妇女和儿童参与了这次创作。图像被合成到一起，并且以真人大小打印，展现了澳大利亚人身高差距的全景。"谁的平均数？"反映的是局部地区人民经常认为首都的政府是以"平均数据"作为依据作决定，而不考虑他们的具体需求。

"数字奥德赛"不仅仅是克雷格·沃尔什在各个通常是偏远的地区的作品或是和社区团体一起合作的作品，这也是一个旅途。项目拥有史诗般的性质、漫长的旅行距离，风景、人和艺术家遇到的经历，使这成为一个促进和协作的丰富源泉。克雷格·沃尔什加入了著名的澳大利亚艺术家团体，包括悉尼·诺兰、汉斯·荷森 和弗雷德·威廉姆斯，他们都曾在澳大利亚偏远乡村地区进行探索并创作。

新媒体时代

李柯臻

"数字奥德赛"是一系列的大型数字媒体艺术。艺术家在旅行中不断创作，向人们诠释了旅行的意义。该项目将大规模计算机合成图像投影到现成的城镇和乡村的环境和公共空间上。用尖端的数码技术，创造出超大尺寸的作品，以体验实地而具象的景象与瞬间来展示艺术；描绘当地历史故事、当前问题。该项目，使乡村和偏远地区不仅能够通过艺术的形式得到展示，还可以在数字媒体提供的优势下，将更高新、先进的技术，积极参与到艺术品的生成和展示中去，创作出更为多样的公共艺术。新媒介的介入，为公共艺术的发展带来了更为深远的意义，为艺术家多样化创作的实现提供了更大的可能性。

Artist: Craig Walsh and regional communities
Location: Touring exhibition/residency at eleven locations: Murray Bridge, SA; Winton, QLD; Cairns, QLD; Mackay, QLD; Woodford, QLD; Gladstone, QLD; Gerringong, NSW; Ballarat, VIC; Hobart, Tasmania; Alice Springs, Northern Territory; and Armidale, NSW, Australia
Media/Type: Igital media
Date: 2010-2011
Commissioner: Contemporary Art Australia
Researcher: Kelly Carmichael

Note: This submission includes the artwork Classi□cation Pending, submitted for the IAPA as a separate work.

Digital Odyssey is an 18-month tour/traveling artist residency undertaken by the Australian artist Craig Walsh. The tour visited eleven locations around remote parts of Australia. The artist adapted and re-executed four major works for this project—Humanature, Cross-reference, Incursion and Classification Pending—and created new collaborative pieces en route. In January 2010, Walsh and his family began the tour in a converted mobile home that doubled as a digital studio. Walsh undertook preplanned residencies that lasted several weeks at a time in regional communities throughout the country. The artist worked in consultation with community groups to create and present digital artworks either adapting one of his existing major pieces or developing new work in response to local environments. A Museum of Contemporary Art Australia (MCA) touring project, Digital Odyssey partnered with local venues at each location to present events including public meetings, talks, workshops and master classes.

Craig Walsh is known for large-scale computer-manipulated imagery projected onto and in response to existing environments and public spaces in both urban and rural settings. These have included buildings, trees, watercourses, and shopping malls. Walsh has a hybrid artistic practice best understood as the exploration of unlikely ideas in unorthodox settings. His practice takes the form of temporary or permanent projection installations using a combination of cutting edge digital technologies to produce monumentally scaled works. Walsh's practice delivers art as experience, as site-specific spectacles and moments in time.

By taking his work on the road via the Digital Odyssey project, Walsh enabled rural and remote communities to not only be exposed to contemporary art, but to also engage actively with the production and presentation of artwork using highly advanced technologies. Alongside recreating major works from the artist's earlier practice, Digital Odyssey allowed the generation of new digital artworks that reflected concerns and interests of particular communities. By consulting and collaborating with each community along the tour, the project fostered a traditional form of cultural exchange, but employed advanced digital technology. Drawing on local history, current concerns or local stories, the project generated works of art that echo back to the community something of itself, creating a site-specific and localized response.

Living in a motor home, the concept of 'home' became fertile ground for a new artwork. Upon meeting recent immigrants from Bhutan, China, Europe, South Korea, and Turkey, a new body of work known as Home in which people from diverse cultures and backgrounds articulate their personal ideas of 'home' came to life. The project was collaboration between Walsh, the artist Hiromi Tango (Walsh's wife), and local communities. A collaborative work titled Who's Average? created on the Digital Odyssey tour offers a community portrait of sorts. Walsh invited people to have their photo taken standing behind two white rods at heights representing Australia's national average: 178.4 centimeters for men and 163.9 centimeters for women. At one location 120 men, women, and children participated. Images are composited and printed life size to create a panorama of Australian difference. Who's Average? reflects the concerns of regional communities who often perceive that governments in city capitals make decisions based on 'statistical averages', without concern for their specific needs.

Digital Odyssey is not simply a presentation of Walsh's work at various, albeit often remote, locations or the creation of new works with communities. The project is the tour. The epic nature of the project and the huge distances traveled, the landscape, people and experiences the artist encountered, which became a rich source of stimulus and collaboration. Walsh joined a group of celebrated Australian artists including Sydney Nolan, Hans Heysen and Fred Williams who had explored and made work in and about Australia's remote rural communities.

Wallumai 风雕塑
Wallumai Wind Sculpture

艺术家：苏珊·米尔恩、格雷格·斯通豪斯、克里斯·托宾
地点：澳大利亚新南威尔士普特尼 Yaralla 路 Bennelong 公园
形式和材料：装置
时间：2011 年
委托人：莱德市地方议会
推荐人：凯莉·卡迈克尔

一个大型的河边再造项目元素 "Wallumai 风雕塑"，把原住民的声音和历史带回到了这片土地上。莱德市任命了一个由苏珊·米尔恩和公共艺术顾问格雷格·斯通豪斯带领的公共艺术团队和土著艺术家克里斯·托宾合作设计河滨路项目的景观，该项目歌颂的是这座城市的土著文化遗产。"Wallumai 风雕塑" 中的 "鲷鱼" 代表着鱼这个意象对于当地土著群体的重要性。这个作品位于帕拉马塔河沿岸，代表的是河被开发前原住民和早期欧洲人享受到的丰饶的自然环境。

三个拔地而起的金属结构矗立在河岸边。鲷鱼的形状富有标志性且容易辨认，鱼的头光滑，亮片鳞片组成的身体闪闪发亮。雕塑的风叶片随着微风转动，闪闪发光的鳞片映射着周围环境和天气变化的情况。把几个雕塑放置在一起，就像一群鱼在不同高度和维度中，它们使水、土地和天空的景象不断地变化。有时鱼看上去像在河里游泳，有时候像在岸边的微风中游动，将土地、水和它们的历史融为一体。

wallumai 的使用来源于在这个地方居住的 Wallumedegal 部落。艺术家根据土著雕刻和绘画来刻画鲷鱼，在土著的雕刻和绘画中，鲷鱼的特点被简化的同时，其骨架结构特征仍保留。该作品代表了该地区的原住民 Wallumedegal 人民的传统图腾。部落的名字很可能来源于 "Wallumai" 和 "matta"，指的是一个地方。那些生活在河的南面，在现在的莱德区域的部落，因此被称为鲷鱼部落。河流和水生生物不仅仅使 Wallumedegal 得到了自然物质上的延续，而且对于他们的文化、身份和生活方式是固有的存在，这反映在他们选择鲷鱼为文化图腾之中。作为一个艺术品 "Wallumai 风雕塑" 成为当地土著居民的身份，他们的捕鱼行为和丰富的海洋生物迁移和生长的象征。

原住民身份和共识是这个作品和它所存在的环境不可或缺的部分。米尔恩领导的公共艺术团队和斯通豪斯、托宾合作，在莱德地方议会采取计划之前，

与当地原著民提出并讨论了这个概念。计划第一个实现的是"Wallumai 风雕塑"。这个雕塑标志着河滨路的总体目标的开端，即通过安装永久性的当代艺术作品反映了历史、故事和当地土著社区的价值观，创建一个沿着海滩的高质量的公共空间。旨在创建更强的原住民社区意识以及他们在莱德社区的土地监管权和存在感。

找寻遗失的文化

李柯臻

"Wallumai 风雕塑"是一个大型的河边再造、永久性的当代艺术作品。该项目旨在建立更强的原住民社区意识、社区土地监管权和存在感。公共艺术参与原始居民社区的再造对于社区物质文明建设和精神文化的形成都具有深远的意义。以艺术的形式将原住民的声音历史重新带回到了这片土地上，在艺术中回味曾经的文化与故事。

艺术家充分发挥和运用了地域文化资源。将当地土著居民的原始图腾，用具有他们生活特色的图案符号表现而出，建立了居民对社区的认同感和归属感。用艺术的方式反映了原始居民的历史、故事和当地土著社区的价值观，再现了这座城市的土著文化。将城市遗失的、原始的、土著的文化和曾经找寻回来，重新带入到新的时代当中。传统与现代的碰撞，将会给城市带来更多未知的可能。

Artists: Susan Milne, Greg Stonehouse, and Chris Tobin
Location: Bennelong Park, Yaralla Road, Putney, NSW, Australia
Media/Type: Device
Date: 2011
Commissioner: Ryde City Council
Researcher: Kelly Carmichael

One element of a greater riverside reinvention project, the Wallumai Wind Sculpture, contributes to bringing Aboriginal voices and history back to the area. The City of Ryde appointed a public art team lead by artists and public art consultants Susan Milne and Greg Stonehouse working with Aboriginal artist Chris Tobin to capture a vision for a river-walk project paying tribute to the city's indigenous heritage. Aboriginal for "snapper," the Wallumai Wind Sculpture celebrates the importance of the fish for local indigenous groups. Located along the Parramatta River, the work refers to the rich natural environment once enjoyed by Aboriginal people and early Europeans before the river was exploited.

Three wallumai sit high off the ground atop metal structures on the riverbank. Taking the iconic and easily identifiable shape of a snapper, the fish have smooth heads and glistening bodies constructed from a formation of sequined scales. The sculptural wind vanes turn with the breeze and their shimmering scales reflect their surroundings and the shifting weather conditions. Placed together as a school of fish at different heights and dimensions, they offer continuously changing views of the water, land, and sky. Sometimes the fish appear to swim in the river, at others they turn inland with the breeze, connecting land and water and their comingling of histories.

The use of the wallumai is derived from the Wallumedegal clan that inhabited this area. The artists drew from the Aboriginal engravings and drawings that simplify the characteristics of the snapper while still retaining its characteristic skeletal structure. The work represents the traditional totem of the area's original inhabitants, the Wallumedegal people. It is likely that the clan's name derives from the word "Wallumai" and "matta," a word denoting a place. Those living on the southern side of the river, in what is now the Ryde area, were therefore the snapper fish clan. The river and its aquatic life not only sustained the Wallumedegal physically. It was intrinsic to their culture, identity, and way of life, and this is reflected in their adoption of the snapper as their cultural totem. As an artwork, Wallumai Wind Sculpture becomes a symbol of the local identity of Aboriginal people, their fishing practices and the rich marine life migrating and feeding here.

Aboriginal identity and consent is integral to this work and the larger context within which it sits. A public art team led by Milne and Stonehouse working with Tobin presented and discussed concepts with local Aboriginal groups before Ryde Council adopted the plan. The first realization of the plan was the Wallumai Wind Sculpture. This sculpture commences the overall aim of the Riverwalk to create a high quality public space along the foreshore through the installation of permanent contemporary art works that reflect the history, stories, and values of local Aboriginal communities. This is intended to create greater community awareness of Indigenous people, their custodianship of the land, and presence in the Ryde community.

两个世界之间
In Between Two Worlds

艺术家：詹森·温
地点：澳大利亚新南威尔士，悉尼唐人街，金贝里
形式和材料：壁画、装置
时间：2011 年至今
委托人：悉尼公共艺术
推荐人：凯利·卡迈克尔

"两个世界之间"是一个较大的公共艺术项目，由澳大利亚艺术家詹森·温发起，其运用深湖蓝色调结合了中国人和土著居民的象征意义。安装物件包括三个主要组件：墙上的壁画、地板上的画、30 个悬浮照亮的精神人物。"两个世界之间"尝试与观众建立友好关系，并探索其场所的肉体性，作品存在于巷道和路面上方的空域，花岗岩上的图案与壁画上的云相辉映。位于悉尼繁华的唐人街的核心区域的金贝里，白天巷子的墙壁被蓝云的壁画和盘旋在头顶的银色人物形象照亮。夜间精神人物以超凡脱俗的蓝光芒照亮巷子，邀请游客探索和深思他们与文化遗产，当代造像和双边文化参考的不寻常的混合。

詹森·温是一位当代艺术家，他在各种媒体工作来检查社会和文化认同。拥有原住民、中国和欧洲体验，并探讨他体验到的多元文化遗产，创建了存在于文化之间的作品。受中国和土著教育双重文化的强烈影响，温对影响他社区的持续的挑战进行探索。作品充满丰富的象征意义和诗歌隐喻，艺术家的中国和土著文化在"两个世界之间"中得到充分的表达。通过这个作品，温力图展示他对悉尼土著卡迪各人的尊重并唤起各种文化——不论新的或旧的——这种文化中称唐人街为家。"我想创造一个体验，就像走在两个世界或在天地之间旅行"艺术家说。

包含风、水、火和土元素的艺术品将中国和土著文化主题带到了悉尼城内。两种文化元素都有它们自己的精神，在中国生肖中所创造的人类带有元素的特点。温用悬浮在金贝里的半人半神灵来反映这一特点，分别代表过去、现在和将来的祖先。暗含天地、元素和对过去与现在祖先的尊敬，温唤起普世主题，他允许工作更开放并提供有关其他文化之间联系的见解。

艺术家的多元文化遗产在场所、历史以及它怎样在悉尼运转上给了他一个独特的视角。作品有时感觉就像一个数字游戏，尤其是夜间被照亮的时候，因为它结合了当代审美和被认可的亚洲图案。"两个世界之间"为其观众

提供了许多连接点，它也是这一地区在公共艺术实践的一种决裂。过去悉尼唐人街的公共艺术被限定在普遍存在和陈旧元素的具象模式中，如灯笼和红光。然而重要的是要识别和定位管辖区华人的社区生活和工作，这种审美没有认识到当代亚洲文化和它的不断变化。"两个世界之间"重新审视了由一个年轻的有动态观的多种族澳大利亚艺术家提供一个引人注目的、受到喜爱的公共艺术作品。

两种文化的融合

李柯臻

"两个世界之间"是一个综合性的公共艺术项目，艺术家用壁画和装置的形式，将中国文明和悉尼土著文化表现出来并融合在一起。

当下，华人遍布世界的各个角落，中华文化在异国的传承、华人在他乡的归属感，在该项目中都会带给人们一些思考和启发。同时，用艺术的方式反映了土著居民的历史、故事和当地土著社区的价值观，歌颂了这座城市的土著文化遗产。艺术家结合了当代审美元素、传统的中国和土著居民的文化，对公共艺术实践进行一种新的尝试。两种文化元素都有它们自己的精神和特点，但它们包容、宽厚、相互接纳，创造了更加震撼心灵的艺术体验，使悉尼的文化更为文明多样，为居住于悉尼的中国人和土著居民带来心灵上的慰藉和归属感。

Artist: Jason Wing

Location: Kimber Lane, Chinatown, Sydney, NSW, Australia

Media/Type: Mural, Device

Date: 2011-present

Commissioner: City of Sydney Public Art

Researcher: Kelly Carmichael

In Between Two Worlds (2011) is a powerful public project by Australian artist Jason Wing meshing Chinese and Aboriginal symbolism in vivid blue hues. The installation consists of three main components: wall murals, floor murals, and 30 suspended illuminated spirit figures. In Between Two Worlds sets out to engage with its audience and also explore the physicality of its site—the work existing in the airspace above the laneway as well as on the road surface, where motifs etched into granite echo the wall mural's cloud design. Situated in Kimber Lane at the heart of Sydney's thriving Chinatown area, by day the walls of the lane are brightened by a mural of blue clouds while silver figures hover overhead. By night the 'spirit' figures illuminate the lane with an otherworldly blue glow, inviting visitors to explore and contemplate their unusual mix of heritage and contemporary iconography and bi-cultural references.

Jason Wing is a contemporary artist working in various mediums to examine social and cultural identity. Of Aboriginal, Chinese, and European heritage Wing owns and explores his experienced multicultural heritage, creating work that exists between cultures. Strongly influenced by his bi-cultural Chinese and Aboriginal upbringing, Wing explores ongoing challenges that impact his community. Both rich in symbolism and poetic metaphor, the artist's Chinese and Aboriginal cultures find full expression in the installation In Between Two Worlds. Through the work, Wing sought

to pay respect to the indigenous Gadigal people of Sydney and evoke the myriad cultures—both old and new—who call Chinatown home. "I wanted to create an experience like walking in between two worlds or travelling between heaven and earth" the artist commented.

Incorporating the elements of wind, water, fire, and earth, the artwork brings important Chinese and Aboriginal cultural motifs to inner city Sydney. In both cultures the elements are said to have their own spirits and in the Chinese Zodiac humans were created with the characteristics of elements. The half human, half spirit figures Wing has suspended above Kimber Lane reflect this, representing past, present, and future ancestors. Suggesting heaven and earth, the elements, and respect for ancestors past and present, Wing evokes universal themes and allows the work to open more widely, offering points of connection between other cultures.

The artist's multicultural heritage gives him a unique perspective on the site, its history and how it functions within the city of Sydney. The work feels like a digital game at times, especially when illuminated at night, as it incorporates a contemporary aesthetic mixed with recognizably Asian motifs. While In Between Two Worlds offers many points of connection for its audience, it also is something of a rupture in public art practice in this area. In the past public art in Sydney's Chinatown has been formulated within a representational mode of ubiquitous and clichéd elements, such as lanterns and red lighting. While it is important to recognize and locate the Chinese community living and working within the precinct, this aesthetic does little to recognize contemporary Asian culture and its constant evolution. In Between Two Worlds readdresses that, offering a striking and much loved public artwork by a young, multi-racial Australian artist embracing a new dynamic.

大魔方箱
The Great Crate

艺术家：Plus One
地点：澳大利亚新南威尔士州悉尼亚历山大区
形式和材料：装置
时间：2012 年
委托人：悉尼市艺术和公共艺术节
推荐人：凯利·卡迈克尔

　　"大魔方箱"是悉尼艺术和公共艺术节的一个临时设施。其形状为一个用256 个塑料箱制成的，10 米长的、巨大的、绿色的立方体，且在该立方体中栽种了数千棵可食用植物。虽然繁茂的绿色植物是临时的，但该项目的目标在于通过强调社区参与，传达可持续的绿色生活方式，从而在该项目结束后延续这种生活方式。该项目，仅从生活和回收材料创建而来，位于绿色广场火车站——澳大利亚最大的城市重建场所，这一地区现已成为悉尼的主要零售、商业和文化中心之一。Plus One 成员 Telly Theodore 评论了该项目："我们希望'大魔方箱'能在几个层面上产生共鸣。比如说，这一选区曾经是这座城市的饭碗—所以我们在该设施安装结束之时，会分发所有可食用植物，供其移植到家里培育，从而鼓励人们记住这一遗产。这是一种在该艺术节结束后继续发展该项目的很好方式。"

　　尽管未来悉尼中心可能永远不会恢复成园艺农田，而且很明显"大魔方箱"的形状取自绿色广场火车站这一名字，但该项目在多个层面上是有效的。该项目——连接了人与地方的平台，唤起了人们对这个场所历史和未来的畅想，使当地人们注意到其社区规划的明显变化，且参与到这一规划中来。此时的"大魔方箱"项目反映了当前的思潮，与更广泛的社会趋势紧密一致。饮食文化发生了变化，推动了当地种植业和可持续种植法。"大魔方箱"项目是一个通过融合农业与激进主义的方式开发类似领域的项目，反映了游击或战术型都市生活在全球范围内发展的趋势。大魔方箱，尽管只是临时的，但它的存在是为了鼓励人们不断讨论如何利用公共空间，如何实践可持续生活政策。

　　"大魔方箱"在本质上是协作式的，Plus One 自身就是创意工作室和个人的协作，包括设计工作室 Telly Theodore 及其同伴，建筑师 Andy Macdonald 以及策展人 Danella Bennett。此外，该项目的核心目标是调动当地人民及其社区积极参与。由于区域背景发生变化，该作品旨在鼓励人们思考绿色广场是一个什么样的村庄以及未来的发展方向。已经有上千包茁壮成长的蚕豆包，通过邮寄，被发送给当地区域的居民和企业。此外，还为阳台或花园中植物的种植提供了种植指南。Plus One 鼓励社区在艺术

& 公共艺术节开始、搜集种子，将其移植到一个巨大的绿色生活广场前，培育种子。其它可食用植物也是可以的，项目鼓励公众在周六将这些种子带至该设施处，参与种植和讨论。在晚上，这一活雕塑和可完全回收的设施通过太阳能 LED 小彩灯照明。

最后，在各方协作和社区参与下，在该项目结束时，还邀请了悉尼居民从"大魔方箱"中选取一株植物，继续在家中培育或食用。项目共收获了 25 000 株可食用植物，包括荷兰芹、鼠尾草、百里香、草莓、山萝卜、小白菜、罗勒、紫罗勒、牛至、莴苣、苹果薄荷、绿薄荷、蜜蜂花、黄豆、越南薄荷、羽衣甘蓝、红叶卷心菜、卷心菜和豆瓣菜。

田园诗意的生活

李柯臻

"大魔方箱"是一个协作式的公共艺术临时设施。该项目的目标在于通过强调社区参与性，传达可持续的绿色生活方式，从而在项目结束后延续这种生活方式。

项目的核心目标是调动当地人民及社区积极参与到其中。鼓励人们去思考绿色社区的概念，以及未来的发展方向。将公共艺术与受众更为紧密的联系在一起，积极热情的互动，相互促进、共同发展。该项目将新的生活理念与生活方式带入到人们的生活当中。给远离田园的都市生活带来绿色天然的色彩，给予快节奏下生活的人们"采菊东篱下，悠然见南山"的田园体验。尽管它只是临时的，但它的存在是为了鼓励人们不断讨论如何利用公共空间，如何实践可持续的生活方式。

Artists: Plus One
Location: Alexandria, Sydney, NWS, Australia
Media/Type: Device
Date: 2012
Commissioner: City of Sydney's Art & About public art festival
Researcher: Kelly Carmichael

The Great Crate was a temporary installation commissioned for Sydney's Art & About public art festival. It took the form of a 10-meter square, giant, green cube made from 256 plastic crates and planted with thousands of edible greens. While the flourish of vibrancy and green plants was temporary, the project aimed to have a life beyond the actual installation period by focusing on community engagement, participation, and communicating sustainable green approaches to living. Created soley from living and recycled materials, the project was situated at Green Square Rail Station, the largest urban renewal site in Australia, which is set to become one of Sydney's key retail, commercial, and cultural hubs. Plus One member Telly Theodore commented on the project:

"We're hoping the crate resonates on a few levels. For example, this precinct was once the city's food bowl—so we' re encouraging people to remember this heritage by giving away all the edible plants at the end of the installation for them to re-pot at home. It's a great way to keep the crate growing after the festival ends."

While central Sydney might not revert to horticultural farmland any time soon—and the physical form of The Great Crate is an obvious play on the name of Green Square train station—the project was effective on multiple levels. A platform for connection between people and place, the project evoked both the history and future of the site and brought the local community's attention to, and engagement with, significant changes planned for their neighborhood. The Great Crate project is also very much of the moment, reflective of current thinking and closely aligned with

trends in wider society. Food culture has shifted to promoting local production and sustainable farming methods and The Great Crate is a project that mines similar territory in the manner it mixes agriculture with activism, reflecting the growing global trend for guerilla or tactical urbanism. While only temporary, the presence of The Great Crate aimed to encourage ongoing discussion about how public spaces are used and how they might incorporate sustainable living policies.

The Great Crate was collaborative in essence—Plus One itself is a collaboration of creative studios and individuals, including design studio Telly Theodore & Associates, architect Andy Macdonald, and curator Danella Bennett. In addition, at the heart of the project was the goal of involving local people and their communities. As an area set for intense change, this work aimed to encourage thinking about what kind of village Green Square was, and what it could become. Thousands of packets of fast-growing broad beans were distributed by mail to residents and businesses in the local area, along with easy to grow instructions to raise the plants on balconies or in gardens. Plus One encouraged the community to nurture the seeds until the Art & About festival began at which point they were gathered and transformed into a giant green living square. Other edible plants were also welcomed for contribution, the public was encouraged to bring them to the installation on Saturdays and participate in planting and discussion. At night, the living sculpture and fully recyclable installation was lit with solar-powered LED fairy lights.

In a final act of collaboration and community involvement, at the end of the project Sydneysiders were invited to take a plant from the cube and continue nurturing it at home, or alternatively eat it. 25,000 edible plants were harvested including parsley, sage, thyme, strawberries, chervil, pakchoi, tatsoi, basil, purple basil, oregano, coral lettuce, applemint, spearmint, lemon balm, pea bounty, yellow beans, vietnamese mint, kale, red cabbage, cabbage, and watercress.

玛丽之灯
A Lamp for Mary

艺术家：米卡拉·德怀尔
地点：澳大利亚新南威尔士州悉尼
形式和材料：装置
时间：2010 年
委托人：悉尼市
推荐人：埃利萨·尤

在 1996 年，一个名叫玛丽的女士被两个男人在当时位于 Surry Hills 叫作 Floods Lane 的地方残忍地强暴了，这是一个在悉尼中心商业区西南方向的郊区。这附近一直是悉尼同性恋大游行的年度聚会场所。在 1997 年，南悉尼理事会和同性恋反暴力项目（AVP）启动了一个活动，来吸引公众对这个悲剧事件的关注，并提高人们针对同性恋暴力行为的了解，形成了一个社区策划和实施委员会，包括艺术家、社区代表和委员会成员。这个委员会努力识别一个暴力防范战略，包括更安全的街道照明、景观美化和街头壁画，作为他们计划的一部分。

1997 年 2 月，在一个仪式中，这个小道被重新命名为 Mary's Place，包括壁画和牌匾的揭幕。在 2006 年，附近贝雷斯福德酒店的装修工作引起的火灾摧毁了壁画。这个社区在 2009 年再次集会，向艺术家公开发出邀请，进行艺术创作，向 14 年前在残忍袭击中存活下来的叫作的玛丽的女子表示敬意。

"玛丽之灯"是由悉尼艺术家卡米拉·德怀尔创建的，于 2010 年安装在 Mary's Place。这个单独的、看起来似乎很传统的路灯是由阳极氧化铝制成的，骄傲地站立在街道边。这个独特的照明器材具有红色的多环和双锥型的外形，会让人想起装饰艺术家庭里面可能用到的传统灯罩。在晚上，路灯会朝路面照射粉红色的灯光，创造一个冥想和回忆的氛围，并且同时提供足够的照明供行人安全通过。艺术家咨询了来自 GLBT（同性恋、双性恋和变性者 和其他组织的代表，包括 Haughton Design、卫斯理传道中心、新南威尔士州警察、ACON 反暴力项目、圣米歇尔圣公会教堂、ACON 年轻女同性恋项目和 Twenty10。一个牌匾上展示着一首诗，社区代表和诗人 Michael Taussig 合作写了这首诗。诗的内容为：

"这条小道的名字和路灯是为了纪念一名妇女，她在这里被暴打和强暴，但是她幸存了下来。她碰巧是女同性恋。当太阳下山时，这盏灯仍然伴随着你，在安静的冥想中阅读这首诗。"这个纪念牌匾上的复原见证了当地 GLBT 社区的成长和力量，城市官员和社区居民保持相关意识，不仅跟 14

年前玛丽的悲剧相关，还包括针对所有倾向女性的暴力行为。卡米拉·德怀尔解释了她这件作品的意图："这个念头是突然出现的，似乎就是应该做的事情，作为保护、治愈和警告措施，同时也用来庆祝玛丽生存下来的力量。"

"玛丽之灯"是公共领域的一件混合艺术作品，作为纪念物以及整合的都市街道设施。这展示了艺术家所领导的社区咨询过程如何用来创建艺术作品，并向社区代表提供一种所有权、骄傲和身份认同感。作品的功能、场所、地点和特定性质也反映了我们社区代表、城市官员、公共艺术管理者以及其他利益相关者越来越希望当代公共艺术在今天所扮演的角色。

公共艺术的人文关怀
李柯臻

公共艺术是一种人文关怀，人文关怀精神是公共艺术的本质要求。以人为本，关心人的生存处境，尊重人的本质，维护人的利益，满足人的需求，表达人性的呼唤，是公众对公共艺术的期待。

"玛丽之灯"是一件公共艺术装置项目，它是以人文关怀为出发点设计而成的，以艺术的方式反对对同性恋的暴力活动。该项目具有与公众产生亲和、积极对话的作用，以艺术的形式传达艺术家想要表达的感悟。立于灯下，粉红色的光芒不仅照亮了前方的路，更向人们诉说着故事，感悟着灯下的心灵，激励着一颗颗勇敢的心。这就是公共艺术的魅力和意义所在，用心灵感悟艺术，用艺术关怀心灵。

Artist: Mikala Dwyer
Location: Sydney, NWS, Australia
Media/Type: Device
Date: 2010
Commissioner: City of Sydney
Researcher: Elisa Yon

In 1996, a lesbian woman named Mary was brutally raped by two men in what was then called Floods Lane in Surry Hills, a suburb located southeast of Sydney's central business district. The neighborhood is well known as a gathering place for the annual Sydney Gay and Lesbian Mardi Gras celebrations. In 1997, the South Sydney Council and the Gay and Lesbian Anti-Violence Project (AVP) initiated a project to bring attention to this tragedy and raise awareness of violence against gays and lesbians. A community planning and implementation committee was formed to include artists, community representatives and council members. The committee worked together to identify a violence prevention strategy that included safer street lighting, landscaping and a street mural as part of their plan.

In February of 1997, the lane was renamed Mary's Place in a ceremony that included the unveiling of a painted mural and plaque. In 2006, a fire at the nearby Beresford Hotel initiated renovation work that caused the destruction of the mural. The community rallied together again in 2009 to develop and issue an open call to artists for an artwork to honor the woman named Mary who survived the brutal attack 14 years earlier.

A Lamp for Mary by Sydney-based artist Mikala Dwyer was commissioned and installed in Mary's Place in 2010. The singular and seemingly traditional street lamp is made from anodized aluminum and stands proudly in the lane. The unique lighting fixture is in the form of a red, multi-ringed and double conical shape, reminiscent perhaps of a traditional lampshade one might

find in an art deco home. At night, the lamp casts a pink glow in the lane creating a meditative sense of place and remembrance, while providing adequate lighting for the safety of passersby. The artist consulted with representatives from the GLBT (Gay Lesbian Bi-Sexual and Transgender) and other community groups including Haughton Design, Wesley Mission, NSW Police Force, ACON Anti-Violence Project, St. Michael's Anglican Church, ACON Young Lesbians Project, and Twenty10. A plaque displays a poem, written in collaboration by community representatives and poet, Michael Taussig. It reads:

"This is a lane with a name and a lamp in memory of the woman who survived being beaten and raped here. She happened to be lesbian. When the sun sets this lamp keeps vigil along with you who reads this in silent meditation."

The reinstatement of this memorial is a testament to the growth and strength of the local GLBT community, city officials and community residents in maintaining awareness, not only of Mary's tragedy 14 years earlier, but also of violence towards all women regardless of sexual orientation. Mikala Dwyer explains her intentions for the work: "The idea came out of nowhere and it just seemed liked the necessary thing to do as a gesture to protect, heal, warn, and also celebrate the power of survival in Mary."

A Lamp for Mary is a hybrid artwork in the public realm that serves as a memorial and as an integrated piece of urban street furniture. It is an example of how a community consultation process, led by an artist, can be used to create an artwork that offers community representatives a sense of ownership, pride and identity of place. The functionality, site and place specific nature of the work also reflects the many roles we, as community representatives, city officials, public art administrators, and other stakeholders, have increasingly come to expect from contemporary public artwork today.

托盘亭子
Pallet Pavilion

艺术家：盖普·菲勒团队
地点：新西兰克赖斯特彻奇市
形式和材料：装置
时间：2012 年 12 月—2014 年 4 月
委托人：盖普·菲勒同克赖斯特彻奇市议会的合作伙伴关系
推荐人：埃丽莎·约恩

从 2010 年 9 月至 2011 年 2 月，克赖斯特彻奇市及其周边地区发生了一系列地震。预计城市中心百分之八十以上的建筑被摧毁，修复成本超过 400 亿新西兰元。作为对 2010 年 9 月地震的回应，艺术协调员 Coralie Win 以及坎特伯雷大学的戏剧与电影研究讲师 Ryan Reynolds 博士组成了 Gap Filler，这是一个由具有创意的专业人士组成的跨学科团队，通过临时的、过渡期的、基于社区的和社会参与的项目，复兴被毁灭性的地震摧毁的多个场所。自 2011 年起，Gap Filler 开始拓展，现在由 Gap Filler 慈善信托管理，董事会由联合创始人 Ryan Reynolds 管理。核心运营团队包括：Coralie Winn（Gap Filler 的联合创始人和创意总监）、Trent Hiles（项目主协调员）、Richard Sewell（项目协调员）、Hannah Airey（管理员）、Sally Airey（教育和外联）和 Claire Cowles（赞助和筹款）。

"托盘亭子"是一个露天社区聚集空间，使用三千多个蓝色的木制 CHEP 托盘和另外捐赠的、借用的和再利用的材料制成。项目位于克赖斯特彻奇市的市中心，皇冠假日酒店的旧址，酒店于 2012 年拆毁。亭子设计用来支持各种文化和社会项目，在其存在期间提供了二百五十个活动。由于严格的场地和建筑规定，一个由设计师、关键建筑专业人士和专注的社区志愿者组成的核心团队领导亭子的设计。预计需要一百多人的熟练和不熟练的劳动力，花上 2 500 小时在 6 周内建好亭子，参与的志愿者年龄在 16 到 65 岁之间，在他们的网站上向约 55 名社区合作伙伴和赞助者致谢。场所设施包括食品和饮料供应、安全、电力、场地维护、视听设备、现场管理团队、厕所和垃圾回收。从 2012 年 12 月到 2013 年 4 月，预计 25000 人以参观者、志愿者、供应商和表演者的身份参与到项目中去。

在六个半月的时间内，在社区场地和聚集场所的建立和准备过程中，Gap Filler 成功地协调和组织多个关键人员、志愿者、熟练劳工、城市代表、社区合作伙伴、赞助商、供应商和观众。在 2013 年 5 月，Gap Filler 开启了一个公众资助的项目来拓展"托盘亭子"的寿命。他们成功筹集了 82 000 新西兰元，作为亭子下一年的运营费用。在 2014 年 4 月，亭子被拆除。这个场所仍然被称为"公用地"，现在是多个地震后组织所在地，包括"闲

置空间生活"（LiVS）、"志愿者军队基金会"（VAF）、"拱廊项目"
以及 FESTA（过度架构节日）所开发的十个层叠木板拱道。通过克赖斯特
彻奇市议会和 LiVS 之间达成的协议，现场仍然支持过渡性项目，甚至对场
所计划了一系列演进的用途。

随着当代公共艺术实践中社会实践的不断演进，可能有必要评估多学科和
跨学科合作项目，以应对涉及都市整建和再生项目的特殊情况。在这种情
况下，"托盘亭子"可以被视为临时的、基于社区的、社会参与的公共艺
术项目的一个例子。该项目把被地震摧毁的主要市中心场所转化为一个迫
切需要的聚集场所，以欢迎居民、城市工作者以及商业回到城市中心。他
们集体的团队技能和经验使他们能够成功地融合艺术、设计、教育和外联
项目，以恢复和培养社区参与、集体和个体创意表达以及被地震摧毁的城
市内的对话。

艺术的多学科、跨学科合作

李柯臻

"托盘亭子"是一个露天社区聚集场所，是一个艺术的多学科、跨学科合
作项目。该项目将被地震摧毁的主要市中心场所转化为一个聚集场所，以
欢迎居民、城市工作者以及商业回到城市中心。通过大量的公众参与，以
恢复和培养社区参与、集体和个体创意表达以及被地震摧毁的城市内的对
话。艺术家更多的关注了作品的广泛参与性与受众性，将公共艺术与受众
更为紧密的联系在一起，积极热情的互动。可以想象，随着当代艺术的发展，
社会实践不断演进、多学科和跨学科合作项目不断增多。文化多样的艺术
形式为城市和人们的生活带来别样的艺术文化氛围，给快节奏的都市生活
带来缤纷的色彩。在未来，公共艺术将以更加包容多样的艺术形式促进城
市的发展，满足公众的需求；反之，多样的文化合作也将为公共艺术拓展
更为广阔和深远的领域创造更大的可能性。

Artists: Gap Filler
Location: Christchurch City, New Zealand
Media/Type: Device
Date: December 2012 - April 2014
Commissioner: Gap Filler in partnership with City of Christchurch Council
Researcher: Elisa Yon

From September 2010 to February 2011, a series of earthquakes struck the city of Christchurch and its surrounding areas. It was estimated that over eighty percent of the buildings in the city center were destroyed, with repair costs over $40 billion. In response to the first earthquake in September 2010, Coralie Winn, an arts coordinator, and Dr. Ryan Reynolds, a lecturer in theatre and film studies at the University of Canterbury, formed Gap Filler, a multi-disciplinary team of creative professionals dedicated to reactivating the growing number of vacant sites left by the devastating earthquake, with temporary, transitional, community-based and socially engaged projects. Since 2011, Gap Filler has expanded and is now administered by the Gap Filler Charitable Trust with a board led by co-founding member Ryan Reynolds. The core operating team includes: Coralie Winn (Gap Filler Co-Founder and Creative Director), Trent Hiles (Lead Project Coordinator), Richard Sewell (Project Coordinator), Hannah Airey (Administration), Sally Airey (Education and Outreach) and Claire Cowles (Sponsorship and Fundraising).

Pallet Pavilion is an open-air community gathering space, built using over 3,000 wooden blue CHEP pallets and additional donated, loaned, and repurposed materials. It is sited in the city center of Christchurch on the former site of the Crown Plaza hotel, which was demolished in 2012. The pavilion was designed to support a variety of cultural and social programs, offering over 250+ events during its time in existence. Due to the strict site and building regulations, a core team of designers, key building professionals, and dedicated community volunteers led the design of the pavilion. It took an estimated 150+ people and 2,500 hours of skilled and unskilled labor to build the pavilion in six weeks, engaging volunteers

aged 16 to 65. Over 55+ community partners and sponsors are acknowledged on their website. Site amenities included food and drink vendors, security, power, site maintenance, audio-visual equipment, on-site administration team, toilets, and waste collection. From December 2012 to April 2013, an estimated 25,000 people participated as visitors, volunteers, vendors, and performers.

In just six and a half months, Gap Filler successfully coordinated and engaged multiple key personnel, volunteers, skilled laborers, city representatives, community partners, sponsors, vendors, and audiences in the making and programming of a community venue and gathering place. In May 2013, Gap Filler launched a crowd-funding campaign to extend the life of the Pallet Pavilion. They successfully raised NZ$82,000 to cover the pavilion's operational costs for an additional year. In April 2014, the pavilion was deconstructed. The site continues to be known as "The Commons" and is now home to a handful of post-quake organizations including Life in Vacant Spaces (LiVS), Volunteer Army Foundation (VAF), The Arcades Project, and a series of ten laminated timber archways developed by FESTA (the Festival of Transitional Architecture). The site continues to support transitional projects through an agreement between Christchurch City Council and (LiVS), and is even governed by an evolving set of aspirations for the site.

As social practice continues to evolve in contemporary public art practice, it is perhaps necessary to begin to evaluate projects initiated by multidisciplinary and interdisciplinary collectives that emerge to respond to unique situations involving urban rehabilitation and regeneration initiatives. In this context, Pallet Pavilion may be understood as an example of a temporary, community-based, socially engaged public art installation. The project transformed a prominent city center site devastated by the earthquakes into a much-needed gathering space to welcome residents, city workers, and businesses back into the urban center. Their collective team skills and experience allowed them to successfully fuse art, design, education and outreach programming together to bring back and nurture community participation, collective and individual creative expression, and dialogue in a city devastated by the earthquakes.

那伽
Naga

艺术家：Leang Seckon
地点：柬埔寨暹粒河畔、尼泊尔加德满都国家动物园
形式和材料：装置
时间：2008 年 10—11 月（柬埔寨），2012 年 11—12 月（尼泊尔）
委托人：柬埔寨、尼泊尔公共艺术组织
推荐人：莱恩·柏格森

"那伽"——这个 5 米高 250 米长的由竹子和再生塑料建造的蛇形的生物——最引人注目的一方面，就是作品所传达的信息。

"通过强调环境开发成本，'那伽'旨在鼓励社区和政府努力保护淡水资源。" Rossi & Rossi 画廊的 Kim-Ling Humphrey 说。

这个作品安装过两次，一次在柬埔寨暹粒河畔，以纪念 2008 年的世界水日；另外一次是安装在加德满都的国家动物园，作为尼泊尔"第二届加德满都艺术节"的一部分，这件作品呼唤人们对全球河道污染的关注。"把塑料和天然设置并列防止—这件作品总是被安装在水体上面—那伽的象征，这传统上代表水，被认为可以保护泉水、水井、河流，并能控制雨水，这使得这个信息非常突出。" Humphrey 补充道。

艺术家 Leang Seckon 更加直接地指出："我希望人们看到这个作品的时候能够理解，这件作品是由垃圾和水污染组成的，如果这种情况继续下去，那伽以及能在河流里找到的其他生物将会消失。"

这个项目还使用本地的信仰和传统来传达其信息。在南亚和东南亚的文化和传统里，那伽是神话中一种具有蛇形的生物，具有变形的能力。在柬埔寨创世神话中，那伽是具有庞大王国的一个蛇类。那伽国王的女儿嫁给了一个印度婆罗门，这个联姻诞生了柬埔寨人民，他们仍然说他们"诞生于那伽"。在南亚和东南亚到处可以看到有关那伽的绘画，尤其是在庙宇和宫殿建筑，强调了他们在当地和每日文化中的重要性。换句话讲，那伽是强大和可辨认的符号，很容易被当地社区理解。

从一开始，在"那伽"的创造和展示过程中，Seckon 就包括了当地社区的参与，这帮助强化和放大了作品所要传达的信息。400 多人参与了"那伽"塑料鳞片的收集、清晰、切削和缝制，而 NGO（非政府组织）参与了作品的安装。

艺术作品的公众参与非常强劲。全白色 "吸引人们过来观看漂浮在暹粒河上的生物"，Seckon 指出。

这件作品在柬埔寨和尼泊尔均有展示，这表明它的象征意义和传达的信息是深远的，而非仅限于单个时空。它明确的文化参考价值可以让观众在本地环境下对信息进行阐释。"这是一件开放和引人注目的艺术作品，"Seckon 说，"它把社区和环境联系在一起"。

装置艺术的意义
李柯臻

"那伽"是一件大型的用于流动展示的开放性公共艺术作品。整个作品的特殊性在于：一方面，它将南亚和东南亚地区都很重视的视觉符号运用于保护水资源的主题中，易被当地社区接受和理解；另一方面，强劲的公众参与性，帮助强化和放大作品所要传达的信息。

它的象征意义和所传达的信息是深远的，而非仅于作品的形象本身；明确的文化参考价值让观众在本地环境下对信息进行阐释。作品所传达的信息是它最引人注目的方面：把社区和环境联系在一起。正是利用这种巧妙地联系，不仅在公共中对作品做了宣传和推广，更使可持续观念在公共因素的影响下，更好的走进公众和社区中去。相信公共艺术在未来的发展中，将不断发挥更多的积极作用，为社区、为城市的发展作出自己特有的贡献。

Artist: Leang Seckon
Location: Siem Reap River, Kampuchea; National Zoo, Jawalkhel, Kathmandu, Nepal
Media/Type: device
Date: Kampuchea: October-November, 2008; Nepal: November-December, 2012
Commissioner: Kampuchea, Nepal, the public art organization
Researcher: Laine Bergeson

One of the most compelling aspects of Naga, a five-meter high and 250-meter long serpent constructed out of bamboo and recycled plastics, is the message behind the work.

"By highlighting the environmental costs of development, Naga aims to encourage community and government efforts in protecting and conserving fresh water sources," says Kim-Ling Humphrey of Rossi & Rossi gallery.

Installed twice—once along the Siem Reap River in Cambodia to mark World Water Day in 2008 and again at the National Zoo, Jāwalākhel, Kathmandu as part of the Second Kathmandu Art Festival in Nepal—the piece calls attention to the world's polluted waterways. "The juxtaposition of the plastic material with the natural setting—it has always been installed above a body of water—and the symbol of the Naga, which traditionally represent water and are believed to protect springs, wells, rivers and control the rain, is also what makes this message so striking," Humphrey adds.

Artist Leang Seckon puts it more directly: "We hope that people, when they see this, understand that this piece was made out of rubbish and water pollution, and that, if this persists, the Naga and all that can be found in the river will disappear."

The project also uses native beliefs and traditions to communicate its message. In South and Southeast Asian cultures and traditions, the Naga is a mythical creature that takes the form of a serpent with the power of transformation. In the Cambodian creation myth, the Naga were a race of serpents that possessed a large empire. The daughter of the Naga King married an Indian Brahmana, and from that union came the Cambodian people, who still say they are "born from the Naga." Depictions of the Naga can be seen throughout South and Southeast Asia, especially in temple and palace architecture, highlighting their importance in local and everyday culture. In other words, the Naga is a powerful and recognizable symbol that can easily be understood by local communities.

Since its inception, Seckon has involved local communities in the creation and display of Naga, which has helped strengthen and amplify its message. Over 400 people were involved in collecting, washing, cutting, and sewing the plastic scales for Naga, and local communities and NGOs were involved in the installation of the work.

Public engagement with the work of art has also been strong. The stark white color "attracted people to come over and have a look at the creature floating above the Siem Reap River," notes Seckon.

Its installation in both Cambodia and Nepal demonstrates how its symbolism and message are wide reaching, not limited to a single time and space. Its clear cultural references also allow the message to be interpreted within a local context by the viewer. "It is a work of art that is open and inviting," says Seckon. "It connects the community to the environment and vice versa."

Rider Spoke
Rider Spoke

艺术家："爆炸理论"艺术团
地点：澳大利亚新威尔士州悉尼
形式和材料：数字媒体
时间：2009年2月11—14日
委托人：英国文化委员会和悉尼市海港管理局联合当代艺术博物馆共同提供
推荐人：凯伦·奥尔森

"Rider Spoke"将剧院与游戏，移动技术和骑自行车相结合，邀请大众来骑装有掌上电脑的自行车，通过穿梭于城市的街道中来反思他们自己的生活，倾听他人的故事。

在该项目中，首先参与者到达一个场所，各参与者带上他们自己的或借来的自行车。在简短的介绍和安全说明后，每位自行车手朝街道驶去，掌上电脑安装在车把上，且有一个耳机提供口头指令和问题。通过使用无线网检测位置，该设备的屏幕会为车手指路，显示附近的"藏身之所"。

一旦车手找到藏身之所——任何其他车手之前未找到的地方——这台设备就会闪光，发出警报，向这名车手提问。然后车手录下问题的回答，这样在后期时其他车手就可收听这一回答。随后，这名车手继续寻找其他车手的藏身之所停下来收听其对相同问题的回答。

"Rider Spoke"这一构思由居住在英国布莱顿的艺术家 Matt Adams, Ju Row Farr 和 Nick Tandavanitj 提出，这些艺术家共同组成了"爆炸理论"（Blast Theory）艺术团。他们的项目即使用交互式媒体来创建开创性的新型表演和互动艺术形式（且这种艺术形式融合了互联网上的观众、现场表演和数字广播），研究社会和政治问题，进行个人通讯时代的表演。在"Rider Spoke"这一项目中，艺术家，即自行车骑手，希望探索私人／公共领域边界。"我们看到了私人和公共领域之间边界的瓦解，"Adams 说。所以他们设计了这一项目，通过让自行车骑手参与公共活动，为他们探索这些边界提供机会。通过让参与者骑自行车，使他们能和这座城市中因行人和城市交通而隔绝的他人更加密切地联系在一起。

"Rider Spoke"最早于2007年出现在伦敦巴比肯，在世界各地巡回演出了七年，共收集了一份有两万多份记录的档案。2009年，该项目作为东南

亚英国文化协会的创意城市项目在悉尼展出。在悉尼，这一项目得到了参与者和评论家的一致喜爱。

"Rider Spoke"是一件如此耀眼的、生动的艺术品。" Metro 的 Lucy Powell 写道。它在投入到纷繁忙乱的城市生活的同时，又强调了个人冷静思考能力的重要性。

"在这个项目结束后，我还有很多想法"。ICON 的 William Wiles 写道："生活在这个世纪，这座联线城市中我们真的是非常幸运，在移动技术的帮助下，我们能重新发现梦想的真谛，能探索一种新型的低技术游戏。"Rider Spoke"有着神奇的魔力。"

倾听他人的故事

李柯臻

"Rider Spoke"是一种以科技化实现新型交流模式的公共艺术作品。艺术家通过让参与者骑自行车，使他们能和这座城市中因行人和城市交通而隔绝的他人更加密切地联系在一起。在未知的空间放空自己，反思自己；在未知的空间与未知的人相互交流，畅聊心声，倾听他人的故事。随着生活节奏的加快，人们自我反思的机会越来越少，人与人之间的沟通与交流越来越缺失。该项目提醒人们，在投入到纷繁忙乱的城市生活的同时，个人冷静思考能力的重要性。使人们在移动技术的帮助下，重新发现梦想的真谛。艺术家将全新的技术注入到公共空间中，GPS 与网络相结合使人们在看不到摸不着的空间中，还能够相互交流相互传递信息，还能够在交流之中与完全不相关的人对话。作品似乎指引人们看到这样一个方向：在未来，公共艺术也许会影响或改变人与人之间的交流方式。

Artists: Blast Theory
Location: Sydney, NSW 2000, Australia
Media/Type: Digital media
Date: February 11 to 14, 2009. Each individual experiencing the project spent up to 60 minutes exploring the city by bicycle.
Commissioner: Presented by the British Council and Sydney Harbour Foreshore Authority in association with the Museum of Contemporary Art (Sydney, Australia)
Researcher: Karen Olson

Combining theater with game play, mobile technology, and cycling, Rider Spoke invites the audience to ride bicycles equipped with a handheld computer through the streets of a city, to reflect on their own lives, and to hear the stories of others.

It begins with participants arriving at a venue, bringing their own bikes or borrowing them. Following a short introduction and a safety briefing, each cyclist heads out into the streets with a handheld computer mounted on the handlebars and an earphone providing verbal instructions and questions. Using wi-fi for detecting location, the screen of the device guides the cyclist and shows nearby "hiding places."

Once a cyclist finds a hiding place—a spot previously undiscovered by any other player—the device flashes an alert and the cyclist is asked a question. The cyclist then records an answer to the question, which can be listened to at a later stage by other riders. The cyclist continues, finding hiding places of other riders where they can stop and listen to other answers to the same questions.

Rider Spoke was conceived by Brighton, UK–based artists Matt Adams, Ju Row Farr and Nick Tandavanitj, who make up the arts

group Blast Theory. Their projects—which use interactive media to create groundbreaking new forms of performance and interactive art that mixes audiences across the internet, live performance, and digital broadcasting—offer inquiry into social and political issues and into performance in the age of personal communication.

In the case of Rider Spoke, the artists, who are all cyclists, wanted to explore the private/public divide. "We're seeing a collapse of the boundary between private and public," says Adams. So they designed the project to give participants the opportunity to explore those boundaries through having cyclists be on their own and engaged in a public activity. The experience is underlined by having participants on bicycles, where they are amongst others in the city yet separate from pedestrians and road traffic.

First presented in 2007 at London Barbican, Rider Spoke has toured the world for seven years and an archive of over 20,000 recordings has been collected. In 2009 it was presented in Sydney as part of the British Council's Creative Cities program in East Asia. There, participants and critics loved the unexpectedness of it.

"Rider Spoke is such a gloriously enlivening piece of theatre," wrote Lucy Powell in Metro. "It manages to embrace the remorseless rush of the city while insisting on the individual's ability to pierce it with quiet reflection."

"As soon as it was over, I wanted more," wrote William Wiles in ICON. "We are truly fortunate in this century, in the wired and anonymous city, to have rediscovered aboriginal notions of songlines and dreamtime, to explore with the aid of mobile technology a new form of strangely low-tech play. Rider Spoke was magical."

Gardensity
Gardensity

艺术家：阿什·基廷
地点：新西兰中部城市克赖斯特彻奇大教堂广场和克赖斯特彻奇美术馆前院
形式和材料：建筑
时间：2010—2011 年
委托人：第六届克赖斯特彻奇公共空间艺术双年展，来自当代艺术博物馆和
澳洲艺术委员会的支持
推荐人：凯伦·奥尔松

"Gardensity" 是一个多方面的项目，旨在吸引新西兰克赖斯特彻奇的公
众参与到一个讨论中，这个讨论是关于如何适应预计的城市市中心人口从
8 000 到 30 000，近乎翻四番的这个问题。本项目由澳大利亚艺术家阿
什·基廷以及合作者建筑师和数字动画师 Dorian Farr、Patrick Gavin 和
Chris Toovey、平面设计师 David Campbell 等一起创立，是为 "第六届
克赖斯特彻奇公共空间艺术双年展" 准备的。

这个项目的主要特点是一个安装在 SCAPE6 内的置顶的建筑动画显示。它
演示了市中心区域如何重建来适应不断增长的人口，是未来的、幻想的、
甚至有点超现实的，这个重新想象的发展——包括一些老建筑的改变——
照耀着白墙，银凸表面，圆顶，是几乎立体化的形状。

在 "Gardensity" 中克赖斯特彻奇的政府生活塔建筑被重塑和转换成一个
聚集圆顶建筑，有雨林和冬季花园，饭店和咖啡馆。Sevicke Jones 被重
新利用包括咖啡馆、书店，设计地面商店、工作室和其上的公寓。Harper
建筑变成了一个空间，可以用作画廊、博物馆、环境研究中心或者城市实
验室。一些相邻空地变成了公共区域，有原始森林和玻璃电梯塔来满足所
有的发展层次。"Gardensity" 的空中建筑有多重用途，如图书馆、天文馆、
办公室、餐厅和休息室。

为了实现目前为止最高绿色星级的目标，"Gardensity" 的所有建筑材料
必须本地采购，包括所有可回收和可再生材料。设计还包括被动式太阳能、
废物回收系统、雨水收集和风能。

由于 2010 年 9 月 4 日在克赖斯特彻奇发生的地震，SCAPE6 被推迟了。
Keating 将他的注意力转向集中管理 "Gardensity" 关于未来克赖斯特彻
奇城的网络讨论。社区参与则是该项目最重要的一个方面。很大程度上因

为地震，在线讨论论坛和 Facebook 讨论变得更为激烈和重要，超出了艺术家的预期。"讨论吸引了许多非专业人士，通常他们不会参与讨论城市发展，并帮助他们理解所述的建筑创新如何帮助解决城市问题。"Keating说道。

当"Gardensity"的最终实体安装开放时——在 2010 年 10 月份，正好在 SCAPE6 之前，SCAPE6 完全延期至 2011 年 8 月份——它不仅包含建筑动画，也包含内部结构的计算机讨论存档。虽然荒谬，"Gardensity"的建筑呈现帮助人们联系到那个时候由克赖斯特彻奇市议会所委托的逼真动画上。"Gardensity"在线论坛的讨论为人们为委员会的"分享一个理想"的倡议铺平道路，这个倡议在 2010 年末推出。

多样的公共艺术资源

李柯臻

"Gardensity"是一个利用了现实资源和网络资源的多方面的公共艺术项目。该项目的目的是为了使公众更好的参与到城市建设的讨论中。为了实现这个目标，项目设计者们利用大空间中动画显示的方式，使公众身临其境去感受这些未来的城市建设的样貌，给公众更多的灵感和激励，使人们更好、更深刻地去参与讨论。在该项目中，工作人员在网络交流平台上发起了有关城市未来规划的讨论。很好地利用了网络的广延性，吸引更多更广泛的公众力量参与到讨论中去，从而极大地丰富了项目的资料收集。该项目作为一个示范，不仅推广了公众参与的理念，而且可以看到在公共建设领域中公众参与开发方面的极大潜力。

Artist: Ash Keating

Location: Cathedral Square and Christchurch Art Gallery Forecourt, Central City Christchurch, New Zealand

Media/Type: Building

Date: 2010-2011

Commissioner: The 6th SCAPE Christchurch Biennial of Art in Public Space, with support from the Museum of Contemporary Art and the Australia Council for the Arts.

Researcher: Karen Olson

Gardensity is a multi-faceted project that aimed to engage the public of Christchurch, New Zealand in a discussion about how to accommodate the expected near-quadrupling of the city's inner-city population, from 8,000 to 30,000. It was created by Australian artist Ash Keating and collaborators—architects/digital animators Dorian Farr, Patrick Gavin, and Chris Toovey, and graphic designer David Campbell—for the sixth Christchurch Biennial of Art in Public Space.

The project's main feature is an over-the-top architectural animation shown inside an installation at SCAPE6. It demonstrates how an area at the heart of the city could be redeveloped to accommodate a growing population. Futuristic, fantastical, and even a bit surreal, the re-imagined development—including the transformation of several older buildings—has gleaming white walls, silver convex surfaces, geodesic domes, and nearly Cubist shapes.

In Gardensity, Christchurch's Government Life Tower building was reimagined and transformed into an agglomeration of geodesic domes housing a rainforest and winter garden, restaurant, and café. The Sevicke Jones was repurposed to include cafes, bookshops, and design stores on the ground floor and studios and apartments above. The Harper building was turned into a space that could be used as a gallery, museum, environmental research center, or urban laboratory. Some adjacent vacant land was transformed

into a public space with a native forest and glass elevator towers to provide access to all levels of the development. The Sky-Loft building of Gardensity was open to several possible uses, such as a library, planetarium, offices, restaurants, and relaxation rooms.

With the aim of achieving the highest Green Star rating to date, all building materials in Gardensity would be required to be sourced locally, including all recycled and renewable materials. The design also included passive solar, waste recycling systems, rainwater harvesting, and wind power.

Because of the September 4, 2010 earthquake in Christchurch, SCAPE6 was delayed. Keating turned his attention to intensively managing Gardensity's internet-based discussion about the future of Christchurch City. Community involvement turned out to be one of the most important aspects of the project. Largely because of the earthquake, the online discussion forum and Facebook discussion became much more intense and important than the artist had anticipated. "The discussion drew in many laypeople who wouldn't ordinarily participate in a discussion about urban development, and helped them to understand how the architectural innovations depicted could help solve the city's problems," says Keating.

When the final physical installation of Gardensity was opened—in October 2010, well before SCAPE6, which was ultimately delayed to August 2011—it contained not only the architectural animation, but also an archive of the discussions on a computer inside the structure. While absurd, the architectural rendering of Gardensity helped connect with people to the very real similar animations that had been commissioned by the Christchurch City Council at that time. And the discussions on Gardensity's online forum paved the way for people to contribute to the Council's "Share An Idea" initiative, which also launched in late 2010.

最后饮品
Last Drinks

艺术家：莎拉·巴恩斯
地点：澳大利亚悉尼马丁 5 号
形式和材料：数字媒体
时间：2012 年 9—10 月
委托人：艺术节
推荐人：里奥·谭

澳洲酒店始建于 1891 年，原址位于悉尼市 Castlereagh 大街，自建造之日起迅速成为城市最迷人的地标性建筑。皇家剧院就坐落在酒店旁边，当时的社会名流喜欢聚集在酒店和宴会大厅，其中最为著名的有 Sarah Bernhardt，Kathryn Hepbur，Marlene Dietric，Laurence Olivie 和 Alfred Hitchcock，他们频繁出入各种上流社会的婚礼宴会、政治活动和商务会议。这里曾经是最受瞩目的地点，而邻近的 Rowe 大街也因沿街遍布的咖啡屋、名品店和书店满足着商务人士的文化娱乐需求。如果说 19 世纪末期，悉尼就已经成为国际化都市，那么澳洲酒店无疑是国际化程度最高的地标性建筑。然而，20 世纪中期，酒店开始走下坡路，由于最新的城市规划，Rowe 大街沿街建筑物尽数拆除，如今这里成为著名的 MLC 中心。

"最后饮品"是一个全方位声音和移动影像装置，当时为庆祝"第 11 届悉尼年度艺术节"而制作。2012 年 9 月至 10 月，艺术家莎拉·巴恩斯将历史音像资料，通过现场移动影像投影仪和声音装置，重现了澳洲酒店和皇家剧院的全盛时期。移动客户端用户也可以通过移动设备欣赏影片。通过建筑物和活动历史图片，投影仪再现了早已不复存在的皇家剧院，而音频资料则反映了当时的人们是如何沉浸于该处建筑的。另外，以播放当代邻居访谈，为装置增添了一份复杂感，历史场景和当代时刻的交织，使人们顿时感觉到时光荏苒，感慨于历史变迁。

"最后饮品"是近期城市艺术项目的延伸。例如，自 2005 年起，德国人 Bremen 制作的"城市荧屏"系列屡次斩获大奖，移动影像和声音装置在全球范围内巡回展览。从艺术历史或技术层面上来说，"最后饮品"并没有展现多少新意，但是从制作水准和重现历史上来看，它可以称作一部佳作。通过生动的影音方式，它能够使当地居民和外来游客感受到这座地标性建筑曾经的辉煌。

毫无疑问，类似于这种暂时性大型装置可以进一步提升公共空间的体验，但是很难准确回答如何提升公共空间。此外，大型广告宣传也可以提升公共空间。那么，如同标题隐含的意思，"最后饮品"能够激发公众的怀旧

情感吗？如果说它可以让人回味那段"美好时光"，那么到底什么是"美好时光"呢？比如说，是社会等级划分严格的那段时期，还是原住民聚居生活的那段时期？"最后饮品"调动现场的情绪了吗？可能是的吧，通过庆祝澳大利亚地标性建筑，如同 Peter Luck 在投影过程中解说的那般，或通过触发观众对悉尼作为国际化大都市的悠久历史的怀旧情绪，从而提高民众自豪感。公共艺术项目的评估工作仍然任重道远。目前来讲，理想的评估工作必须以可用信息和严谨的推测为前提条件。

公共空间的提升与发展

李柯臻

　　"最后饮品"是一个全方位声音和移动影像装置，通过影像的展示，使公众感受到公共空间昔日的风采。该项目的可借鉴之处在于：在构思方面，制作者利用公众的怀旧情结，带人们回到最鼎盛时期，从而提升该公共空间在观众心中的魅力值。在传播方面，制作者利用移动客户端，让公众可以随时随地的进行观影，从而提高了项目的传播覆盖面。在制作方面，该项目从制作水准和重现历史上来看，可以称作一部佳作。毋庸置疑，该项目可以进一步提升公共空间。但是，它对于公共空间的提升度到底有多高尚不明朗。况且，该公共空间和项目所重现的时候的模样也大不相同，作为并不了解这段历史的群体来说，对该公共空间的历史感知也许并不理想。

Artist: Sarah Barns
Location: 5 Martin Place, Sydney, Australia
Media/Type: Digital media
Date: September-October 2012
Commissioner: Art & About Festival
Researcher: Leon Tan

Hotel Australia opened in 1891 on Castlereagh Street in Sydney, and rapidly acquired a reputation as the city's most glamorous and iconic social venue. Situated directly across the backstage entrance to the Theatre Royal, the hotel and its banquet hall hosted celebrities such as Sarah Bernhardt, Kathryn Hepburn, Marlene Dietrich, Laurence Olivier, and Alfred Hitchcock, alongside high society wedding receptions, political events, and business meetings. It was the place to be seen, and its immediate vicinity (Rowe Street) catered to the cultural and entertainment needs of those whose business or aspiration was to be seen, with cafés, galleries and bookstores. If Sydney had become a world city in the late nineteenth-century, Hotel Australia was an important node in the flow of an international class. By the mid-twentieth century, however, the hotel's fortunes turned, and the building and Rowe Street were demolished in 1971 to make way for a new development, the current MLC Centre.

Last Drinks was a multi-site sound and moving image installation commissioned for Sydney's 11th annual Art & About Festival. Over a one-month period (September to October 2012), artist Sarah Barns staged on-site moving image projections and sound installations drawing on archival footage and audio recordings from the heyday of Hotel Australia and the Theatre Royal. The audio was also available to users through their mobile devices. Projections recreated a sense of the long gone Theatre Royal through footage of the building and its activities, while the soundscapes heightened

audience immersion in the site's past. Recordings of interviews with members of the neighborhood today added a layer of complexity to the installation, weaving history and the contemporary moment together in such a way as to refute the conventional understanding of time as the chronological succession and displacement of the past by the present.

Last Drinks belongs to a recent proliferation of urban projection artworks. As an example, the UrbanScreen collective from Bremen, Germany have, since 2005, executed several award-winning, moving image and sound installations in cities around the world. While Last Drinks is not charting new ground from an art historical or technical point of view, it is nevertheless commendable for being well executed and highly attuned to the historical specificity of the site. It is an imaginative way of engaging local residents and visitors in the history of an iconic place.

There can be no doubt that large-scale installations such as this temporarily transform the experience of a public space. The question of how precisely it transforms the public space is rather more complicated to answer. After all, large-scale advertising campaigns also transform the experience of public space. Does Last Drinks create a sense of nostalgia, as the title implies? If it does generate a longing for the 'good old days,' what exactly was good about those days? Were the wide class divisions of those days, or the treatment of Aboriginal communities at the time good, for example? Does Last Drinks encourage emotional investment in the site? Perhaps it does, by celebrating Australian icons such as broadcaster Peter Luck in the projections, or by reminding audiences of Sydney's long history as an international city, and thereby contributing to civic pride. Rigorous frameworks for the evaluation of public art projects remain to be developed and implemented. For now, a great deal of evaluation must rest on available information and careful conjecture.

红坊区和滑铁卢区的美丽之旅
Redfern Waterloo Tour of Beauty

艺术家：Squatspace
地点：澳大利亚悉尼红坊区和滑铁卢区
形式和材料：公共活动
时间：自 2005 年起，3—4 小时的旅程
委托人：艺术家发起
推荐人：里奥·谭

红坊区和滑铁卢区是悉尼两大郊区，紧邻中心城市商务区，每个区域都聚居着大量原住民，房屋排列紧凑。每当提起红坊区，众多澳大利亚人就会联系到 T.J. Hickey 和红坊暴动。某天，Thomas J. Hickey 为逃避警察追铺，骑着自行车一路飞奔，但是在一个拐弯路口处失去控制，他重重地从自行车上摔下，不幸地被路边栅栏刺死。验尸官报告使警察免除责任，但是 Hickey 的家人和支持者对结果表示抗议，他们一再强调警察开车猛追死者，至少这是导致他死亡的一个间接原因。2004 年 2 月 14 日，Hickey 的死亡和权力机关对案情的处置不当彻底激怒了民众，暴动持续九个小时，悉尼各个地方的原住民青年聚集在红坊区，将燃烧弹和石块扔向警察。最终，警察使用水枪制服了暴动人群。悉尼红坊区和滑铁卢区经常成为人们宣泄不满情绪的公共场所。

"红坊区和滑铁卢区的美丽之旅"将社会紧张情绪作为素材来源。Squatspace 艺术家 Lucas Ihlein 解释道：项目标题对应的双关语是"责任之旅"，因为来到悉尼红坊区就如同进入战区一样。通过审视社会和生存空间问题，作品拷问着当前政府对待红坊区和滑铁卢区未来发展的态度，这体现在政府规划之中，两个区域将被推倒重建，许多低收入居民或原住民将因此流离失所。这还体现在外来游客的郊区之旅，美丽的景观值得他们流连忘返，但是自上而下的区域规划将重新改造区域布局。

自 2005 年项目启动以来，开展了大量旅游活动。2005 年至 2009 年间，开展了约 15 次旅游。每次旅游持续三到五小时，期间 Squatspace 艺术家们被分成每组最多 30 人，他们以小组为单位骑或乘坐巴士到达预定地点，尤其是那些将要重建的区域。在每个站点，游客们和当地居民交谈，内容涉及区域历史和未来发展。在这一过程中，许多游客得知布洛克小区聚居着大量原住民，只有联系到殖民历史原因和当代非殖民化运动才能让他们理解存在的问题和挑战。

不禁要问：在城市变迁的过程中，什么在失去，而人们又能够得到什么？怎样才能更好地进行？另外，它是如何影响居民的？"红坊区和滑铁卢区的美丽之旅"就是艺术史学家和理论家 Grant Kester 描述的"风俗画"，其利用人们的对话体现社会艺术，作为变迁中的刺激或媒介。Squatspace 牵头的项目虽然没有解决澳大利亚原住民和非原住民之间的冲突和问题，但是它成为两个群体相互联系的纽带。此外，它使公众了解到红坊区和滑铁卢区空间动态和紧张关系，同时使他们带有批判性的思维直面由政府牵头的大型重建计划。

呐喊

李柯臻

　　"红坊区和滑铁卢区的美丽之旅"是一场短时的公共参与活动。该活动的主题是通过旅行的形式来反对政府的城市规划计划。该项目运用艺术的形式进行公共活动，带领人们宣泄心中的呐喊。项目采用短时旅游的形式，提高公众参与的积极性和趣味性，使游客们在旅行途中，了解了城市历史以及思考城市规划中的问题。在几年间中，艺术家策划开展了十几次之多，将其铸成一场持续的公共活动。显然它比单一的公共活动更具影响力和说服力。"有时候仍不免呐喊几声，聊以慰藉那在寂寞里奔驰的猛士，使他不惮于前驱。"艺术家以这样的形式，来鼓舞许多将要因城市改造而流离失所的低收入居民或原住民，为他们谋取利益，使他们继续安稳的生活。

Artist: Squatspace

Location: Redfern and Waterloo, Sydney, Australia

Media/Type: Public events

Date: 3-4 hour tours, occurring sporadically since 2005

Commissioner: Artist initiated

Researcher: Leon Tan

Redfern and Waterloo are two Sydney suburbs in the vicinity of the central business district, each with extensive public housing and a large aboriginal or indigenous population. Redfern is also, for many Australians, associated with the name T.J. Hickey and the Redfern riots. Thomas J. Hickey apparently fled from police on a bicycle one day, and lost control around a corner. He was thrown off the bike and impaled himself on a fence. The coroner's inquest absolved the police of any responsibility, but Hickey's family and their supporters disagreed, insisting that the police were at least partly responsible for the death since they were chasing him in a car. Hickey's death and the handling of the case by authorities ignited a nine-hour riot on February 14, 2004 in which aboriginal youth from across Sydney gathered in Redfern and attacked police with Molotov cocktails and bricks. They were eventually subdued with water hoses. Not surprisingly, Redfern and Waterloo are parts of Sydney in which public spaces are often filled with social tension.

Tour of Beauty took this social tension as its source material. Squatspace artist Lucas Ihlein explains that the project title puns on 'tour of duty,' since going to Redfern is like going into a war zone. As an initiative focused on critically thinking through social and spatial problems, Tour of Beauty questioned prevailing government attitudes toward the future of Redfern and Waterloo, attitudes embodied in plans to bulldoze and revitalize the area, which would

consequently displace many residents on low incomes and from aboriginal backgrounds. It did so by taking visitors on tours of the suburbs highlighting beautiful aspects of the place that were worth preserving, against top-down plans to redevelop and gentrify the district.

Numerous tours have taken place since the project launched in 2005. Between 2005 and 2009, approximately 15 tours were conducted. Each tour lasted for between four and five hours, during which time Squatspace artists would take groups of up to 30 'tourists' on bicycle or by bus to predetermined sites, typically those earmarked for redevelopment. At each of the stops, tourists encountered locals who would share their perspectives on the area, its history and its future. In this process, many of the tourists became palpably aware of the existence of The Block, a dense space inhabited largely by indigenous peoples, rife with problems that can only be understood in terms of colonization and the contemporary challenges of decolonization.

The tour asked: what might be lost and what might be gained through the process of urban transformation? How could it be done better? And how would it affect actual people on the ground? Tour of Beauty is what the art historian and theorist Grant Kester would characterize as a 'conversation piece,' an instance of socially engaged art using dialogue as a stimulus/medium for change. While it cannot be said to have solved the problematic and conflicted relations between indigenous and non-indigenous Australians, Squatspace's project did broker encounters between the two groups. It also raised awareness of the spatial dynamics and relational tensions in Redfern and Waterloo, while simultaneously confronting lofty, government-led revitalization initiatives with a critical dose of cynicism.

向失落空间朝圣
Homage to the Lost Spaces

艺术家：迈克·休森
地点：新西兰基督城蒙特利尔大街 350 号
形式和材料：壁画
时间：2011 年 2 月 22 日—2012 年 10 月
委托人：艺术家发起
推荐人：里奥·谭

克兰默法院始建于 1873 至 1876 年，隶属于基督城师范学院。现如今，该建筑物已经成为基督城最具代表性的建筑遗产。因 2011 年地震受到毁损，该建筑与其他建筑物印有红色标记"待拆"。除学院以外，曾有一群艺术家生活工作在这个哥特式建筑中。"向失落空间朝圣"是一个临时性（游击型）装置，由土木工程师、艺术家迈克·休森花费数月完成，他是此次拆迁项目的负责人。该作品旨在唤起人们对于因地震受损的基督城建筑群的历史回忆，赞美生命和创新型建筑。正如迈克·休森所说："我想通过这个作品唤起人们对于克兰默法院相同的记忆，在它消逝之前帮助人们回忆它曾经热闹的景象，美满快乐，家庭和睦。"

公共艺术作品占地 120—130 平方米，规模较大。它由大型木夹板组合而成，上面绘有图画，矗立在地震受损旧址。响应早期的空间幻觉艺术尝试，以古罗马（庞培）壁画作品的"第二"或"建筑"风格为典型，迈克·休森绘制的图画使人们产生强有力的幻觉，审视着过去居民的生活和工作空间。其中一幅图画横跨克兰默法院的渗透孔，描绘着一个男人骑行在建筑物内，另外一幅图画展示着一位艺术家的工作室和生活空间，还有一幅图画则让我们看到一位男士正在接电话，好像在跟观众诉说着什么。

"向失落空间朝圣"能够吸引基督城居民的关注。事实上，一些项目经费由当地商业团体资助。迈克·休森艺术也能够唤起国际观众的关注，通过大众在线媒体、杂志和博客，包括 Design Boom, Juxtapoz 以及 The Cool Hunter 等刊物。通过这种方式，迈克·休森从客观角度出发，将绘画作品联系过去的生活和地标性建筑的精神内涵，让观众不禁审视城市遗产建筑的拆迁项目。此外，艺术家透过作品表达哀思。正如他说的那般："重要的是，我们能够向对待其他失落空间那样看待这些地方。这就好比我们在准备一场葬礼，穿上正服，打扮自己，最后伤感于永别。"

考虑到地震带来的诸多破坏，基督城居民在心理上尚未治愈，一系列自然灾害更增加了社会不稳定因素。迈克·休森的直觉很尖锐，本地居民仍然满怀伤感。可以预见的是，"向失落空间朝圣"通过阻扰日常重建工作进一步增进了伤感情怀，使得人们停留在过去的回忆之中，失落空间的情感维系被赋予更多的空间去发泄、去传播，而最终消逝。迈克·休森项目是一个很好的范例，体现了艺术家们是如何起到社会"症状学家"的作用，在毫无征兆的破坏和严重的自然灾害过后，帮助社区居民重新恢复意识。

情感的寄托与希望

李柯臻

"向失落空间朝圣"是一个临时性装置，以壁画的形式展现。作品旨在通过公共艺术的展示勾起人们对于灾害发生前生活的回忆，寻找失去的意识。在表现形式中，艺术家尝试空间幻觉艺术，在受损的建筑上表现出日常生活的空间错觉，使人们产生强烈的幻觉，审视过去的生活和工作空间。在受众群体中感悟，作品不仅针对当地居民，而且利用线上和线下等多种形式，让更多的人感受到这份对于过去的哀思。更重要的是，该作品指出了公共艺术一个新的方向。艺术家可以通过公共艺术作品表达对于社会心理学的关注。类似于在严重的破坏或自然灾害过后，通过便于公众接受的艺术表现形式来帮助社区居民找到情感的寄托，给予希望。

Artist: Mike Hewson
Location: 350 Montreal St., Christchurch, New Zealand
Media/Type: Mural
Date: February 22, 2011-October 2012
Commissioner: Artist initiated
Researcher: Leon Tan

Cranmer Court was constructed between 1873 and 1876 to house the Christchurch Normal School. The building was, until recently, one of Christchurch's most significant heritage buildings. Damaged by the 2011 earthquake, it was 'red-stickered,' along with many others, for demolition. Apart from the school, Cranmer Court at one point housed a community of artists, who both lived and worked within the Gothic style architecture. Homage to the Lost Spaces was a temporary (guerilla) installation by civil engineer turned artist Mike Hewson in the months leading up to the building's demolition. It was intended to evoke the recent histories of quake damaged Christchurch buildings, and pay tribute to the lives and creative practices the buildings accommodated. As Hewson remarked,
  "I intended this work to project the same spirit of life back into Cranmer Courts, to help people remember, before it is gone, that this building, too, was once full of community, fun, and family."

Ambitious in scale, the public artwork occupied 120-130 square meters. It consisted of large-scale plywood mounted images erected on-site amidst the quake damage. Echoing early experiments in spatial illusionism typical of the so-called 'second' or 'architectural' style in ancient Roman (Pompeian) mural painting, Hewson's images produced powerful illusions of looking into the living and working spaces of former building residents. In one of the prints, installed across a gaping hole in Cranmer Court, we see a man bicycling

inside the building. In another, we see the studio and living space of an artist-resident. In yet another, we see a man talking on the cellphone, oblivious to his audience.

Homage to the Lost Spaces appeared to capture the attention of Christchurch residents. Some of the project costs were, in fact, underwritten by local businesses. Hewson's art also captured the attention of international audiences, receiving reviews in popular online media, magazines and blogs, including Design Boom, Juxtapoz, and The Cool Hunter, among many others. In this way, Hewson succeeded in his objective of drawing attention to the former life and spirit of an iconic place, and in asking audiences to carefully consider the demolition of the city's material heritage. The artist also intended the project as a kind of collective mourning. As he put it, "It is important we acknowledge these places in the same way we do other losses. It's almost like we are preparing it for burial—dressing it up, grooming it and saying a final goodbye."

Given the extent of the quake damage, and the fact that Christchurch city is still recovering from a psychologically traumatic and socially destabilizing series of natural disasters, Hewson's intuition seems accurate, that local communities have much grieving to do. Homage to the Lost Spaces conceivably facilitated this grieving process by interrupting the everyday routines of reconstruction, creating reflective pauses in which memories of the past and feelings associated with loss were given room to surface, circulate, and perhaps dissipate. Hewson's project is a good example of how artists can function as 'symptomatologists' of society, helping communities, in this case, to make sense amidst and after the senseless disruption and violence of natural disaster.

SCAPE 6
SCAPE 6

艺术家：黛博拉·麦考密克
地点：新西兰克赖斯特彻奇和奥克兰
形式和材料：装置
时间：2011年中的6个月
委托人：第六届克赖斯特彻奇公共空间艺术双年展
推荐人：里奥·谭

"SCAPE 6"是新西兰克赖斯特彻奇的公共艺术双年展，通过改变城市的永久和临时公共艺术品赢得了良好的声誉。考虑第二次国际公共艺术奖项的项目是第六届SCAPE，原计划于2010年举办，由于地震，延期了两次，在2011年才实现。克赖斯特彻奇的地震致使市中心破坏严重，促使"SCAPE 6"组织者改变事件形式，将该项目第一次划分在克赖斯特彻奇和奥克兰之间。"SCAPE 6"在两个城市间，还同时举行了两个其他大事：奥克兰艺术博览会和克赖斯特彻奇艺术节。2011年项目包括一系列媒介的公共艺术品，涉及到的参与艺术家是 Anton Parsons, Darryn George, Ahmet Ogut, Ash Keating, Joanna Langford, Héctor Zamora, Ruth Watson 和 Richard Maloy。

Parsons 的 Passing Time 也许是最传统的作品：一个圆形雕塑，由编号是1906年—2010年的砖块组成。1906年克赖斯特彻奇理工学院建立(CPIT)；CPIT是委员会的主要支持者。2010年"SCAPE 6"启动。

土耳其艺术家 Ahmet Ogut 的 Waiting for a Bus 更具互动性维度，尤其受到观众的好评。它是由装饰着彩灯的公共汽车候车亭的旋转木马组成。路人可以坐在候车亭，在旋转木马内部考虑安装位置（在坎特伯雷博物馆外）。

Keating 的 Gardensity 可以被认为是一个相关工作，目前它的目的是刺激和保持"关于克赖斯特彻奇市中心未来的交互式在线对话"。Gardensity 涉及建立目的网站和论坛（不再活跃）以及一个建筑动画（一个虚构的未来的克赖斯特彻奇）来主办和引发讨论。在城内，在2011年地震后的三年，中心地区依然在建筑，这个作品似乎尤其相关。

墨西哥艺术家 Hector Zamora 的 Muegano 是另一个圆形雕塑，高7米，长20米，在赖斯特彻奇的亭湖植物园的 Kiosk 湖。由26个镀锌钢温室框架，它引发了对传统格式和几何图形的反思。考虑到极端轴有26个温室框架，Muegano 也表达了地震引发的主办城市的混乱（文字颠覆和商业住宅建筑的破坏）。

在"SCAPE 6"中，所有的项目的讨论没有范围，但是以上示例的目的是提供一种整体管理者项目的感觉。委托艺术家在公共空间实现他们的作品，其中的想象力和活动可以在集体创伤后构建社区弹性，因为这些文化活动展示了受影响人群的团结。给出的许多作品也可以，至少对于一些观众来说，反映了一种生活在在建城市的生存压力。"SCAPE 6"值得称道的是，在两种挫折和逆境下，成功实现了公共艺术委员会的连贯计划。双年展的广泛媒体报道，很大程度上是积极的，证明了它在当地和国家的重要意义。

公共艺术双年展

李柯臻

"SCAPE 6"是一次公共艺术双年展。在"SCAPE 6"中展出了众多的公共艺术作品。这些公共艺术作品贴近公众生活，加强了公众的体验感和参与性，体现了艺术越来越多的人文意识与社会关怀。由于此次双年展是在灾后重建的氛围下举行的，其中的展示作品和一些参与活动使人们在集体创伤后重构公共意识，进行地方重塑项目。许多作品反映当下公众在城市重建时的生存压力，更让人们找到一种共鸣。"SCAPE 6"实现了公共艺术于现实生活的实际意义。将艺术带入生活，不仅仅有文艺的感受，更有对生活的向往与期望，有深刻的感悟与启发。这将是公共艺术在我们的生活中真正的意义。

Artist: Deborah McCormick
Location: Christchurch and Auckland, New Zealand
Media/Type: Device
Date: Six months of 2011
Commissioner: SCAPE 6 Public Art Christchurch Biennial
Researcher: Leon Tan

SCAPE is a public art biennial based in Christchurch, New Zealand, which has earned a reputation for transforming the cityscape through the commissioning of permanent and temporary public artworks. The project under consideration for the second International Award for Public Art is the sixth edition of SCAPE, planned originally for 2010, but realized only in 2011, after two postponements due to earthquakes. The Christchurch quakes, which left much of the inner city devastated, prompted SCAPE 6 organizers to change the event format, dividing the program for the first time between Christchurch and Auckland. SCAPE 6 took place in two cities in conjunction with two other events, the Auckland Art Fair and the Christchurch Arts Festival. The 2011 program consisted of public works in a range of mediums, and involved the participation of artists Anton Parsons, Darryn George, Ahmet Ogut, Ash Keating, Joanna Langford, Héctor Zamora, Ruth Watson, and Richard Maloy.

Passing Time by Parsons was perhaps the most traditional work: a sculpture in the round made up of numbered blocks indexing the years 1906-2010 inclusive. 1906 was the year in which Christchurch Polytechnic Institute of Technology (CPIT) was founded; CPIT was a major supporter of the commission. 2010 was the year that SCAPE 6 should have launched.

Waiting for a Bus by Turkish artist Ahmet Ogut had a more interactive dimension and was especially well received by audiences. It consisted of a carousel of bus shelters adorned with colored lights. Passersby could sit in the shelters and contemplate the installation site (outside the Canterbury museum) from within the revolving carousel.

Gardensity by Keating could be considered a relational work, insofar as its object was to stimulate and maintain "an interactive online dialogue about the future of inner-city living in Christchurch." Gardensity involved a purpose built website and forum (no longer active) as well as an architectural animation (an imaginary future Christchurch) to host and provoke discussion. In the context of a city whose central district is still largely under construction three years after the 2011 quake, this work seemed particularly relevant, although the extent to which consequential conversations took place remains unclear.

Muegano by Mexican artist Hector Zamora was another sculpture in the round, standing seven meters high and 20 meters long above the Kiosk Lake in the Christchurch Botanic Gardens. Made of 26 galvanized steel greenhouse frames, it invited reflection on traditional house formats and geometries. Given the extreme axes along which the 26 greenhouse frames were arranged, Muegano also expressed something of the quake-induced chaos of its host city (the literal upending and destruction of commercial and residential buildings).

There is not the scope to discuss all the projects in SCAPE 6, but the examples above are intended to provide a sense of the curatorial program as a whole. It is possible that the imagination and activity on the part of commissioned artists realizing their works in public spaces contributed to building community resilience after a collective trauma, since these cultural activities demonstrated solidarity with the affected population. It is also likely that many of the works provided, at least to some audiences, a kind of existential respite from the stresses of life in a city under construction. SCAPE 6 is commendable for successfully realizing a coherent program of public art commissions in the face of two setbacks and highly adverse circumstances. The extensive media coverage of the biennial, largely positive, attests to its local and national significance.

中南美洲
South & Central America

中南美洲

"在远方山坡之上，古老的圣安东尼奥城的灯光隐约可见。" 1951 年，凯鲁亚克的《在路上》描写新墨西哥州圣安东尼奥城的大街小巷，空气是芬芳和温柔的，微风习习的金秋里充满了神秘的气氛。一个身穿白色印花绸衫的少女的影子在充满生气的黑夜里出现，"噢，她真是美得让人不敢相信。"

南方在世人的眼中是什么样子呢？古老陈旧潮湿粘滞慵懒和生命力。今天再去看，绚丽的色彩被厚实的灰褐调子压得密密实实，却依旧能感受到浓烈。在美洲中南部，色彩绚丽不再是看出来的，而是被感受。绵长历史中固守，让出现在中南美洲的公共艺术，在现代媒介的表现形式下，依然透出古老的秩序与原始的风情。这些创意的初衷各有不同，有关乎生死无常，如 "几何的意识" "意义（Sentido）" "Konbit 住房"；有关怀天地自然，如 "北极星公园" "绿色汽车"；有美化社区生活，如 "Luz nas vielas 社区改造" "圣米格尔的露天博物馆" "找寻我的梦（野性的呼唤）" "贫民区现代艺术节" "呼叫的游行"；有希望也有反思，如 "El Bibliobandido 项目" "玻璃迷宫" "巴别塔" "沙滩怪兽" "Leões 之屋"；有新媒体创造新境界，如 "雷达" "7700 万件画作"。许多主题的诉求都是公共艺术的主题,投射于不同地域与文化后，展现出别样的独特性与多元化，这也正是公共艺术的魅力所在。中南美洲的瑰丽与落魄、文化与贫穷、传统交流与新新媒介,势必为这片神秘的土地酝酿出不竭的灵感源泉。新的表达形式伴随历史与文化的长河，更迭、融合，引导着人们不禁期待那 "美得让人不敢相信" 的明天。（吴昉）

Luz Nas Vielas 社区改造
Luz Nas Vielas

艺术家：博阿·米斯图拉团队
地点：巴西圣保罗贫民社区
形式和材料：壁画
时间：2012 年
委托人：西班牙驻巴西大使馆
推荐人：格雷戈里·多尔

"Luz Nas Vielas 社区改造"（位于光明巷中）是对圣保罗市附近的 Vila Brasilândia 狭窄人行道选定墙壁进行改造。这个项目由总部在马德里的艺术家团体发起，博阿·米斯图拉与附近的家庭密切合作，尤其是孩子们。"Luz Nas Vielas 社区改造"不同于具象的壁画工程，在紧邻人行道处使用复杂、重叠的建筑物平面，创建一个聪明又俏皮的光学画面：在合适的角度，其中一个字似乎跃入空间，悬停在人行道的水平线上。

博阿·米斯图拉由一个跨学科的团队组成，包括建筑、视觉艺术、营销、设计。该团队活动频繁，虽然不是专门创作壁画，在西班牙驻巴西大使馆邀请下，和 Vila Brasilândia 低收入社区成员一起协作，这个社区是圣保罗城里许多"贫民区"之一。

"我们在创造作品的过程中积极参与到社区中去，"博阿·米斯图拉的成员 Pablo Ferreiro 说，"影响的强大是令人难以置信的，因为他们成为这件作品共同的作者。"与社区的家庭一起合作，艺术家很快着迷于这些用于步行的蜿蜒"小巷"。研究小组发现那些街道成为了社区的定义元素——不仅是因为它们的主要建筑功能，还因为它们构成了地区的主要公共生活。

在与社区成员的对话中，几个词成为居民共同的重要的价值观：Beleza、Firmeza、Amor、Docura、Orgulho，意为：美、力量、爱、甜蜜和骄傲。突出的墙体呈一个个复杂的几何体相连在一起，而不像城市一样呈方格状，这就造就了一种创作方法，利用平面创造一种光学错觉。

在数以百计的当地儿童帮助下，团队第一次在墙上和人行道上画明亮、艳丽的颜色。画成的文字只有在特定的位置才能清晰可辨。从其他角度来看，该项目形成简单、大胆的画。

"这个项目给了贫民区的街道新生，"Ferreiro 说，"这给社区的居民提供一个新的视角，加强了邻居之间的团体感。"对于那些贫民窟之外的人们，

此项目提醒着人们，一个看似"破败"的社区有着"强大和积极的价值观。"最后，或许最重要的是，该项目在数百名儿童的帮助下，"鼓励社区孩子追逐自己的梦想并实现它们。"

项目的审美暗示着作品有这样一个追求，通过避免传统具象壁画，巧妙地将文字漂浮于颜色上，集体的力量赋予了贫民区蜿蜒街道新的意义。没有拥挤、危险、困惑或肮脏，街道洋溢着创造力和惊喜。项目暗示着，实现强大的第一步，仅仅是采取正确的视角。

决定性视角

吴昉

蜿蜒曲折的石阶小巷，闪身穿越其间的孩童，让人联想起亨利·卡蒂埃·布勒松的某一幅摄影作品，那位"决定性瞬间"大师认为摄影暗藏辨识功能，能把真实事物世界的节奏辨识出来，而形式间的联系是关键。博阿米斯图拉这项改造贫困社区的城市艺术项目，则赋予圣保罗落魄街道另一种神奇的"决定性瞬间"。一个特定的视角、一刻停滞的时光，拼合出美好的释义。从三维世界穿越至二维平面，以歧途不平的墙和路构筑标准完美的字与词，"Luz Nas Vielas 社区改造"项目在琐碎、纷乱的环境中，创造值得聚焦的主体。或许，在人们捕获决定性视角的那一瞥中，望见的才是人与生活最真实的面貌。

Artist: Boa Mistura
Location: Vila Brasilândia neighborhood, São Paulo, Brazil
Media/Type: Mural (Painting)
Date: 2012
Commissioner: Spanish Embassy in Brazil
Researcher: Gregory Door

Luz Nas Vielas (In Light Alleys) transformed selected walls in the narrow walkways of the Vila Brasilândia neighborhood of São Paulo. The project was initiated by the Madrid-based artist collective Boa Mistura, in close collaboration with families, especially children, in the neighborhood. Instead of a traditional, representational mural project, Luz Nas Vielas, used the complicated, overlapping planes of the buildings adjacent to the walking paths to create a clever and playful optical perspective-painting: at just the right angle, one of several words appears to pop into space, hovering at the horizon line of the walking path.

Boa Mistura consists of a multidisciplinary team drawing from architecture, visual arts, and marketing/design. The group works frequently, though not exclusively, with paint on walls. In this case, the group was invited by the Spanish Embassy in Brazil to work with community members in the low-income neighborhood of Vila Brasilândia, one of many "favelas" that ring the massive city.

"We actively involved the community in the process of creation of the works," says Pablo Ferreiro of Boa Mistura. "The impact is incredibly strong, because they became co-authors of the piece." Working with families in the neighborhood, the artists soon became fascinated by the winding "alleys," designed for foot traffic, that wend crookedly through the neighborhood. The team found them to be defining elements of the neighborhood—not only because they are a primary architectural feature, but because they figure largely in the communal life of the district.

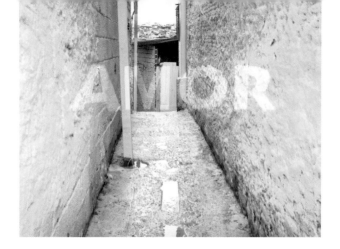

In conversations with community members, several words emerged as important values shared by residents: Beleza, Firmeza, Amor, Doçura, and Orgulho, which translate to beauty, strength, love, sweetness, and pride. The overhanging walls, which interconnect in a complicated geometry unlike a city built on a grid, suggested an approach that capitalized on the planes to create an optical illusion.

Enlisting hundreds of local children to help, the team first painted bright, flamboyant colors on walls and walkways. Then the words were painted in such a way that only at certain locations would they be legible. While complicated in terms of perspective, the project resulted in simple, bold paintings.

 "The project gave a new identity for the favela's streets," says Ferreiro. "It gave the inhabitants a new perspective of their community and reinforced the sense of group identity among the neighbors." For those outside of the favelas, Ferreiro says, the project serves as a reminder that a seemingly "run-down" community has "strong and positive values." Finally, and perhaps most importantly, the project, in enlisting hundreds of children to help, "encouraged the community kids to chase their dreams and make them true."

The aesthetic of the project hints at the work required in such a chase. By avoiding traditional representational murals in favor of cleverly situated words floating in color, the collective forces new meaning from the crooked streets of the favelas. Instead of cramped, dangerous, confusing, or unsanitary, the streets reveal ingenuity and surprise. The first step in achieving strength, the project implies, is to simply adopt the right perspective.

几何的意识
Geometry of Conscience

艺术家：阿尔弗雷多·伽
地点：智力圣地亚哥纪念和人权博物馆
形式和材料：装置
时间：2010 年
委托人：纪念和人权博物馆
推荐人：格雷戈里·多尔

在为饱受 17 年皮诺切特军事独裁折磨的受害者（在世或去世）创作纪念作品时，艺术家阿尔弗雷多·伽彻底背离了典型的充满损失和悲伤的纪念碑模式。在圣地亚哥的纪念和人权博物馆，该纪念碑位于其地下。

纪念碑每次仅限 10 人访问，浏览大约需要三分钟。游客拾级而下 33 个台阶，来到了绝对的黑暗之中，在黑暗的一分钟后，500 个影子，每一个代表一个政体的受害者，在一个墙上慢慢被照亮，在两侧墙上相对的镜子里无限地反射。灯光达到最大强度后被关掉，让观众陷入黑暗，强烈的后像留在了他们的视网膜上。

皮诺切特独裁统治在 24 年前才结束，这至今仍然使智利人民记忆犹新。成千上万的市民被政府杀害，成千上万的人因政治原因被折磨或监禁。在 2006 年去世的时候，皮诺切特面对着成百上千的反人类罪的指控。确实，独裁统治触动着每个现存智利人的生命。

阿尔弗雷多·伽的纪念碑上刻画的这 500 个影子的一半属于活着的智利人，另一半属于的那些"消失"的人。因此，作品"纪念的不仅仅是受害者，还是 1700 万智利今天还活着并努力追溯他们共同历史的智利人"，阿尔弗雷多·伽的工作室负责人 Capucine Gros 说。

在公共纪念碑中，"几何的意识"脱颖而出。大多数纪念馆是默祷和追忆的地方。但像这件作品，他们在记忆中的过去和鲜活的现在之间插入了一个空隙。在阿尔弗雷多·伽的作品中，悲恸的人和那些令人悲恸的人之间的距离被缩短。在一定程度上，这是因为纪念馆使用了混合的照片，并体现出独特的生物体特性。

这个作品不仅是一个令人不安的作品，因为它强加了一个相对隔离环境，使观众陷于黑暗；它也构建了体验功能的核心角色"后像效应在视网膜上留下了 100 万个光点，从生理上把影子的图像嵌入到了观众的视觉记忆中，"Gros 说，"因此，工作的成功依赖于对作品的概念的理解（观众的智力和情感）和观众的不可避免的生理信号（观众的身体）。"

"几何的意识"完成于 2010 年。智利人阿尔弗雷多·伽凭借他在电影、摄影、公共媒介的作品广受赞誉，这其中频繁地使用到了灯箱。他代表智利出席了"第 55 届威尼斯双年展"。

灯光穿透身体

吴昉

自然界的生命周而复始、新新相续，而自然间的人，有生之涯却稍纵即逝，生命因短暂而可贵，也因此，无妄之灾的刻意戕害，才是无法容忍的罪。杀戮和迫害，是红色的凶器和白色的恐惧，时间滤去刺痛，留下沉默，但圣地亚哥纪念和人权博物馆地下的黑暗，却并非遗忘。上帝创造了光和生命，阿尔弗雷多·伽用灯光和无穷的镜面创造出生命的幻象，参观者的影像与受难者的光，都被重重复制、永无终结。苏珊·桑塔格在《重生》中絮叨"人应该总是预料生活中最糟的事情——生活是一种长期的凄凄惨惨和碌碌无为——个人不该抗议，应该抽身而不是投身。"但显然，阿尔弗雷多·伽用地下的暗室剥夺了这种权利，穿透黑暗的灯光，在寂灭后才灼烧眼帘，好像受害者在受害后才为人知晓那样，短暂的视觉记忆，也可以看作重生的一种。

Artist: Alfredo Jaar
Location: Museum of Memory and Human Rights, Santiago, Chile
Media/Type: Installation art
Date: 2010
Commissioner: Museum of Memory and Human Rights
Researcher: Gregory Door

In creating a memorial for the victims, living and dead, of the 17-year Pinochet military dictatorship, artist Alfredo Jaar made a radical departure from the typical monumentalization of loss and grief. Housed on the grounds of Santiago's Museum of Memory and Human Rights, the memorial is situated underground.

Access to the memorial is restricted to only ten people at a time, and viewing the piece takes approximately three minutes. Visitors descend by 33 steps into absolute darkness. After a full minute in the dark, 500 silhouettes—each representing a victim of the regime—slowly brighten on one wall, reflected infinitely in two facing, mirrored side-walls. After the lights reach their full intensity, they snap off, plunging the viewers into darkness, with an intense afterimage left on their retinas.

The legacy of the Pinochet dictatorship, which ended only 24 years ago, is still fresh in the lives of Chileans today. Thousands of citizens were murdered by his government, and tens of thousands were tortured or imprisoned for political reasons. At the time of his death in 2006, Pinochet faced hundreds of charges of crimes against humanity. Truly, the dictatorship touches the lives of everyone alive today in Chile.

It is fitting then, that half of the 500 silhouettes depicted in Jaar's memorial belong to living Chileans; the other half are those of the "disappeared." As a result, the work is "not a memorial not for victims only but rather for the 17 million Chileans who are alive today and trying to retrace their common history," according to Capucine Gros, Jaar's studio manager.

In the universe of public memorials, The Geometry of Conscience stands out. Most memorials are static places of contemplation and memory. As such, they insert a gap between the past that is remembered and the living present. This distance between those who mourn and those for whom they mourn is shortened in Jaar's piece. In part this is because of the mixture of photographs used for the memorial, and in part it's because of the piece's unique physicality.

Not only is the piece profoundly unsettling as it imposes a relative isolation and deprives the viewers of light; it also builds physicality into a central role of the function of the experience. "The after-image effect imprints their retinas with a million dots of light and physically embeds the silhouettes into the audience's visual memory," says Gros. "The work's success, therefore, relies on both the conceptual understanding of the work (the viewer's intellect and feelings) and the audience's inevitable physiological reception to it (the viewer's body)."

The Geometry of Consciousness was completed in 2010. Jaar, a Chilean, is highly acclaimed for his work in film, photography, public interventions, which frequently include light boxes. He represents Chile in the 55th Venice Biennale.

圣米格尔的露天博物馆

Museo de Cielo Abierto en San Miguel
(Open Air Museum of San Miguel)

艺术家：Centro Cultural Mixar 和居民
地点：智利圣地亚哥圣米格尔社区
形式和材料：壁画
时间：2010 年
委托人：Centro Cultural Mixart、智利的国家艺术和文化委员会
推荐人：格雷戈里·多尔

"圣米格尔的露天博物馆"建立在壁画艺术的基础上，正如作品名称所示，是一座无"防护栏，警卫或参观时间限制"巨大的、高品质的对全民开放的博物馆。这个项目的先锋是 Centro Cultural Mixart，由 Villa San Miguel 社区的工人阶级居民开发和管理；迄今为止，壁画家沿着这一社区的街道，在建筑物上共完成了 4 000 平方米的壁画。

Villa San Miguel 是智利圣地亚哥的一个低收入社区。该社区开发于 20 世纪 60 年代，经历了不同程度的失修和遗弃。壁画项目的一个重要方面是向该社区居民提供壁画的布置和题材。

"所有壁画在画之前，都得到了邻里的同意。" Roberto Hermandez，项目经理说，"这一简单事实创建了一种归属感和对其社区的特殊自豪感。由于所有居民都直接或间接参与了壁画的创作，因而圣米格尔的街头艺术得到了居住在该区域的居民的照顾和保护，而不需要任何形式的警卫或护栏。"

壁画中的意象反映，进一步论述了该区域丰富的壁画传统。在参与该项目的 70 多名艺术家中有来自智利和其他国家的经验丰富的壁画家和年轻的、白手起家的街头艺术家。这些艺术作品是创新的、富于表现力的、当代的，而基于社区的创作流程则使艺术家与附近的居民建立更加密切的联系，从而丰富双方的体验。

该项目具有明确的社会实践目的。据 Hermandez 所说，附近的建筑物和公共场所经历了多次失修和遗弃。他说，壁画项目"实现了唤醒社区，将其变革成重新激活社会、改善居住在该区域内人们的生活质量的有力例证这一奇迹。公共艺术是家庭和 Centro Cultural Mixart 采取具体的、有规划的系统行动的工具。"

据 Hermandez 所说，这一社会议程既有有形，也有无形的成分。露天博物馆，具有强大的基层组织，增强了附近居民的归属感以及对其周围环境的自豪感。同时，公共艺术作为属于社区的、独特的无形集体资产。最终，该项目利用了公共艺术来资助社区的改进，它吸引了该社区外部人士的注意和投资。"这标志着在该社区中创建文化性和艺术性更强的作品的开始，其目的是使 Villa San Miguel 社区成为一个文化和旅游的标志。" Hernandez 说。

私我也

冯正龙

博物馆随处可见，露天博物馆也并不稀奇，对于博物馆的官方保护与建设更是理所当然，而"圣米格尔露天博物馆"的奇妙之处却在于超脱官方的维持，形成一座无"防护栏，警卫或参观时间限制"对全民开放的高品质博物馆。项目的创建以追求归属感与自豪感为目标，通过对所有周边居民的参与邀约，在完成壁画的一点一画中建立起居民心中属于自己的博物馆。较之金碧辉煌的大型博物馆，警卫森严，肃然起敬却难免隔阂于无形。圣米格尔的街头博物馆，墙面五彩放任自由，居民参观如同归家，对自家的保护与照顾，自然无需警卫或护栏去横生枝节。拥有是参与的最高形式，因为，城北徐公再美，总不敌私我也。

Artist: Centro Cultural Mixart & Residents
Location: Villa San Miguel neighborhood, Santiago, Chile
Media/Type: Mural (Painting)
Date: 2010
Commissioner: Centro Cultural Mixart and Chile's National Council of Art and Culture
Researcher: Gregory Door

The Museo de Cielo Abierto en San Miguel (the Open Air Museum of San Miguel) builds upon the mural-arts tradition to create, as the project title suggests, a huge, high-quality museum open to all without "protection bars, security guards, or visiting time restrictions." Spearheaded by the group Centro Cultural Mixart and developed and managed by the residents of the working-class neighborhood of Villa San Miguel, the muralists have to date completed some 4,000 square meters of artwork on buildings along the heavily traversed thoroughfares of the community.

Villa San Miguel is a lower-income neighborhood of Santiago, Chile, developed in the 1960s and suffering, to varying degrees, from disrepair and abandonment. An important aspect of the mural project is giving residents of the community agency in the placement and subject matter of the murals.

"All the murals were approved by the neighbors before they were painted," says Roberto Hernández, project manager. "This simple fact creates a sense of belonging and special pride for their neighborhood. Since all residents have participated directly or indirectly in the making of the murals, the street art of San Miguel is taken care of and protected by the people that live in the area, without the need of guards or fences of any kind."

The imagery in the murals reflects and expands upon the rich mural tradition of the region. Among the more than 70 artists who have participated in the project are experienced muralists as well as younger self-made street artists from Chile and beyond its borders. The artworks are innovative, expressive, and contemporary, and the community-based process brings the artists into close contact with neighborhood residents, enriching the experience on both sides.

The project has an explicit social activist purpose. The buildings and public areas in the neighborhood faced severe deterioration and abandonment, according to Hernández. The mural project, he says, "has achieved the miracle of waking up the community, transforming it into a powerful example of social reactivation and improvement of quality of life for the people who inhabit this area. Public art is the tool for making specific and systematic actions planned among the families and Centro Cultural Mixart."

This social agenda has both intangible and tangible components, according to Hernández. With a strong, grassroots organization, the open-air museum has given residents of the neighborhood an "improved sense of belonging" and pride in their surroundings. At the same time, the public art itself serves as a tangible and unique asset belonging to the community as a collective. Finally, the project has leveraged public-art financing for neighborhood improvements, and it has drawn attention and investment from outside the neighborhood. "It's a start for creating more cultural and artistic work and initiatives in the neighborhood, designed to turn the Villa San Miguel into a cultural and touristic icon," says Hernández.

El Bibliobandido
El Bibliobandido

艺术家：玛丽莎·雅恩
地点：洪都拉斯 El Pital 村庄及其周边乡村
形式和材料：互动行为
时间：2010 年
委托人：非政府组织
推荐人：格雷戈里·多尔

"El Bibliobandido" 项 目 利 用 艺 术 和 表 演 来 提 高 文 化 素 养。
Bibliobandido 或 Book Bandit 乔装打扮，"带着对故事的渴望，他漫游
在丛林中，恐吓小孩子，除非他们能写一些故事来满足他贪得无厌的胃口"。
该项目，由艺术家玛丽莎·雅恩和洪都拉斯 El Pital （La Ceiba 附近）的
村民合作启动，通过志愿"图书馆委员会"提供书籍制作、讲故事和基础
文化素养支持。项目由 15 位成人和青少年构成。

每个月，志愿者们会上演 "El Bibliobandido" 干预，戴面具的人物会骑马
或骡子进城，孩子要专门编造一些故事来牵制这个强盗。"孩子和成人要
么相信 "El Bibliobandido" 是真实的，要么愿意相信这个人物是真实的。
所以每个人都会在这个传说中扮演一个角色，虽然他们只是假装相信这个
人物"，这位艺术家说。

"El Bibliobandido" 项目使 El Pital 及周边社区的识字率提高到了约
80%。该项目，用高可见性、艺术干预代替传统模式的教学，以亲切友好
和令人愉快的方式提高了识字意识。这些支持强盗装扮的工作室教授了实
地读写能力，同时较长的故事情节吸引了参与者，使他们急于了解下面的
故事情节。

该项目的志愿部分也很重要。许多参与者告诉艺术家，志愿活动是一次转
变的体验。"他们通过教学学到知识，他们可以看到新社区，了解历史、
文化和政治（很多人很少离开他们的村庄），发展他们可加以实践的领导
能力。"玛丽莎·雅恩说。

"El Bibliobandido" 项目是两年工作的巅峰，在此期间，玛丽莎·雅恩
在该社区内建立了关系。"需通过现场组织来创建具有深远影响的艺术作
品。"她说。影响已经扩散。尽管玛丽莎·雅恩已经转向其他项目，但青
少年和成年人的"图书馆委员会"接管了该项目且扩展了该项目。现在"El
Bibliobandido"在 18 个邻近村庄的儿童达到了五百多名，而其他社区也
采纳了这一构思，把它变成了自己的项目。

玛丽莎·雅恩继续成立了立足于纽约市的艺术团 REV "艺术家、制媒者，低收入工人、移民和年轻人结合大胆的想法和良好的研究，从而对我们面临的问题创造创意媒体"的非盈利工作室。她的艺术实践根植于"艺术能够在培养实际影响中发挥作用"这一概念。

但是，艺术家玛丽莎·雅恩很快指出：她认为自己仅仅是从洪都拉斯的协作者中汲取创意的项目发起人。"这种形式的可见性不仅令人振奋，也能帮助他们更加了解他们面对的问题，筹集资金使其工作获得可持续发展。"

文化胃口养成记

吴昉

尽管人生识字忧患始，但文字带来的愉快和安慰倒也无可否认。"El Bibliobandido"项目有着咒语般的魔力，将孩童引入文字编撰的故事世界。意识流创作、自发式写作，这些成人文学的书写技巧，在洪都拉斯乡村道边孩子们的心中，只是游荡丛林对故事有着贪婪胃口的强盗，教化与惊喜同行，对付强盗的过程实在讨人欢喜。据说，该项目使周边社区的识字率提高到了约80%，每月盛装出行的"Bibliobandido"，十足一场文化恣意的狂欢，孩子们伸长脖颈、翘首以盼，酝酿出故事的起伏，抛下情节线索的接力棒。随着候场、登台、转场、再登台，故事的延展似乎永无止境，每一个结局都是新的开始，所有结尾的故事都是完整的，所有接近结尾的创作过程都是完美的。欢腾嬉笑中，文字创作的魅力在幼年心中生根发芽、影响深远。在这个从不结束的故事中，强盗们正在月行一善。

Artist: Marisa Jahn
Location: Village of El Pital and surrounding countryside, Honduras
Media/Type: Various interventions
Date: 2010
Commissioner: NGO Un Mundo
Researcher: Gregory Door

The Bibliobandido project uses art and performance to boost literacy. The Bibliobandido, or Book Bandit, wears a disguise and "ravenous for stories, roves the jungles terrorizing little kids until they write stories to nourish his insatiable appetite." Launched by artist Marisa Jahn in collaboration with villagers in El Pital, Honduras (near La Ceiba), the program provides book-making, story-telling, and basic literacy support through a volunteer "library committee," consisting of about 15 adults and pre-teens.

Each month, the volunteers stage a Bibliobandido intervention with a masked character that rides into town on a horse or mule to demand a quota of stories made especially to keep the bandit at bay. "The kids and adults either believe El Bibliobandido is real or they like to believe the character is real. So there's a role for everyone to play in the legend, even if it's pretending as if they believe," says the artist.

The Bibliobandido project raises awareness about the 80 percent illiteracy rate in El Pital and surrounding communities. By replacing the traditional models of teaching/tutoring with a high-visibility, artistic intervention the project raises awareness while remaining approachable and delightful. The workshops supporting the bandit's appearance teach hands-on literacy skills, while the larger storyline keeps the participants engaged and excited for the next installment.

The project's volunteer component is important, too. Many participants tell the artist that volunteering is a transformative experience. "They get to learn by teaching, they get to see new communities and learn about history, culture, and politics (many would otherwise rarely leave their village), and they get to develop leadership skills that they put into practice," says Jahn.

The Bibliobandido project is the culmination of two years of work, during which Jahn built relationships within the community. "It takes on-the-ground organizing to create artwork with deep impact," she says. That impact has spread. While Jahn has moved on to other projects, the "Library Committee" of teens and adults has taken up the project—and expanded it. The Bibliobandido now reaches more than 500 children in 18 neighboring villages, and other communities are adopting the idea and making it their own.

Jahn, went on to found the art collective REV, based in New York City, a "non-profit studio where artists, media makers, low-wage workers, immigrants, and youth combine bold ideas and sound research to produce creative media about the issues we face." Her artistic practice is rooted in the notion that "art can play a role in fostering tangible impact."

The artist is quick to point out, however, that she considers herself solely the initiator of a project designed to draw out creativity from her collaborators in Honduras. The IAPA nomination is shared, she says, with those who worked with her. "Visibility in forms like this is not only heartening, but also helps them increase visibility towards the issues they face and raise funds to make their work sustainable."

北极星公园
Estrella del Norte Park

艺术家：巴勃罗·卢比奥
地点：波多黎各巴亚蒙 Bobby Capó 大道
形式和材料：空间艺术
时间：2010 年
委托人：巴亚蒙政府
推荐人：格雷戈里·多尔

　　"北极星公园"是一个规模宏大的改造项目，这个项目将原来位于波多黎各的第二大城市巴亚蒙中心的一处废墟，改造成一个全新的公园。对于巴亚蒙政府来说，成立该项目的目的并非单纯出于刺激经济发展的需要。由于还考虑到社会犯罪和被弃用的风险，巴亚蒙政府希望通过这次的改造，让这片废墟成为一个具有地标性意义的公共艺术空间，它能够象征和展示这座城市为了发展所作的努力。

这里交通网发达，建有大学，是许多旅客和购物者的目的地。可惜的是，此前这些优势资源并没有得到充分的重视和利用。改造前的很长一段时间里，这片废墟仅仅是吸毒者的集聚地。"这里是一个犯罪率很高的地方。在重建之前，我们收集了上百个吸毒用的针筒。"巴勃罗·卢比奥告诉我们。巴勃罗·卢比奥是"北极星公园"设计团队的一员，他的团队从五个实力相当的竞争队伍中脱颖而出，负责废墟的改造。他们的使命不仅是重建废墟，更重要的是使之成为巴亚蒙居民喜爱的文化中心。

项目的成果就是我们今天看到的"北极星公园"。公园设计的灵感源自宇宙中的星体，公园主体由一些圆形结构构成，包括一面弧状的墙。400 盏太阳能灯，象征着宇宙中的 400 颗星星，而位于中心的是北极星。一个大舞台和固定的座位构成了表演空间，附近还新建了一个餐馆供游客用餐。

巴勃罗·卢比奥最伟大的贡献是创作了一个纪念性雕塑，共 616 英尺长，横跨了整个公园。其中一部分是一个立体三维的球体，25 英尺高，利用太阳能持续发亮和旋转。他希望人们能够和他的这件作品产生互动，"以某种方式，与我们自身以及我们身处的环境建立联系，感受空间、时间和光"。

新的公园里不再是吸毒者聚集，巴亚蒙市民纷纷被新公园吸引前来。"改造这片沉闷的荒地是一件非常有趣的事，我们赋予了它更有意义的社会文化价值。"巴勃罗·卢比奥说，"现在，'北极星公园'已经成为一个名副其实的文化园。"改造项目的内容丰富多样，公园里还有演唱会、艺术展、

艺术工作室，共同构成一个开放的、创意的空间。波多黎各缺少的正是这样的一个空间，他表示，因此这个公园才能把在地铁附近的人们吸引过来，更何况它就在一个重要的交通中转站对面。

作为一个"社会文化的符号"，这片空间非常成功——就像所有优秀的公共艺术作品一样，它是一个独特的当代艺术品。这座新雕塑，传达着这座城市再次繁华的美好愿景。

将"北极星公园"与一般改造项目区分开来的一个重要特征是巴勃罗·卢比奥对社会正义的投入。作为改造的内容之一，他为狱友们开创了一个艺术项目，来帮助将犯罪分子重新融入社会。"他们和市民一起和谐地工作"，巴勃罗·卢比奥说，"他们中的多数已经用他们出色的表现和成就证明了他们已经改过自新，并得到市民的理解和认可。"

肖申克之光

吴昉

北极星是天空北部的一颗亮星，靠近正北方位，千百年来，迷航的人们靠它的光亮来找回方向。"北极星公园"原来只是一片瘾君子的荒芜地，颓靡与病毒是弥漫在这个巴亚蒙城中角落挥之不去的气息。对政府而言，犯罪率的控制远比开发经济、降低废弃风险来得重要。巴勃罗·卢比奥通过空间与装置形成的艺术之光，将罪恶危险的废墟照亮，成为如今公众喜爱的文化中心之一。公园设计的灵感来自包容万象的宇宙，大型装置勾勒出银河中随日光持续发亮旋转的北极星，与四百多束星光一起，带领往日的迷途羔羊们重返"智性"。整个项目的亮点还在于协助那些昔日囚禁于图圄的失足之人，帮助他们再次融入社会，回归自我。"北极星公园"的闪耀星空映出肖申克永恒的拯救之光。

Artist: Pablo Rubio
Location: Bobby Capó Avenue, Bayamón, Puerto Rico
Media/Type: Space art
Date: 2010
Commissioner: City of Bayamón
Researcher: Gregory Door

Estrella del Norte was a large-scale redevelopment project that turned a centrally located abandoned lot in Puerto Rico' s second-largest city, Bayamón, into a new park. Worried about crime and declining use, Bayamón officials sought something more than an economic revitalization project; they desired an iconic, marquee-ready piece of public art that could help define and galvanize the city's revitalization efforts.

Although situated in a prime location, with easy access to mass transit, universities, and destinations for tourists and shoppers, the abandoned lot chosen for redevelopment had long been left to addicts. "The space was a high crime area where hundreds of syringes were collected before construction," reports Pablo Rubio, the artist on the team that designed Estrella Del Norte Park to replace the vacant lot. Rubio's team was selected from five competing groups and charged with creating a park setting that not only transformed the space, but also could serve as a "socio-cultural icon" for the residents of Bayamón.

The result is Estrella del Norte Park. Inspired by universal shapes, the park is comprised of several circular shapes, including a curved wall. A solar-powered lighting feature marks the location of 400 stars with the North Star (Estrella del Norte) at the center. A large stage and permanent seating provides a performance space, and the design includes a new restaurant.

Rubio's contribution is a monumental sculpture that spans the entire park with its 616-foot length. A section of the sculpture includes a three-dimensional, 25-foot-tall star that both rotates and is lit by solar-powered lights. The sculpture is part of Rubio's broader artistic agenda to encourage those who interact with his work to "get in touch, in one way or another, with our own being and the position we occupy in our environment in relation to space, time, and light."

Instead of drug users, the new park attracts a broad segment of Bayamón's population. "Designing for barren and idle spaces is one of my interests because we can appropriate and repurpose the space for a better socio-cultural urban use," he says. "Today, Estrella Del Norte is a cultural space." The variety of programing includes concerts, art festivals, arts workshops—and open, recreational space. Puerto Rico lacks similar spaces, he says, so the park draws from surrounding areas in the metro, especially since its location is directly across from a main mass transit stop.

As a "socio-cultural icon," the piece has all the marks of success—it is a distinctive, contemporary artwork that is as recognizable as any good "logo" public artwork. Publicity and urban branding has begun the process of associating the city and its revitalization dreams with the aspirational new sculpture.

One feature that sets Estrella Del Norte apart from similar large-scale transformative placemaking projects is Rubio's very practical commitment to social justice. As part of the project, he created Art for Inmates, a rehabilitation program that brought prisoners into the community. "They worked and interacted in harmony with the civil population," says Rubio. "Many of them obtained pardons for their excellent behavior and achievements."

Konbit 住房
Konbit Shelter

艺术家：加勒多尼亚·柯里、Kt·蒂尔尼、本·沃尔夫
地点：海地科米尔镇
形式和材料：空间艺术
时间：2010 年
委托人：尚不明确
推荐人：格雷戈里·多尔

为响应 2010 年海地地震，美国艺术家发起了"Konbit 住房"项目，其成员包括加勒多尼亚·柯里（亦称 Swoon）等。艺术家们与海地 Komye 镇的居民合作，建造了一座防震建筑作为城市社区中心，该建筑由当地材料做成。从那时起，该项目持续进行，其宗旨已经由原来建造基础设施扩展为实现多样的社会目标和艺术目标。

1 月 12 日海地地震，达到灾难性的 7.0 级，夺去了超过 10 万人的生命，严重损毁了国家基础设施，同时也加重了这个发展中国家的负担。Komye 镇位于震中附近，其建筑大量毁坏，亟待重建。

艺术家们在该项目的网站上，对"如何给处在危机时期人们产生积极的影响"很感兴趣。为此，他们联系了的 Komye 镇活动家，这些活动家曾促进该项目的启动。

经过集体讨论认为，当地居民最迫切的需要是建造公共社区，从而便于居民交流，接受课堂教育以及其他的项目开展。通过合作，成员们组织设计施工了一个"地球包"建筑，并进行了装饰。该建筑的材料由当地土与混凝土混合而成，简称"超级土"。项目成员之一的建筑师 Nader Khalili 发明了该建造工艺。自社区中心建造完成之日起，成员们又建造了两栋建筑，为单亲家庭提供住宿。

因为存在实际困难和文化困难，开展这类工作是让人担忧的。 即使是第一世界到第三世界的善意介入，也会导致经济发展的角色问题，世界资源的不平等分配问题，还有文化力量的差异问题。在"Konbit 住房"项目中，即使没有明确公开说明这些问题，也在执行过程中说明了这些。组员们一开始就和居民通力协作，包括与当地农场建立合作伙伴关系，使用在美国筹集的资金，并提供有意义的工作岗位。

这是地方决策充分运用的好例子。柯里说道："我们建造的社区中心和房子已经投入使用，并深受百姓喜爱"，并指出这些建筑只是其中一部分。"该

项目仍在继续开展，我们认为这是和 Komye 镇合作以来取得的最大成功。此外，我们也很有信心积极进行海地灾后建筑重建工作。"

在现阶段"Konbit 住房"项目开始从事成人和儿童的工作，关注住宅建设，建筑风格以及工程建设。柯里说道，"我们希望提出有用的解决方案和想法，用来指导当地其他住宅建造项目……该阶段是学习阶段，为我们下一步工作奠定了基础，指明了方向。"

柯里的工作，连同其他艺术家一起，由原来办公室的小打小闹发展到公众参与，从原来墙上的蓝图规划发展为自制舰队一样的宏大实践。她已然从她工作中蜕变为一名艺术家，一名活动家。"我认为，人们可以从事一些工作来弥补政府工作的不足……我必须依靠自身的力量去实践，这才是艺术家的本色。因此，即使只是刚刚起步，就像我现在从事可持续发展住宅建筑一样，我也会满怀兴趣，并且将其作为我的工作核心。我是艺术家，来自艺术家团队……我想，正如一名艺术家一样，如果搞不清自己，又何谈能取得骄人成绩呢？"（摘自"艺术之声"杂志）

后灾难建设中审美
吴昉
在人类与自然共处的过程中，总是祈求能一如既往、风调雨顺，但往往伴随着不期而至的天灾与人祸，天灾是一个临界点，不可预测又无法避免。灾害过后的重建与规划，是人类智慧与自然和好妥协的产物。"Konbit 住房"项目于 2010 年启动，原本只是一项应对自然灾害作出的快速救助措施，为海地灾害区的人们搭建家园与社区空间，而现已演变成艺术家 Swoon（原名加勒多尼亚·柯里）及其团队对当地的长期支持项目。从"淹没的祖国"、新奥尔良市音乐房到布拉多克废弃教堂改造艺术中心以及"漂浮的城市"，Swoon 的作品正如 MoMA 策展人莎拉·铃木（Sarah Suzuki）所言：出发点不在于对商业文化的冷嘲热讽或批判，而具有情感核心。"Konbit 住房"项目中堆叠构成的建筑空间、色彩缤纷的墙绘、精细雕琢的木板花纹，显示了项目设定的目标不仅是现实环境的应急救灾，更是抚慰居民心灵创痕的艺术创作，其本质是爱、温暖与美。

Artist: Caledonia Curry, Kt Tierney, and Ben Wolf
Location: Town of Cormiers, Haiti
Media/Type: Space Art
Date: 2010
Commissioner: Unknown
Researcher: Gregory Door

Konbit Shelter began as a response by U.S. artists, including Caledonia Curry (a.k.a. Swoon), to the Haitian earthquake of 2010. In collaboration with residents in the town of Komye, Haiti, the artists developed an earthquake-proof structure made from local materials to serve as the town community center. The project has since evolved into an ongoing relationship, and its mission has expanded beyond the built infrastructure to include a wide variety of social and artistic goals.

Haiti's earthquake of January 12, with a catastrophic magnitude of 7.0, killed upwards of 100,000 people and unleashed unimaginable devastation on the country's infrastructure, already taxed by the ongoing struggles of a developing economy. Located near the epicenter, the town of Komye lost much of its built environment and had immediate needs for rebuilding.

Artists Caledonia Curry, Kt Tierney, and Ben Wolf shared an interest in "how the creative process might positively impact people's lives in times of crisis" according to the project's website. Together, they made contact with activists in the Komye community who helped facilitate the launch of the project.

It was collectively decided that the foremost needs of the townspeople was a public community space where citizens could meet, classes could be taught, and other planning and programming could take place. Working through a collaborative process, the group designed, built, and adorned an earth-bag structure from local dirt mixed with concrete, or "Super-Adobe," a process developed by architect Nader Khalili, one of the team of artists on the project. Since the community center was completed, the group has built two additional structures to serve as single-family homes.

Work of this sort is fraught with practical and cultural difficulties. Even well intentioned interventions from the "first" world into the "third" can call into question the role of economic development and the unequal distribution of world resources, as well as cultural power differentials. The Konbit Shelter project, while not explicitly addressing these matters in its publicity, has addressed them structurally. The group has worked collaboratively with residents from the beginning, including forming partnerships with local farm organizations and using funds raised in the United States to provide meaningful employment on the ground.

The result has been a dramatic example of creative placemaking. "The community center and houses we have constructed are also used and loved," says Curry, adding that the project was only partly about the built environment. "The project is ongoing, and we consider its biggest success to be its relationship to Komye, as well as its ability to creatively evolve with the changing circumstances of post-earthquake Haiti."

In its current phase, Konbit Shelter is taking up education for both children and adults, focusing on home building, architecture, and engineering. Curry says, "Our hope is to generate creative solutions and ideas within the community, and to let those ideas inform the next series of homes created in the village... This period is one of learning and listening, and from here we will take direction for the next steps of our work."

Curry's artwork has evolved from a studio-based print practice to public interventions from wheat-pasted prints on public walls to a flotilla of homemade vessels built collaboratively with other artists. She has evolved an arts/activist rationale for her work: "I think that people can pick up where governments are failing... I have to do it from my place of strength, which is as an artist. So even if it seems a point of departure—like now I'm interested in sustainable home building—I take that interest and bring it back to my core and say, okay, I'm an artist, I come from an artist community... I think like an artist, how do I achieve something unique without losing that sense of who I am?" (Art Voices Magazine).

Leoes 之屋
Casa dos Leoes

艺术家：亨里克·奥利维拉
地点：巴西阿雷格里港 Andradas 507 街道
形式和材料：雕塑
时间：2009 年
委托人：第七届南方共同体双年展
推荐人：格雷戈里·多尔

艺术家亨里克·奥利维拉的 "Leoes 之屋" 这项雕塑作品用建筑物拆除过程中所产生的胶合板为载体，通过胶水粘结，形成一个外形隐约可辨的有机形状体。作为博物馆现代雕塑作品，奥利维拉的这项作品极具写实生活气息，与画廊传统整洁无菌的 "白色大箱子" 形象大相径庭。 在巴西圣保罗 "第七届南方共同体双年展" 上，艺术家亨里克·奥利维拉的作品选材为二层废弃已久的住宅，配以雕塑外形而形成。该外形突破建筑结构的门窗，直直逼向公共街道。

与亨里克·奥利维拉以往画廊展览作品不同，这项现代雕塑作品位于市中心繁忙街道上，一经展出即引起了公众注意——尽管不乏负面评价。该作品，在施工之初，就已引起了大众的关注，且时常有人群驻留针对该作品发表自己的看法，这位艺术家如此说道。 许多路人感到非常激动，因为这常年废旧的建筑物终于又引起了公众注意。但人们更为好奇的是这建筑物到底最终会用来做什么呢？然而，随着施工的进行，该作品引发了全城范围内的对于公共艺术品价值更深层次的热烈讨论——同时也伴随着对社会腐败现象的评论。

据奥利维拉说， 该作品的施工过程也为热议话题之一。奥利维拉从垃圾堆或者建筑物装卸卡车处收集胶合板围栏碎片，从中抽离出一片片胶合板条，然后进行粘结，最后形成具有独特风格的有机形状体。奥利维拉将这样的过程描述为 "让材料回到它的来源处——大街"。这些材料可以与建筑物具有年代感的外墙完美融合，从而造成 "该雕塑是从内部生长出来" 的假象。这座废弃的建筑用一层围栏挡住，围栏所用的材料不是其他而是同雕塑一样的材料——随处可见的胶合板围栏。这是艺术家奥利维拉特意为之，意图为该作品增加一层新的含义。 "这项作品是用胶合板这种材料制成，这些材料一直都在，它就是组成那些围栏的材料。但现在它们不同了，更加有血有肉了。" 奥利维拉说道。

这项作品产生的影响远远超出了奥利维拉的预料。 "这项作品从最初就没有设想能成为可以跟公众广泛互动的作品。" 奥利维拉补充道。 "作品并

没有特别指向这个或那个当地社区，也从没有想过可以为人们提供哪些帮助，并没有想教给人们什么，不论这项作品美或丑，只是想做一种呈现，表达事物的存在。"尽管最初这项作品没有这些预想，该作品的的确确在这次双年展上引起了大众对于公共艺术的热烈讨论。一位杰出却有些保守的哲学家批评该作品为怪兽之屋，是这次双年展留下的最坏的公共艺术品，正是该评价把对该作品的讨论推向了高潮。

公众的热烈讨论引发了更为尖锐问题的讨论，不仅仅关于公共艺术，也涉及枯萎荒芜、城市美化等美学问题，奥利维拉提到：这项雕塑作品的造型是以人体肿瘤的医学成像为基础的，意在比喻"人类身体失调与巴西大城市社会发展失调有共通之处"。奥利维拉这样说道："作品主要描述巴西社会是如何对待贫穷问题以及边缘化人群的。"文字意义上的"社会问题"，借助于贫民窟老旧的、腐烂的树木，来表达"衰老的器官及已损坏的皮肤"。借助于"衰老器官"这一明晰概念，奥利维拉对于不断城市化世界的公平性及美学性进行了讨论。

老树著花无丑枝？

吴昉

丑是否有内涵？丑是否能引起比美更强烈的震撼？丑的历史究竟是相对于美的历史而言，还是具备独立的审视价值？在巴西圣保罗"第七届南方共同体双年展"上，两层废弃的老旧住宅，从垃圾堆和建筑废料中收集的胶合板围栏碎片，粘合而成畸形扭曲的有机形态附着于墙面。突兀、畸变、生物体增殖的即视感引起视觉心理直观的不适，艺术家奥利维拉声称以此代表肉身的腐坏，指涉社会发展的失调，的确形象。据称该作品本意并不为引发观者的交流，只是将作品呈现，逼近街道、迫人直视、拒绝外在交流却又强行侵占行人意识，姿态傲娇如一针见血，确诊社会内分泌失调的病灶。通过公众行为表现社会发展暴露的问题，并不鲜见，将丑陋赤裸裸地摆在日光之下，就好比肿瘤切除后的端盘检验，刺激神经，令人印象深刻。只不过插上艺术的小红旗，是否就真能掩住恶意并激发审美的好感，却还有待时间来验证。

Artist: Henrique Oliviera
Location: Rua dos Andradas 507, Porto Alegre, Brazil
Media/Type: Sculpture
Date: 2009
Commissioner: VII Bienal do Mercosul
Researcher: Gregory Door

The sculptural works of artist Henrique Oliviera are built from scraps of deconstructed plywood glued into vaguely organic forms. In museum installations, his works achieve a life-like form that contrasts with the 'white box' sterility of the gallery. For São Paulo, Brazil's VII Bienal do Mercosul, Oliviera took over an abandoned two-story dwelling and built a sculptural form that seemed to bulge out of the structure's windows and doors and leak into the public street.

Unlike Oliviera's gallery works, this installation near a busy thoroughfare in the city drew immediate public attention—and not all of it positive. The interest began during construction, according to the artist, when crowds began to gather and comment on the work. Many of these passersby were excited that the long-abandoned property was at last receiving some attention. Their curiosity focused on what use the building would be put to. As the work progressed, however, the piece raised a larger, citywide discussion about the value of public artworks—and served as a commentary on social decay.

The process of construction is an important aspect of this larger conversation, according to Oliviera. The artist gathers scraps of plywood fencing from dumpsters and other trash sites. The layers of plywood are peeled apart and then re-glued into his distinctive organic forms. Oliviera describes this as a process of "returning the material to its original place—the street." Instead of the "aseptic environment of galleries and museums," he says, "in this project, the materials had a harmonic integration with the building's aged walls," which supported the illusion that the sculpture "grew from its interior."

An added layer of meaning came from the fact that this abandoned property had been boarded up with the very same materials—the ubiquitous plywood fencing—that Oliviera repurposed in the sculpture. "This work was then made of the same kind of material that was already there. It was the same fence, but with its 'flesh' modified," he says.

The impact of the piece far exceeded Oliviera's expectations. "The work was never conceived as an interactive public project," he says. "It was never directed for this or that local community, never pretended to help people, nor to teach anyone, nor to be beautiful nor ugly—but just to be a presence." In spite of these lack of pretensions, the piece became a magnet for a larger discussion about public art and the Bienal. This conversation reached its peak when a prominent, conservative philosopher focused his ire on "the monster house" as an example of bad public art left in the wake of the Bienal.

This public debate helped raise critical issues, not only about public art, but also about blight, urban renewal, and aesthetics generally, the artist says. The form of his sculpture was based on medical images of human tumors. The piece drew a parallel between "the human body disorders and social disorders common in the big cities of Brazil," says the artist. "The work gave vision to the way Brazilian society has treated the problem of poverty and marginalized populations. Taking the term 'social tissue' to its literal significance, the layers of aged and rotten wood, commonly seen in the construction of shanty towns, play in this work the role of degenerated organs and diseased skin." By making explicit these "degenerated organs," Oliviera contributed to an important discussion about the fairness and aesthetics of our increasingly urbanized world.

巴黎广场最深沉的静默

Máximo Silêncio em Paris

艺术家：詹卡洛·内里
地点：巴西里约热内卢巴黎广场
形式和材料：LED 装置
时间：2012 年
委托人：阿尔贝托·萨莱瓦、OI·富图罗·弗拉门戈
推荐人：格雷戈里·多尔

意大利艺术家詹卡洛·内里已经在几个地方推出了 Máximo Silêncio，罗马大竞技场也不例外。这些表面上看起来很简单。成千上万的 LED 球形灯，直径大约有 10 英寸，它们全都铺设在地毯上。夜空中，不同颜色的灯光以各种图案闪烁，并以其柔和的光芒变化着景观。

这个项目带来了无与伦比的感官体验。灯光巧妙地改变着景观，造成一种错位感，这种可辨识、备受詹卡洛·内里推崇的地标景观就是这样奇妙变幻着。

一般来说，詹卡洛·内里的工作台展现是无可非议的，它主要强调灯光的感官体验。然而，在里约热内卢，该项目却以一个社会活动家的色调呈现。艺术家推荐将 Praça Paris 这个 20 世纪 20 年代的美丽公园置于城市中心。但是，在赞助商看来，这个公园的位置太危险，晚上会招致一些袭击。所以他们建议将公园建在更高档的地方。然而，詹卡洛·内里却坚持保留 Praça Paris 的建筑地点。

"但结果却令人大吃一惊：每天晚上，成千上万的人都会从城市前来，于是晚上公园的每个地方就变成了拥挤、欢快、安全的地方，这种场景在多年来还是第一次"艺术家说："人们或是乘着出租车和豪华轿车从 Zona Sul（富裕的"南部"地方）前来,或是乘着公共汽车和火车从 Zona Norte（贫穷的"北部"地方）前来。而当地居民则步行前来。很多人说，他们从来没踏进过公园，尽管他们只住在几百码远的地方。"

事实上，景观本身是让人叹为观止的。装置包括 9 000 个灯，覆盖近一万平方英尺的种植草坪，还包括公园独特的装饰艺术装置。LED 灯来源于廉价的非数字设备，却能发出耀眼的光芒，一时间，灯光的随机脉冲能够聚合成有趣的图案，甚是壮观。

詹卡洛·内里出生于那不勒斯,并在纽约度过了他早期的职业生涯。在这里,他创造了大型雕塑和装置,并赢得了广泛声誉。通常来说,他的作品稍微抽象并具有影响特质。其中,他最知名的不朽桌子和椅子被称为"作家"。(在 2010 装置中,他运用大型照明椅子来重复这种修辞。)

由于它的位置备受争议,人们认为这个公共艺术公园不太安全。感觉上,"巴黎广场最深沉的静默"是背离詹卡洛·内里更抽象、更注重感官的项目。作为通过场所营造公共项目的示范能力,它的一个重大贡献就是打破了社区之间的传统障碍。在审美上,该项目为许多不同的观众所亲近,同时也保持了转换的可能性。

值得肯定的是,除了詹卡洛·内里拒绝将此项目转移到"更安全"的地方外,"巴黎广场最深沉的静默"更多是偶发事件的结果,而非规划当中的事情。然而,当无法预测出席观众的反响、批评和欣赏力时,这个偶发事件就成了公共艺术项目的重要组成部分。其他拥有社会转型的直接议程项目可能已经获得了更多可衡量的目标。然而,詹卡洛·内里的光束展示却给了观众一个措手不及,产生了"偷袭"性的震惊效果。

地光
吴昉

巴黎广场公园所处的位置在当地居民与赞助商看来,不够安全且缺乏足够的吸引力,很难想象城中心的夜晚会由于某些原因而冷清无人气,这一切在艺术家詹卡洛·内里点亮九千余盏 LED 彩灯后,发生了转折性改变。上城区与下城区的人们蜂拥而至,原本冷落的地界在一夜间焕发华彩,成为众人热议的焦点。这或许是公共艺术的美妙之处,通过人为营建,改变地区环境的命运。人们希冀星光而不得,就在地上造就一片星空,所谓无与伦比的感官体验,是不知天上人间今夕何夕的沉醉与放松,暗夜里的光会有神性。项目命名为"巴黎广场最深沉的静默",实则以无声的彩灯吸引众人的聚拢,为原本悄然的夜晚增添人间的欢声和笑语。

Artist: Giancarlo Neri
Location: Praça Paris, Rio de Janeiro, Brazil
Media/Type: LED Installation
Date: 2012
Commissioner: Alberto Saraiva, OI Futuro Flamengo
Researcher: Gregory Door

The Italian-based Giancarlo Neri has staged Máximo Silêncio in several locations, including in Rome's Circus Maximus. The installation is deceptively simple. Thousands of LED globe lights, approximately ten inches in diameter, are laid in a carpet on the ground. The lights blink different colors in random patterns through the night, transforming the landscape with their soft glow.

Engaging with the project is a sensory experience. The lights subtly alter the landscape and cause a sense of dislocation—the recognizable landmarks that Neri prefers are transformed and made unfamiliar.

Neri's stagings of the piece are frequently uncontroversial, with the emphasis on the sensory experience of the lights. In Rio de Janeiro, however, the project took on a social-activist hue. The artist proposed the location, Praça Paris, a beautiful park designed in the 1920s in the city center. Sponsors of the work, however, worried that the location was too dangerous, with numerous assaults at night, and suggested another park in a more upscale location. Neri, however, insisted the location remain the Praca Paris.

"The result was phenomenal: thousands of people came every night from every part of town and the park turned into a crowded, cheerful and safe place at night for the first time in many years," says the artist. "People came in their taxis or fancy cars from the Zona Sul (the rich "South Side") or with buses and trains from the Zona Norte (the poor "North Side") while local residents just came walking, many saying they had never set foot in the park though they lived a few hundred yards away."

The spectacle itself was breathtaking. The installation included 9,000 lights and covered a rolling parkland of nearly 10,000 square feet and including the park's distinctive Art Deco fixtures. The LED lights, constructed from inexpensive, non-digital equipment, cast a pale glow, while the random pulses of light converged into interesting patterns over time.

Born in Naples, Neri spent his early career in New York City, and has gained fame for his large-scale sculptures and installations, which often possess a slightly abstract and somewhat threatening quality—his best-known piece is a monument-sized table and chair called The Writer. (He repeated the trope with large illuminated chairs in a 2010 installation.)

Because of its location in a contested space, a park considered too unsafe for public art, Máximo Silêncio em Paris feels like a departure from Neri's more abstract, sensory projects. It makes a significant contribution as a demonstration of the ability of public placemaking projects to break down traditional barriers between communities. While being aesthetically approachable to many different viewers, the work held the potential for transformation.

The primary impact of Máximo Silêncio em Paris is admittedly more a result of happenstance than intention, apart from Neri's refusal to move the work to a "safer" location. Yet this happenstance is a vital component of public-art projects—when it becomes impossible to predict the responses, critical and appreciative, of the attending audience. Other projects with a direct agenda of social transformation might have accomplished more measurable aims, yet Neri's light display caught the viewer off guard and provided a "sneak attack" on convention.

雷达
The Radar

艺术家：池田亮司
地点：巴西里约热内卢
形式和材料：装置
时间：2012 年
委托人：Outras Idéias para o Rio
推荐人：格雷戈里·多尔

"雷达"是一个临时安装在特定场所的光投射和声音装置。在里约热内卢的迪亚波罗沙滩上，艺术家池田亮司把一个大的网格投射到巨大的沙滩上。池田亮司以他提供额外的音乐背景而出名，投影加上模仿老派声纳的声道。横扫投影区的条状和点状灯光好像声纳装置的回声。

池田亮司的作品作为里约两年一次的 Outras Ideias para o Rio（OiR）活动的一部分，在 2012 年秋季举办。这位在日本出生的法国艺术家是几名被邀请在里约标志性景观进行临时特定场所艺术创作的艺术家之一。

跟大多数视觉艺术相比，"雷达"在更大程度上反对将作品翻译成文字。观看该作品的视频能提供一定程度的体验。这件作品中光和声的复杂相互作用，加上风和海浪的声音，创造了一个完全包围的感知体验。

使用 NASA（美国国家航空和宇宙航行局）数据来确定光点模式，并在星星出现在天空时，从现场指出成千上万颗星星的位置。艺术家指出，将星星精确地投影到沙滩上是他试图"让人们感到一丝崇高。"

这件作品的力量部分来自光投影和音景自然派生的节奏以及天然声音／海浪及其拍打沙滩声音之间的对比。"它的成功来自于自然完美的结合，" OiR 的制作人 Marcelo Dantas 说，"波浪和无线电波；灯光和天上的星星；沙滩上的人们。"雷达"创造了一种诗意，通过灯光、节奏、沙滩和水进入里约的灵魂。"

这个作品出人意料的一个方面就是它变得越来越互动。每晚，亲朋好友等待投射开始，并在开始时鼓掌。在作品展出的几周内，人们学会如何和灯光互动，在沙滩上照明下跑来跑去。其他人则站在一旁观看灯光和"表演者"。

对池田亮司而言，互动"成为这个项目最漂亮的一部分"。"当地人自发地参与到项目中去，这些人不一定都是艺术爱好者，"这位艺术家说，"他们只是下午在沙滩上消遣而已。"

在这个作品的展览期限，池田亮司把这些参与者视为真正的合作者。"他们开始认可这件艺术作品，找到各种方式参与/玩耍。最后，我强烈地感觉到没有人的话，这件艺术作品就一点意义都没有。可以这么说，我只完成了一半的作品，这些人完成剩余的部分。"

合拍

吴昉

池田亮司（Ryoji Ikeda）是日本电子音乐界和数码多媒体创作界的领军人物，他尝试各种形式的创作，包括静态影像、雕塑、声响和新媒体等，经常从细微处入手研究超声波学、频率学和声音本身的基本特性。而超越科技这层面纱，艺术家的内心将声音视为一种"感觉"而非物理现象，试图展现这种感觉与人类感知之间的关系。作品"雷达"捕捉的是一种人、自然、科技三者间和谐的节拍，光投影下自然派生的音乐节拍、海浪拍岸水打沙滩的节拍、互动人群脚踩浪花的节拍，"雷达"把握住来自各处节拍契合的一瞬，来迎合参与者心脏的跳动。诚如制作者所言，将一种诗意注入里约的灵魂。

Artist: Ryoji Ikeda
Location: Rio de Janeiro, Brazil
Media/Type: Installation
Date: 2012
Commissioner: Outras Idéias para o Rio
Researcher: Gregory Door

The Radar was a temporary, site-specific light projection and sound installation. Artist Ryoji Ikeda projected a large grid over a gigantic swath of sand on Diablo Beach in Rio de Janeiro. Ikeda, known for his spare soundscapes, paired the projection with a soundtrack that mimics old-fashioned sonar beeps. The various striped and dotted lights that swept across the projection area likewise seemed an echo of a sonar device.

Ikeda's piece was part of Rio's biennale, Outras Ideias para o Rio (OiR), which took place in fall 2012. The Japanese-born French artist was one of several invited to create site-specific temporary works on Rio' s iconic landscape.

To a greater degree than most visual art, The Radar defies translation into words. Viewing a video of the piece provides some sense of the experience. The complex interplay of the light and sound in the piece, added to the sounds of the wind and surf, create a fully enveloping sensory experience.

The patterns of points of light were determined using NASA data and indicate the positions of thousands of stars as they appear in the sky from the earth at the site. The artist said this precision, in projecting the stars onto the sand, was his attempt "to make people feel a little bit of the sublime."

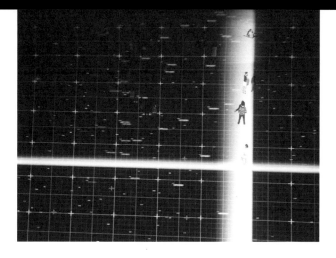

Part of the power of the piece comes from the contrast between the naturally derived rhythms of the light projection and soundscape, and the natural sound/waves of the ocean and its rhythmic pulse on the beach. "Its success came from this well-woven integration with nature, says Marcelo Dantas, producer of OiR. "The waves and the radio waves; the light and the stars above; the people in the sand. The Radar produced a poetic immersion into the soul of Rio with bits, bytes, sand, and water."

An unanticipated aspect of the piece is that it became highly interactive. Families and friends awaited the start of the projection each night, applauding when it began. During the weeks that the piece was active, people learned how to interact with the light, running back and forth across the sand to play in and out of the illumination. Others stood by watching both lights and "performers."

This interaction became "the most beautiful part of the project" for Ikeda. "It was the spontaneous engagement by local people who were not specifically art-lovers," the artist says. "They were just there on the beach to spend their own time all afternoon."

Over the duration of the piece, Ikeda came to see these participants as true collaborators. "They started to recognize the work and to invent ways to engage/play with it. Eventually I strongly felt that the work didn't make any sense without the people. Let's say, I made half of it and the people completed the rest of it."

玻璃迷宫
Glass Labyrinth

艺术家：罗伯特·莫里斯
地点：巴西里约热内卢，市政府和市立剧院之间的广场
形式和材料：装置
时间：2011 年
委托人：里约热内卢
推荐人：凯莉·克里斯坦森

罗伯特·莫里斯创造了"玻璃迷宫"，2012 年，作为公共艺术国际展览的一部分，它被临时安放在里约热内卢。玻璃迷宫在 Cinelândia 地区的里约热内卢历史中心，沿着其历史性城市轴线展览了两个月，它吸引人们前来，到这里探险。三角形的玻璃迷宫位于 Câmara dos Vereadores（市政府）和市立剧院之间的广场。

罗伯特·莫里斯，1931 年出生于堪萨斯城，是美国最著名的在世艺术家之一。他以使用铝、钢和毡制品等工业材料而著称。20 世纪 60 年代和 70 年代，在概念艺术、极简主义和地球艺术发展过程中他发挥着重要作用。

此作品是 Other Ideas for Rio (OiR) 节的一部分，展览时间从 2012 年 9 月持续到 2012 年 11 月。两年一次的项目 OiR 以罗伯特·莫里斯的作品开始，贯穿 2016 年奥运会。同时将里约热内卢办成一个露天画廊，促进城市位置的原始介入。以前从未参与此城市创作的国际著名艺术家也被邀请过来，为里约热内卢城市景观的主要工作献计献策。该项目旨在促进文化民主化。

将名称 OiR 倒拼成"Rio"，具体指的是，以不同的方式看城市。第一阶段的六个作品包括英国艺术家 Andy Goldsworthy (Cais do Porto) 和 Brian Eno (Arcos da Lapa)，西班牙艺术家 Jaume Plensa (Enseada de Botafogo)、美国艺术家 Robert Morris (Cinelândia)、日本艺术家 Ryoji Ikeda (Arpoador) 和巴西艺术家 Henrique Oliveira (Parque Madureira)。这个项目由汇丰银行、里约热内卢、里约热内卢州的政府赞助，以及 Oi Futuro 和文化部 (MinC) 的支持。Marcello Dantas 为馆长。

"玻璃迷宫"推动了公共艺术的边缘化发展，使观察者成为参与者，他们在通过迷宫的同时，既能观察又能参与。这并不像一个迷宫，但却是一个

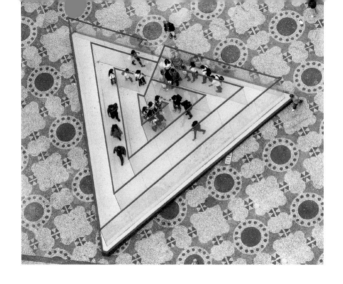

真实的迷宫，只有一种方法可以进来或同样的方式出去，而没有其他选择路径。在安放期间，项目被破坏了，部分被粉碎，但它仍被视为是成功的作品。迷宫是由透明的玻璃墙、金属帽和玻璃的顶部和底部边缘组成。地平面是浅灰色豆砾石，灌浇混凝土，这会使参与者在里面走的时候很舒服。雕塑持续管理是防止玻璃痕迹和划痕的一个重大考虑决策。

最近，罗伯特·莫里斯被委托在堪萨斯城创建一个类似的项目，同时他也受到了大厅家庭基金会资助的 Nelson-Atkins 博物馆的邀请。与在里约热内卢的工作不同，堪萨斯城的版本将会具有永久性，成为美国第一个永久性莫里斯迷宫。

透过玻璃照见的自己
吴昉

罗伯特·莫里斯的玻璃迷宫装置，像极了放大的万花筒底，一个个行走其间的人就如同构成变化无穷的彩色玻璃片，不可预测却又殊途同归。从整体观赏性而言，玻璃迷宫是美的，剔透晶莹，结构稳定，光线穿过迷宫，改变自身方向，这一切赋予作品朦胧的幻象。作为公众艺术，玻璃迷宫还另有一种诠释，当人走入迷宫，隔着玻璃看见的，正是来路与去路上的自己，因为别无其他路径选择，擦身而过重叠的身影就如同一次次与自己的相遇和别离。从入口到出口，前进的路线被预设了，而欲望和决策的驱动，创造出一个生动的世界。在看得见内外的构造中移动，却有着无法触及的隔阂，循序依次向前，无法后退，不能折返，只能在下一个转角处，望见另一个曾经，抑或未来的自己。透过层层玻璃，罗伯特·莫里斯似乎在向世人传递着禅宗的奥义。

Artist: Robert Morris
Location: The plaza between the Câmara dos Vereadores
 (City Council) and the Teatro Municipal, Cinelândia,
Rio de Janeiro, Brazil
Media/Type: Installation
Date: 2011
Commissioner: Inside Out Rio
Researcher: Carrie Christensen

Robert Morris created the Glass Labyrinth, temporarily installed in Rio de Janeiro in 2012 as part of an international exhibition of public art. Displayed for two months in Rio's historic center district of Cinelândia, on Rio's historic urban axis, Morris' Labyrinth compelled observers to participate in the work, finding their way through the glass maze. The triangular glass labyrinth was located in the plaza between the Câmara dos Vereadores (City Council) and the Teatro Municipal.

Morris, born in Kansas City in 1931, is one of America's most eminent living artists, known for his use of industrial materials, including aluminum, steel and felt. In the 1960s and 70s, he played a leading role in the development of process art, minimalism, and earth art.

The installation was part of the Other Ideas for Rio (OiR) festival and was on display from September 2012–November 2012. The biennial project OiR, which began with Morris's piece, runs through the 2016 Olympics and considers Rio de Janeiro an open air art gallery, promoting original interventions in locations around the city. Internationally prestigious artists who had never before created for the city were invited to propose major works for Rio de Janeiro's urban landscape. The project aims to democratize access to culture.

The name OiR— "Rio" spelled backwards—refers precisely to the idea of thinking about the city in a different way. The six works in the first phase included British artists Andy Goldsworthy (Cais do Porto) and Brian Eno (Arcos da Lapa), Spanish artist Jaume Plensa (Enseada de Botafogo), American Robert Morris (Cinelândia), Japanese Ryoji Ikeda (Arpoador), and Brazilian Henrique Oliveira (Parque Madureira). The project was sponsored by HSBC bank, Oi, the Government of Rio de Janeiro State, and the City Government of Rio de Janeiro, with cultural support from Oi Futuro and the Ministry of Culture (MinC). Marcello Dantas is the curator.

The Glass Labyrinth pushed the boundaries of public art in that it made the observer into a participant—passing through the labyrinth one simultaneously observed and participated in it. Unlike a maze, this is a true labyrinth, with one way in and the same way out; there are no choices to make except following the path. The project was vandalized and part of the labyrinth was shattered for a period during its installation, but it was still deemed a success. The labyrinth was composed of transparent glass walls with a metal cap and base at the top and bottom edges of the glass. The ground plane was a light gray pea gravel on poured concrete that produced a sound underfoot for participants. Ongoing management of the sculpture was an important consideration to keep the glass from having marks and scratches.

Morris was recently commissioned to create a similar project in Kansas City, commissioned by the Nelson-Atkins Museum with funding from the Hall Family Foundation. Unlike the work in Rio, the Kansas City version will be permanent—the first permanent Morris labyrinth in the U.S.

意义
Sentido

艺术家：杰泽贝尔·斯托雷
地点：阿根廷布宜诺斯艾利斯，Arroyo 和 Suipacha 十字路口
形式和材料：多媒体
时间：2012 年
委托人：未提供
推荐人：格雷戈里·多尔

为了纪念在布宜诺斯艾利斯以色列大使馆旧址的自杀式炸弹中丧生的公民，艺术家杰泽贝尔·斯托雷将视频投影和现场钢琴表演结合。从黄昏直到午夜一直亮着的投影，描述出一组蜡烛的画面，一只蜡烛代表在 1992 遇袭事件中失去生命的一位公民。在此次现场表演中，Alfredo Corral 演奏了俄罗斯作曲家穆索尔斯基的作品"图画展览会"。

1992 年，一名自杀式炸弹袭击者开着一卡车的炸药冲进了位于 Arroyo 和 Suipacha 十字路口的以色列大使馆。使馆大楼以及附近的天主教堂和学校被毁坏。240 人受伤，29 人死亡，这次袭击事件也成为对大使馆做出的最致命的攻击。一个与伊朗有着所谓关系的以色列组织承担了责任。在此次事件中，未宣判任何人有罪。这一遗址现在是一座小型纪念公园，以色列 Embajada de 广场的所在地。

为了纪念此次爆炸中的受害者，受过画家培训，正在攻读艺术治疗硕士学位的杰泽贝尔·斯托雷，提出了和平祝福这一概念，"不需要用任何言语来感动人们的灵魂。反之，我们可以在唤起恐惧和绝望的这个地方献上美丽与和平"。据这位艺术家所说，有 200 多人参加了守夜仪式，包括附近的拉比和天主教神父。广场附近的邻居打开了他们的窗户，聚精会神地看着。爆炸事件中的幸存者和死伤者的亲人也参加了这一仪式。

"在场人士被音乐和图像吸引住了，"杰泽贝尔·斯托雷说，"爆炸事件的幸存者和死者家属感动得热泪盈眶，表达了他们无限的感激之情。"没有人会拒绝如此大规模灾难中的受害者家属公开哀悼、纪念他们失去的至亲，而"意义"针对艺术、纪念仪式和悲痛提出了一些有趣的问题。

这一临时艺术干预包括蜡烛的大荧幕投影（该艺术家的资料声称一个蜡烛代表一名受害者，共 22 名受害者，但实际上是 29 名）以及另一位作曲家复制的一段音乐。但是，如果要将这种小型的临时干预与资金充足的永久性纪念仪式进行对比是不公平的。

不过，在其他临时性纪念干预措施这一背景下考虑"意义"是值得的。一个很好的范例是"逝者9000（Fallen 9000）"项目，该项目招募了志愿者在诺曼底海滩的沙土中刻上9 000个蜡印轮廓，来纪念战争中双方的受害者。这一临时干预有几个特征是"意义"缺乏的。它适于这一场所，动员广大人民积极协作，而不是旁观；它有着独特的、强大的美学效果；反驳了人们对诺曼底登陆历史叙事的认识。

该项目，不再是严肃投影中呈现的悲伤的普遍象征，如果杰泽贝尔·斯托雷追求的是这一事件和现场的复杂性，那么该项目可以多深入、多有趣呢？在呈现的三大宗教这一背景下，该事件的意义是什么？这一罪行又是如何谈论宗教和政治仇恨的？什么能解释恐怖主义——尤其是自杀性爆炸呢？地方当局和此次罪行中未受到惩罚的公民又是如何串通一气的？艺术作品如"意义"或"Fallen 9000"，可能都没有回答这种复杂的问题，但至少艺术应提出质疑。

艺术治疗法

吴昉

艺术治疗又称艺术疗法，是心理治疗的一种，有别于一般心理治疗多以语言为主要沟通媒介，艺术治疗通过提供艺术素材、活动经验等作为治疗的途径。"意义"创作者斯托雷的灵感核心即艺术疗法，从听觉艺术与视觉艺术——两种人类智慧表现形式中最直接和耐人寻味的方式，帮助人们寻找战胜身心困境的出路与方法。投射在墙面上的蜡烛呈冷色调，冰蓝的火焰与周围暖黄的灯光形成对比，强化视觉印象，也揭示生死两相隔的残酷。相比公共艺术，"意义"更像是一场小规模的私人悼念会，感动来源于切身之痛，与他人无关，忽略了以艺术唤起情感同生并发的意图。

Artist: Jezabel Storey

Location: Arroyo and Suipacha, Buenos Aires, Argentina

Media/Type: Multimedia

Date: 2012

Commissioner: None provided

Researcher: Gregory Door

To commemorate the civilians killed by a suicide bomber at the former site of Buenos Aires Israeli Embassy, artist Jezabel Storey combined video projection and a live piano performance. The projection, which remained lit from dusk until midnight, depicted a group of candles, one for every citizen who lost a life in the 1992 attack. The live performance featured Alfredo Corral performing Pictures at an Exhibition by Russian composer Modest Mussorgsky.

In 1992, a suicide bomber drove a truck of explosives into the Israeli Embassy, at the intersection of Arroyo and Suipacha. The building was destroyed, along with a nearby Catholic church and a school. 240 people were injured and 29 were killed, making it the deadliest attack on an embassy, and one of Argentina's worst terrorist attacks. An Islamic group with alleged ties to Iran took responsibility. No one was convicted in the crime. The location is now the site of a small commemorative park, the Plaza Embajada de Israel.

In seeking to commemorate the victims of the bombing, Storey, who is trained as a painter and is pursuing a master's degree in art therapy, invokes the notion of a peaceful offering. "No words were needed to move the soul of the people. Instead, beauty and peace were offered in a place that evokes terror and desperation." More than 200 people attended the vigil, according to the artist, including both rabbis and Catholic priests from the neighborhood. Neighbors to the plaza opened their windows and looked on. Survivors and relatives of the dead and wounded attended the ceremony as well.

"The audience was captivated by the music and the image," says Storey. "Survivors of the attempt and the families of the deceased were moved to tears and expressed their enormous gratitude."

While no one would deny the families of victims of a tragedy of this

magnitude the chance to grieve publically in commemoration of their lost loved ones, Sentido does raise interesting questions about the role of art, memorials, and grief.

This temporary intervention consisted of a large-screen projection of candles (the artist's materials claimed one for each of the "22 victims;" in fact there were 29) and the reproduction of a piece of music by another composer. It would be unfair to compare such a small-scale, temporary intervention to well-funded, permanent memorials.

Yet it is worth considering Sentido in the context of other temporary, commemorative interventions. A good example is the Fallen 9,000 project, which enlisted volunteers to etch 9,000 stenciled silhouettes in the sand of Normandy Beach to commemorate victims from both sides of the war. That temporary intervention had several qualities lacking in Sentido: It was appropriate to the site; it enlisted the active collaboration, rather than spectatorship, of a mass of people; it made a unique and powerful aesthetic statement; and it challenged assumptions about the historical narrative of the D-Day invasion.

Instead of a pervasive symbol of grief presented in an un-ironic projection and accompanied by a well-worn piece of the Western musical cannon, how much deeper and more interesting could the project have been had Storey pursued the complexity of the event and the site? What is the significance of this event in the context of the three major religions represented? What does the crime say about religious and/or political hatred? What can explain terrorism—especially suicide bombing? How complicit were local authorities and citizens for this unpunished crime? An art piece such as Sentido or Fallen 9,000 may not answer such difficult questions, but serious, worthwhile art owes us a responsibility to at least ask them.

巴别塔
Torre de Babel

艺术家：玛尔塔·米努欣
地点：阿根廷布宜诺斯艾利斯市圣马丁广场
形式和材料：装置
时间：2011 年
委托人：布宜诺斯艾利斯市政府
推荐人：艾希莉·金敦

2011 年 5 月 13—28 日，为庆祝布宜诺斯艾利斯市被列入世界首都名录，玛尔塔·米努欣的个人作品"巴别塔"临时安置在圣马丁广场。该装置由一个七层 25 米高的金属脚手架和铁丝网组成，陈列着超过八万册图书。这些图书流派各异，有 40 多种语言，均由世界各地的使领馆捐赠。在装置安装期间，100 多人的团体参观了该作品。在"巴别塔"中，公众可以近距离查看书名，同时聆听不断反复播放的多语言版本单词"图书"，有时如音乐般悦耳，有时如同低声吟唱。在装置撤离前的最后一天，许多人受邀领取钟爱的图书，限每人一本。剩余的图书作为"巴别图书馆"项目的一部分，被一并捐赠给 Gálvez 公立图书馆。

玛尔塔·米努欣是一位阿根廷艺术家，因个人理念化、行为化和参与化作品通常涉及政治话题，表达极度激进主义以及作品融入幽默和荒诞的元素（代表作："奶酪版米洛的维纳斯"），而闻名于世。玛尔塔·米努欣的个人作品中还体现出时下盛行的解构主义观点。例如："詹姆斯·乔伊斯的面包之塔（Dublin，1980）"，由 5 000 条面包构成，但不久就被公众"破坏"，因为他们把面包拿回家。需要提及的是"图书状帕特农神庙"这部作品，它与"巴别塔"有重要的渊源。这部作品由三万册图书构成，按照著名的帕特农神庙堆砌，但是在 20 世纪 80 年代初，阿根廷独裁政府对它发出禁令。在这些作品中，我们发现建筑材料具有以下三大功能：构成作品结构，评论政治或文化时事以及公众可以利用、分配。

玛尔塔·米努欣个人作品标题常常参考了巴别塔神话，这可以追溯到《创世纪》一书中的犹太和基督教传统，或者其他古典著作。这则神话讲述的是一群大力士想要建造一座通天塔。上帝通过发明语言向他们发难，这群大力士因沟通不畅而分崩离析，最终项目成了"烂尾楼"，他们分散到世界各地。这则故事意在抨击傲慢，而神话则讲述语言的来源。

纵观艺术史，"巴别塔"是一个普遍的主题，也是西方文化的一个重要的语言知识标志。在这部作品中，玛尔塔·米努欣展现了这些主题，但是从公共领域上来讲，它也发挥着一个更大的作用，即突出人类与生俱来的本性，有一种想了解以及被了解的欲望以及当无法沟通和学习他人而产生的绝望。所有参观"巴别塔"的游客得到了宝贵的机会，了解和被了解。那些拿走图书的参观者再现了神话场景，破坏"巴别塔"，向世界各地散播知识和语言。他们因艺术这个共同的目标而走到一起，正如玛尔塔·米努欣所说，艺术无国界，最终不需要翻译语言就能够产生共鸣。

巴别塔的瓦解

吴昉

人类互相理解，故可齐心协力，建巴别塔通天，而上帝迁怒人类的通天之欲，令世间有千万种语言，巴别塔因此顷刻瓦解。"圣经"描述的这个故事成为现实生活中人际沟通的困境渊薮，反复被提及，巴别塔俨然成为创作的母题。玛尔塔·米努欣的"巴别塔"装置艺术，将各种语言的书籍堆砌成塔，参观者可沿螺旋扶梯拾级而上，同时还可听到配乐单词朗诵，由米努欣本人亲自作曲，其他艺术家用各种不同的语言念诵"书—书—书"，让观者从视听感官上真切感受巴别塔困境的寓意，身临其境。一般的装置作品在撤展时即告终结，而"巴别塔"借助读者抽取书籍这一动作，在观者的协助下，最终实现了巴别塔现实意义的瓦解。如此完整的一次艺术创作，使神话也具备了可实现的操作性。"当人们听说图书馆已经收集齐全所有的书籍时，首先得到的是一种奇特的幸福感。"阿根廷作家博尔赫斯在其"通天塔图书馆"中，如是说。

Artist: Marta Minujín
Location: Plaza San Martin, Buenos Aires, Argentina
Media/Type: Installation
Date: 2011
Commissioner: City of Buenos Aires
Researcher: Ashley Guindon

Marta Minujín's Torre de Babel was temporarily installed in Buenos Aires' Plaza San Martin from May 13-28, 2011 in celebration of the city being named the Book World Capital. The installation consisted of a seven-story, twenty-five meter metal scaffolding hung with wire netting, which displayed over 80,000 books. The books were donated from embassies all over the world and represent all genres and over 40 languages. Groups of 100 people toured the work over the course of its installation period. Once inside Torre de Babel, the public could peruse the book titles and listen a recording of the word "book" spoken repetitively, like music or chanting, in many languages. On the last day of the installation, individuals were invited to take a book of their choosing. Leftover books are part of an archive entitled The Library of Babel, which was donated to the Gálvez Public Library.

Marta Minujín, an Argentinian artist, is most known for her conceptual, performative, and participatory work, which often deals with political issues in a monumental and radical fashion, incorporating elements of humor and the ridiculous (for example, a Venus de Milo cut from cheese). The notion of destruction is also a prevalent motif in Minujín's work. For example, The James Joyce Tower in Bread (Dublin, 1980), made of 5,000 loaves, was made to be "destroyed" by the public as they took the bread home with them. This work is important to note in relation to the Torre de

Babel, as is the Parthenon of Books (1983), which consisted of a Parthenon-shaped construction of 30,000 books that had been banned by the Argentinian dictatorship during the early 1980s. In each of these pieces, the building materials have three functions: to make up the structure of the work, to comment on a political or cultural situation, and to be utilized and distributed to the public.

The title of Minujín's work refers to the myth of the Tower of Babel, which has its origins in the Judeo-Christian tradition in the book of Genesis, but is also referenced by other ancient traditions in similar stories. The myth is of a powerful group of people striving to build a tower to heaven. God confuses them by inventing languages, thus ending their ability to communicate and forcing them to abandon the project and scatter over the earth. The moral of the story is about pride, and the myth points to the origins of languages.

The Tower of Babel has been a common motif throughout art history and a symbol of language and knowledge for Western culture. In this work, Minujín played on these themes, but the work also served a greater function in the public realm. It referenced the innate human desire to understand and be understood, as well as the frustration that results when there is an inability to communicate and learn from one another. Participants who came to Torre de Babel were given the opportunity to understand and be understood. Those who took away the books are re-enacting the myth, destroying the tower and scattering the knowledge and languages out into the world. They came together with a common purpose for the sake of art, which, as Minujín says, ultimately needs no translation.

找寻我的梦（野性的呼唤）

Olhar nos meus sonhos (Awilda)

艺术家：乔玛·帕兰萨
地点：巴西里约热内卢市博塔弗戈湾
形式和材料：雕塑
时间：2012 年
委托人：Other Ideas for Rio 双年艺术展
推荐人：艾希莉·金敦

2012 年 7 月至 10 月，乔玛·帕兰萨个人作品"找寻我的梦（野性的呼唤）"
在巴西里约热内卢市博塔弗戈湾临时安装。该作品被"Other Ideas For
Rio"双年艺术展选为六大公共艺术作品之一，该双年展将在 2016 年巴西
奥运会期间备受瞩目。该作品是一块雪白的 40 英尺高女性的头部到颈部的
身体部位。它由玻璃纤维、大理石粉和聚酯树脂制成，安装在博塔弗戈湾，
临近海岸线，整个形体就好像从水中浮现。

乔玛·帕兰萨是一位加泰罗尼亚雕塑家，公共作品寓意丰富，富有整体感。
他试图通过作品探索个人与集体之间的联系性。比如另一件个人代表作"皇
冠喷泉"（2004），位于芝加哥市千禧年公园，刚好传达了这种理念。

在创作里约热内卢的"找寻我的梦（野性的呼唤）"作品之前，帕兰萨就已
经运用一个白色、不朽的女性头部比如：临时安装在纽约市麦迪逊花园的
作品"回音"(2011)，圣海伦斯的"梦"（2009），以及另一部永久安装
在萨尔斯堡的"野性的呼唤"(2010)。每次帕兰萨都会使用一个光滑的空
白人脸来展现居住的地点。例如，"回音"指代的是希腊神话中的女神，
她具有复制其他人的思想的能力，所以说纽约闹市中的雕塑好比一个安静
聆听并储存千百万行人的记忆库。通常情况下，头部创造指代一条特定的
信息。例如，萨尔斯堡的"野性的呼唤"由多层大理石板层层叠加而成。
而这个多层结构具有一定的含义，它代表了过去几个世纪以来欧洲移民的
人文情怀、梦想和希望。不论何处，帕兰萨使用这种简单却优雅的意向反
映所在地的独特之处。

通过里约热内卢的"找寻我的梦（野性的呼唤）"，帕兰萨试图"让大家做梦"，
因此将雕塑打造成梦境。"野性的呼唤"一词意思是"野性"，这源自一
段历史传奇，曾经有一位斯堪的纳维亚公主为逃避逼婚，最终成为一名海盗，
这段历史代表着主人公强大的意志力和丰富的内心世界。帕兰萨的雕塑充
分体现了这些人物特质。雕塑是寂静和神秘的延伸，极具女性魅力，浮沉
俗世，直面内心世界，因此成为里约热内卢市的地标性雕塑，正对应了矗

立在基督山顶的里约热内卢基督像，他张开双手，似乎在拥抱着城市（甚至全世界）。而她是如此地平易近人，充分向公众展示自己，但是从内心上来看，她又是一块空白画布，等待被诠释、解读和梦想。她让观众合上双眼，挖掘内心。

帕兰萨的雕塑具有独到之处，向公众展现充满魅力和直击内心的事物，足以引起他人的注意，同时给予他们解读、审视和反思的空间。这部作品为个人和集体之间的互动创造了空间，意义非凡。帕兰萨并没有赋予含义，而是充分给予公众自由，按照自己的理解进行解读。

在海的远处，水是那么蓝

吴昉

"她是一个古怪的孩子，不大爱讲话，总是静静地想什么事情。当别的姊妹们用她们从沉船里所得到的最奇异的东西来装饰她们的花园的时候，她除了喜欢像高空的太阳一样艳红的花朵以外，只愿意有一个美丽的大理石像。"在巴西里约热内卢的博塔弗戈湾，帕兰萨用 12 米高的雕塑点缀海面，洁白的少女，静谧阖眼，好像安徒生笔下海的女儿。艺术家说作品是为了让人们有梦。沿着蔚蓝的海湾，一张光滑的脸庞，与伫立山顶的里约热内卢基督像遥相呼应，构成奇妙的对照。"找寻我的梦"是一件融入环境的作品，雕塑宁静的白与海洋遥远的蓝，投射入人心，产生美好的愿望。安徒生童话世界中海的女儿，年满十五，就能浮上海面，可以在月光下、石头上，看巨大的船只身边驶过，可以看到树林，看到城市。博塔弗戈湾的 14 岁少女，是否迫不及待了呢？

Artist: Jaume Plensa
Location: Botafogo Bay, Rio de Janeiro, Brazil
Media/Type: Sculpture
Date: 2012
Commissioner: Other Ideas for Rio
Researcher: Ashley Guindon

Jaume Plensa's Olhar nos meus sonhos (Awilda) was temporarily installed in Botafogo Bay in Rio de Janeiro, Brazil from July to October of 2012. The work was selected as one of six public artworks for Other Ideas For Rio, a biannual art event that will culminate in the 2016 Olympic Games. The work consists of a white, forty-foot bust of a woman from the neck up. It is made of fiberglass and marble dust mixed with polyester resin. It was installed in the bay, close to the shoreline, so that the figure would appear to be emerging from the depths of the water.

Jaume Plensa is a Catalan sculptor whose public work is often figurative and monolithic. He tends to explore community, citizenship, and the play between the individual and the collective through his artwork. For example, he is well known for his Crown Fountain (2004) in Chicago's Millennium Park.

Plensa has utilized the motif of a white, monumental, female head in works before Rio's Awilda. Echo (2011) was installed temporarily in Madison Square Park, New York, Dream (2009) can be seen in St. Helens, and another Awilda (2010) is permanently installed in Salzburg. Each time, Plensa uses a smooth, blank face as a representation of the place in which it rests. For example, Echo alluded to the Greek myth of a nymph who could only repeat the thoughts of others, so the sculpture in busy New York acted as a tranquil reservoir for the thoughts of millions of passersby. Often,

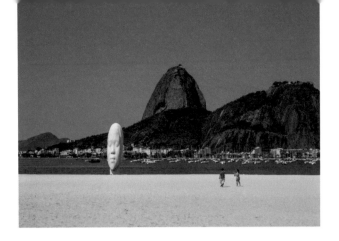

the fabrication of the head points to a specific message. The Salzburg Awilda, for instance, is made of layered slabs of marble stacked on top of each other. This layered configuration points to the particular meaning of this work, which represents the hopes and dreams of the layers of humanity that have immigrated to Europe over the centuries. Regardless of the location, Plensa uses this simple and elegant motif to highlight unique aspects of the site.

For Rio's Awilda, Plensa wanted to "make the people dream," and so he placed the figure in a dream-like setting. The name Awilda means "untamed" and is historically associated with a legend about a Scandinavian princess who became a pirate rather than being forced to marry. This history, combined with the rest of the title, Olhar nos meus sonhos / To see my dreams, alludes to a character of strong will and vast inner space. Plensa's sculpture captures these attributes fully. Awilda is a monumental figure of serenity and mystery. Rooted in water and framed by the iconic landscape of Sugarloaf Mountain, the sculpture acts as the female, earthly, inward-facing counterpoint to Rio's more iconic monument, Cristo Redentor (Christ the Redeemer), which embraces the city (and the world) from his heavenly perch on Corcovado Hill. She is approachable and touchable, fully displayed for the public but looking inward from it, a blank canvas for interpretations, thoughts, and dreams. She encourages the viewer to close their eyes and see within.

With this sculpture, Plensa demonstrates a unique ability to present a public with something that is beautiful and compelling enough to demand attention, while simultaneously open to interpretation, critical conversation, and personal reflection. The work creates space for both individual and collective interaction and meaning. Rather than assign a meaning to a place, Plensa allows the place—and its people—to arrive at their own meaning.

绿色汽车
Green Cars

艺术家：毕加·理
地点：巴西圣保罗市以及其他城市，奥地利格拉茨市
形式和材料：装置
时间：2009 年
委托人：毕加·理
推荐人：艾希莉·金敦

"绿色汽车"由毕加·理首创，是一个涉及多领域的创新收藏，位于巴西圣保罗市。作为该系列作品的一部分，专业创意人员安装了 11 件作品，地点遍布圣保罗市以及其他巴西城市，奥地利格拉茨市。每辆汽车内部都覆盖着植被，放置于公共场所。一些汽车甚至装有绿色灯，在晚上会发出亮光。

2000 年以来，毕加·理的许多作品以影音、新媒体、动画和数字化技术呈现。主要客户有可口可乐，福特汽车，Trident 口香糖和美国 HBO 电视台，他们在舞台、人行桥和建筑物上呈现出绚丽多彩的灯光和影像画面，令人印象深刻。为庆祝圣保罗建市 425 周年，一部影像制作呈现在喷泉之上（伊比拉布埃拉公园喷泉，2013 年），专为 Absolut Vodka 酒业公司（2011）打造的互动式影音作品，使参与者可以通过动态激活，在白色画布上"随性发挥"以及 2011 年高达 262 英尺的 LED 显示屏展现的 Planeta Terra 盛事。这些宏大的作品仅仅是 BijaRi 创作的一小部分，具有科技含量高，艺术创新表现力强等特点。

除了这些高技术含量作品外，也有一些简单但优雅的公共艺术作品，寓意幽默，通常涉及一些社会和政治热点。其中，"鸡"（2002）描述的是两只鸡被放生到两处公共地点，一个是上流社会的聚居地，而紧挨着就是另一个工人阶级聚居地，但是这打破了两处公共地点的日常规律。与之相比，"绿色汽车"虽然没有引起这么大的争议，但是寓意相似。该项目是大型制作系列"城市与自然"（2007 年至今）的组成部分，一些城市结构，包括广告牌、垃圾桶甚至公交车都被改造成绿色花园。

从许多方面来看，"绿色汽车"都堪称一部佳作。首先，它反映了城市和自然空间的冲突。当旧汽车不再依靠于人工技术时，他们被覆上一层植被，绿意盎然，充满活力，不再是过去那般死气沉沉。行人路过时都会驻足停留领略人造建筑和自然之间的和谐统一。从这一点上来说，这些作品极力展现公共艺术，供人们观察、欣赏和反思。其次，当属公共领域起到的作用。

这些作品安装结束后，当地媒体集中报道了一些装置，但是不久后，大部分作品被城市职能部门拆除，提醒人们关于艺术和政治之间的冲突。一些报道过的作品被保留数周或数月，有时提醒着公众市政系统存在一些弊端。此外，一些汽车被邻居充分利用起来，他们负责保养汽车，有时也在里面聚会闲聊。最后，这部作品让我们可以直接接触公众艺术领域。有的人认为，"绿色汽车"被视作累赘，占据着街道大部分空间，或者占了一些停车位。有些车主也表示支持，他们认为应该发挥汽车最大的作用。每当出现一些公共艺术，人们的反应不尽相同，有的持功利主义观点，而有的则持自由主义观点。该作品风趣幽默、创意十足，且寓意深刻，值得相关研究人员进行案例分析。

自然时代
吴昉

工业革命以机器取代人力，机器的发明及运用成为那个时代的标志，因此被历史学家称为"机器时代"。白云苍狗，当社会发展与人类生活无法承受机器迅猛的超负荷，加上机器时代以自然环境为代价的无节制开发，一种更紧迫的需求开始滋长，那就是"绿色汽车"作品中对自然的再次期待。废弃不用的汽车，像从工业革命战场退下的老兵，奄奄一息又百无一用，却经由毕加·理的敏锐视角，发现了其中有关城市工业与自然环境的议题。如同过去的人们无法预计自然时代的意义，今天的我们看待机器时代也无法一言蔽之。耗尽的汽车可以完成历史使命在机器时代，也可以延续使命。

"绿色汽车"一路从过去驶往将来，这株城市的移动盆景，用新奇的形态引发议论与思考，同时也为城市带来一份超现实的趣味。

Artist: BijaRi
Location: Multiple sites in São Paulo, multiple sites in other cities in Brazil, and Graz, Austria
Media/Type: Installation
Date: 2009
Commissioner: BijaRi
Researcher: Ashley Guindon

Green Cars is an initiative by BijaRi, a multidisciplinary creative collective of São Paulo, Brazil. As part of this series, the group of creative professionals has installed eleven works throughout São Paulo, in other Brazilian cities, and in the European city of Graz in Austria. Each work consists of a car stripped of its insides, planted with vegetation, and left in a public place. Some cars also include green lights to illuminate them at night.

Since 2000, much of BijaRi's work has taken the form of video, new media, motion graphics, and digital technology. Their client list includes Coca-Cola, Ford, Trident, and HBO, and they have produced impressive public light and video displays on stages, bridges, and buildings. These works have included a video production to celebrate São Paulo's 425th anniversary, which was projected onto a water fountain (Ibirapuera Park Fountain, 2013), an interactive video work for Absolut Vodka (2011) in which participants could "draw" on a blank canvas through motion activation, and a 262-foot LED video screen for the Planeta Terra events of 2011. These stunning displays are just a small sample of the many technologically advanced and artistically innovative works in BijaRi' s portfolio.

In addition to these astounding and visually appealing feats of technology, the group has also produced simple, elegant, and often somewhat humorous public art interventions aimed at social and political commentary. An example in this vein is Chicken (2002), in which two live chickens were released in two busy public spaces —

one upper-class area, one working-class neighborhood—as a means of interrupting the daily routines and patterns of both publics. Green Cars is similar in its aim, though perhaps less controversial. The project is part of a larger body of work called Urban Nature (2007-present), in which urban structures—such as billboards, garbage bins, and even a bus—are turned into gardens of lush greens.

Green Cars is compelling in several ways. The first is the most apparent: the tension between urban and natural space. As the old cars are stripped of their technology and man-made parts, then filled with plants, they become green, living, breathing organisms in contrast to their original purpose. They are there for the contemplation of passersby, highlighting the dichotomy of the built and the natural. In this sense they act as much public art does—as a monument to an idea to be observed, considered, and appreciated. Beyond this, however, is their function within the public realm. Some of the installations are reported by locals and removed by the city almost immediately after installation, making them a symbolic reminder of the tension between art and politics. Some were reported but not removed for weeks or months—reminders of a sometimes broken municipal system. Still other cars were adopted by people in their neighborhood, who take responsibility for their maintenance and use them as a gathering space. The work, then, offers us insight into art in the public realm. For some members of the public, Green Cars was seen as an unnecessary blight on their street, or, perhaps, as unnecessary usage of a parking space. Others took ownership of the works, embraced their message, and put them to good use. These reactions—from the utilitarian to the libertarian—take place whenever art enters into the public realm. The work's layers make for an interesting, innovative, and poetic case study.

7 700 万件画作
77 Million Paintings

艺术家：布莱恩·伊诺
地点：巴西里约热内卢市拉帕拱桥
形式和材料：新媒体技术
时间：2012 年
委托人：Other Ideas for Rio 双年艺术展
推荐人：艾希莉·金敦

影音作品"7 700 万件画作"由布莱恩·伊诺呈现，于 2012 年 10 月 19 日至 22 日安装在巴西里约热内卢市。该作品被"Other Ideas For Rio"选为六大公共艺术作品之一，该双年展将在 2016 年巴西奥运会期间备受瞩目。作品分为影像和音频部分，投影在 Lapa Arches 之上。

布莱恩·伊诺是一名著名的氛围音乐家，也是一位流行音乐艺术家和演唱会灯光特效师，其代表作有"U2"和"Coldplay"。2006 年"7 700 万件画作"曾以在线音乐的方式传播。该作品由近 300 首音乐、原创数字艺术片组成，所有这些元素经自由组合，展现了一种无穷无尽的光、色、声的循环效果。该作品由个人收藏，已经在世界各大画廊以及艺术科技中心展出。此外，它也曾经安装在城市地标性公共空间，最著名的当属 2009 年在悉尼歌剧院展出。布莱恩·伊诺经常加入一些新鲜元素，使得作品呈现多变的风格，互不雷同。

2012 年 10 月 19 至 21 日，2012 年版"7700 万件画作"临时性安装在拉帕拱桥。拉帕拱桥是 Carioca Aqueduct 的城内建筑，是 18 世纪的遗产建筑，横跨中心城区，如今成为仅存的交通高架桥。虽然 Arches 周边设施陈旧，满目疮痍，但是它富有文化底蕴，街头遍布音乐表演和俱乐部。

布莱恩·伊诺尝试通过"7 700 万件画作"吸引路人驻足欣赏。里约热内卢市内，他创造了许多新型设计，添加更多细节。为回应城市噪音，他还将声音融入现场播放的"音乐带"。光影动态极具视觉冲击，使人们情不自禁地驻足欣赏，并完全融入周围氛围。慢镜头和作品变化需要慢节奏的欣赏，一瞥而过只能让人们错过这部杰作。当然，布莱恩·伊诺做到了这些，他"俘获"了观众的目光。不管天气情况多么恶劣，总会发现人群驻足在装置周围，欣赏着光影交织，享受音乐魅力。由于作品的可参与度高，观众们也愿意为它驻足停留。通过光影特效，作品投影于拱桥之上，他们可以融入到作品画面，这在以前是无法想象的。

拉帕拱桥为"7 700万件画作"提供了一个崭新有趣的背景。一方面，拱桥历史久远，标志着深厚的文化底蕴，与新媒体技术的华美效果相比，形成了强烈的反差。另一方面，两种元素的完美融合充分展现了人类技术的先进性，一新一旧，但是都代表了当时的前沿科技。布莱恩·伊诺将个人作品与历史名作联系起来，因此作品可以被视作深厚历史底蕴的进一步延伸。

氛围中的延迟

吴昉

布莱恩·伊诺是著名音乐制作人，对电子元素的应用有着创造性的发明和实验精神，引领了受人瞩目的"氛围音乐"，即一种依靠合成器创作出某种环境的音乐类型，"7 700万件画作"正是这种结合环境、光、影、音乐、人，应运而生的新媒体作品。布莱恩·伊诺自称作品是一种影像绘画，18世纪古风荡漾的拉帕拱桥随电音节拍幻出绮丽色彩，熟悉的景观在夜幕中成功猎奇，吸引过路行人的驻足，如同伊诺曾经评论合作过的U2乐队音乐理念：每个人都有来到派对的感觉，不打算让任何人觉得它是不可接近的。20世纪70年代，伊诺在专辑制作过程中，发明了"磁带延迟"技法，将吉他音乐做循环延迟，通过现代录音技术另辟蹊径。"7 700万件画作"的拉帕拱桥也在特殊的音乐氛围中，循环往返在各色图形的光影里，时间在这一刻获得延迟，而行人思绪的火花却被多重感官带动，节奏摇摆中慢慢飞扬。

Artist: Brian Eno
Location: Lapa Arches, Rio de Janeiro, Brazil
Media/Type: Software/DVD combination
Date: 2012
Commissioner: Other Ideas for Rio
Researcher: Ashley Guindon

The audiovisual work 77 Million Paintings by Brian Eno was installed in Rio de Janeiro, Brazil, on October 19-21, 2012. The work was selected as one of six public artworks for Other Ideas For Rio, a biannual art event that will culminate in the 2016 Olympic Games. The work consists of a video and sound work projected on the Arcos Da Lapa (Lapa Arches).

Eno is perhaps most well-known for his contributions as an ambient musician and for his work with pop music artists such as U2 and Coldplay, for whom he creates concert lightscapes, but he also works as an artist. 77 Million Paintings was originally produced as a video and software work in 2006. It contains music and approximately 300 abstract, original digital artworks, all of which are cycled through randomly, thus creating a seemingly endless display of light, color, and sound. The piece can be bought for individual use, and has been exhibited in various configurations at galleries and art and technology centers worldwide. It has also been installed in iconic public spaces before, most notably the Sydney Opera House in 2009. It is an ever-changing spectacle, never the same twice, and Eno adds new content often. He refers to his work as "video painting."

The 2012 version of 77 Million Paintings was installed at Lapa Arches for a short duration—October 19-21, 2012. The Lapa Arches are part of the Carioca Aqueduct, an eighteenth-century relic that runs through the middle of the city and now serves as a bridge for the

only remaining active tram line. The area surrounding the Arches could be considered old and run-down, but it is also culturally rich, with a lively live music and nightclub scene. It is a bustling place, with plenty of sound and traffic.

Eno's aim with 77 Million Paintings seems to be to bring that traffic to a halt. For the Rio installation, he created many new designs with extra details. He also mixed sounds into the existing ambient "music-scape" live in response to the noise of the city. The captivating light and movement aimed to arrest the viewer and immerse them in an ambient atmosphere. The slow motions and changes of the work must be taken in slowly; they cannot be simply glanced at or they might be missed. Eno certainly succeeded in arresting the viewer. The crowd that gathered to watch the performance was so transfixed that they stayed even through a rainstorm. The audience also remained with the work because of its participatory nature. By using the light and their bodies to cast shadows on the arches, they embedded themselves in the work and interacted with the Arches in a way they may never have been able to otherwise.

The installation at the Lapa Arches gives 77 Million Paintings a new and interesting context. On one hand, the Arches, so old and steeped in history and culture, contrast nicely with this spectacular display of new media and technology. On the other hand, both elements represent human-made technological advancement—one old and one new, but both cutting-edge for their time. Eno sees his work in relation to the history of painting, not that of video art, and so it is fitting for the work to be paired with a historically rich setting.

沙滩怪兽
Strandbeest

艺术家：泰奥·扬森
地点：许多地点，包括阿根廷布宜诺斯艾利斯
形式和材料：装置
时间：2012 年
委托人：Tecnopolis
推荐人：里奥·谭

"沙滩怪兽"是荷兰物理学家和艺术家泰奥·扬森长期从事的项目，研究他称之为"新型生命"的逼真动态雕塑的创造与演化过程。这些怪兽最早出现于 1990 年，其外观特别采用骨架结构，制作材料为常见的塑料管。它们形体巨大（其中许多对人类还而言就是庞然大物），由风力驱动，其行动依赖于海岸边的气流，这样它们完全能够正常运转，而无需借助任何形式的电子学外力。值得注意的是，经过大量实验以后，它们已经具备了"原始"的感官功能，可以探测水、躲避障碍及在风暴来临前自行固定在地上。这些怪兽的演化源自于它们"DNA"的变化，如改变塑料管长度以及添加泵、阀和迎风面。

每头怪兽的制作都需历时一年左右，在此过程中扬森会根据遇到的困难（如风暴抵抗）对它们加以修改，使得新版的怪兽在旧版的基础上都有所提高——这是一种辅助进化。扬森这一系列项目迭代出现的部分是科技园区 2012 展示的不断拓展的恶魔图鉴中两种怪兽的安装，该活动是拉美地区最大的艺术、科学、技术盛会。这两种被采用的怪兽是 Animaris Umerus 和 Animaris Ordis Mutantis；前者是长 12 米、宽 4 米的大怪物，后者的体形更接近于人类。

"沙滩怪兽"最初只是在海滩走动，当多足一起移动时，它们或从容优雅，或踌躇不前。一不小心碰到了它们，就好像亲眼目睹了初级智能正试图去搞懂它们存在的世界并力图去适应所在的气候条件一样。说到场所营造，这些怪兽有着毋庸置疑的使一个地方面目一新的能力，它们可以创造意外的相遇，让人顿时感到既熟悉又陌生。感到熟悉，是因为它们唤起了人们关于发育中的婴儿或儿童探步行走的回忆，而且它们的制作材料是塑料，这在当今消费社会中无处不在；感到陌生，是因为它们的外形奇异或怪诞，但却与人类有相通的地方，同时也由于它们的行动方式、感官和对环境的反应。

"沙滩怪兽"给我们的提醒是，生命就在我们身边，也存于我们内心；人类并不是唯一值得认真对待的生命形态。这一提醒减弱了人本主义和理性主义思想的过分要求，该思想错误地把人类置于进化的最高点，并使我们相信生命和意识是有机体特别是人类独有的品质。泰奥·扬森开始将怪兽用于增强气候变化和海平面上升的意识，或许这不是一个巧合。它们让我们反思无节制的人类生活对其他生命形态的影响，思考人类工业化和大众消费于与我们生活在同一个星球上的海洋生命及其他物种造成的后果。怪兽多足行走在海滩上，就好像是来自不明未来的幽灵，在那时人类生活也将变得越来越不稳定。

该项目的卓越之处在于它提醒了人类要关注其他生命和未来，同时也在于培育怪兽所需的技术天赋和毅力。包括许多自发项目的"沙滩怪兽"给了世界以灵感（人们根据泰奥·扬森证明了的原则制作他们自己的"沙滩怪兽"），展示了该项目是如何抓住世界各地的想象力的。

"新型生命"
吴昉

泰奥·扬森创作的"沙滩怪兽"并不恐怖，精细的结构搭建，更像是海滩边扩张版的立体骨架构成，停止的时候它就是静态装置，运动的时候则是具备仿生特性的动态雕塑，动静皆宜。纽约时报在评论"沙滩怪兽"时，以"为什么没有东西是真正活着的？"为标题，其实，"沙滩怪兽"作为公共艺术，给予观者，或者说"传统生命"的是一种反思的意义，人类并不是唯一现存的生物体，而公众也可以参与创作来孕育更多的"新型生命"，通过自我反省来重新看待世界。据说，泰奥·扬森已与日本学研社（Gakken）合作，提供"怪兽们"的全尺寸设计微缩版本，迷你的家庭版"新型生命"已经诞生，一派生机蓬勃的奇妙景象。

Artist: Theo Jansen
Location: Multiple, including Buenos Aires, Argentina
Media/Type: Installation
Date: 2012
Commissioner: Tecnopolis
Researcher: Leon Tan

Strandbeest is a long-term project by Dutch physicist/artist Theo Jansen, involving the creation and evolution of what the artist calls "new forms of life," uncannily life-like kinetic sculptures. Started in 1990, Jansen' s strandbeests typically take on a skeletal appearance, and are constructed out of everyday plastic tubes. Monumental in scale (many of them dwarf the human body), they are animated by wind, living off air currents along the coast, and have the capacity to propel themselves without the aid of electronics of any kind. Remarkably, they have, through numerous experiments, developed 'primitive' senses enabling them to detect water, evade obstacles and anchor themselves into the ground in anticipation of storms. The evolution of Jansen's creatures depends on varying their 'DNA,' namely, the length of the plastic tubes, as well as the addition of pumps, valves and wind catching surfaces.

Each of Jansen's beasts takes approximately a year to create, and each new strandbeest improves on the previous version, with the artist implementing modifications in response to the difficulties encountered (e.g. withstanding storms)—a kind of assisted evolution. One iteration of Jansen' s strandbeest projects consisted of the installation of two creatures from a growing bestiary at Tecnopolis 2012, the largest art/science/technology event in Latin America. The two animals were Animaris Umerus and Animaris Ordis Mutantis, the former, a gigantic 12-meter in length and 4-meter in height, and the latter, closer to human scale.

Originally designed for roaming beaches, the strandbeests move gracefully at times, and hesitantly at others on their multiple feet.

Encountering them, it feels as though one is in the presence of a rudimentary intelligence attempting to make sense of its world, and to adapt to its climatic conditions. As far as placemaking goes, these beasts have an unquestionable ability to transform a site, generating encounters that are at once familiar and alien. They are familiar in that they recall the exploratory movements a developing infant or child makes, and insofar as they are constructed of plastic, a ubiquitous material in today's consumer society. They are alien in that they incarnate forms that are strange or bizarre, that nevertheless share with us certain possibilities, that of movement, sensation and reaction to the environment.

Strandbeest reminds us that life is all around as well as inside us, that humans are not the only lifeform worthy of serious consideration. This reminder erodes strains of humanistic and rationalist thought, which have placed humanity erroneously at the apex of evolution, and convinced us that life and consciousness are exclusive qualities of the organic, and especially human, body. Perhaps it is not a coincidence that Jansen began making them in response to growing awareness of climate change and the rising sea levels. The beasts make us reflect on the impact of unrestrained human life on other forms of life. They make us consider the consequences of human industrialization and mass consumption on the life of the oceans and the other species with whom we share the planet. Walking on its many feet, the beasts of the beaches appear like specters from an uncertain future, in which human life has become increasingly precarious.

The project's excellence consists of these reminders of other lives and futures, as well as the technical genius and perseverance it has taken to cultivate the strandbeests. The many self initiated projects Strandbeest has inspired around the world (people making their own strandbeests based on the principles demonstrated by Jansen's) provides a measure of how the project has captured the imagination of communities around the world.

贫民区现代艺术节
Ghetto Biennale

艺术家：安德烈·欧仁、利亚·戈登
地点：海地太子港附近的大街上
形式和材料：互动行为
时间：2009 年（后续分别在 2011 年和 2013 年）
委托人：原始资金来源于参与人员，部分来源于赞助支持
推荐人：杰西卡·菲亚拉

"当第一世界艺术遇到第三世界艺术会发生什么呢？会引发流血冲突吗？"带着这一发人深省的问题，艺术家团体 Atis Rezistans 于 2009 年在海地太子港附近的大街上举办了第一届"贫民区现代艺术节"。艺术节打出了莉亚·安莎杜娃将美国和墨西哥边境比作创伤口的原创标语，充分展露了第一世界与第三世界的明显差异，并在此二元文化间创造了一个"第三空间"，使得背景不同的艺术家都能通力合作，一同工作以及进行辩论和讨论。

当代艺术市场日益全球化却未能很好地处理艺术家遇到的不平等待遇，面对这种情况，艺术家安德烈·欧仁和馆长利亚·戈登发起了贫民区现代艺术节。旅行签证遭拒，Atis Rezistans 的艺术家发现，尽管他们的工作可以穿越国境，他们自己却不能。他们已被边缘化，失去了社交网络的机会和在展会上的声音。面对这种流动性的缺失，"贫民区现代艺术节"扭转了流向，将国际艺术社团带到了海地，此举削弱了参与其中的障碍，提供了新观点的展示平台，促进了跨越障碍的对话，提高了街道艺术家的知名度，给出了改善海地负面媒体形象的办法。

到访参加第一届"贫民区现代艺术节"的艺术家于 2009 年 11 月底开始陆续抵达，他们有三周的时间来布置现场的项目，与当地艺术家和社团保持通力合作，充分利用当地可利用的材料。完成的 35 个项目在形式和范围上各不相同。卡罗尔·伦格·弗朗西斯和泰勒·乔纳斯·拉巴泽创立了"海地制造"——由慈善机构捐赠的二手服装造就的流行款式。杰西·达林与附近的儿童和青少年成立了"垃圾教堂"，一个由废弃材料建造的用于举行音乐会、展会及电影放映的场所。塞图·琼斯与当地医疗机构合作制作并分发了 1 000 个内含花草种子的粘土球来绿化城市。当地的青少年则扮演起了外国记者的角色，进行"贫民区电视采访"，严肃地瞪大了眼睛注视着整个艺术节，这时候"外国视角"常常聚焦于海地。

定期按时举行的"贫民区现代艺术节"（2009 年、2011 年和 2013 年）给当地带来了 60 000 美元以上的收入，尤其重要的是，该地区已被指定为"红区"，外国工人及非政府组织不能进入。艺术节衍生了一批正在进

行中的合作、项目和机会，包括 Atis Rezistans 至少 15 名成员参加海外展会及在海外居留的机会。该活动除了吸引到访的艺术家和学者，也吸引了大批当地观众和海地以外的艺术家前来参加。

与 Atis Rezistans 重新目的化的文化碎片的审美理念一致，"贫民区现代艺术节"是该理念的具体实践。作为艺术节的文化机构，Atis Rezistans 给显著不同的世界——艺术节与贫民区——搭建了互相进入桥梁，如此来破除潜在的"问题化"，增加双方对彼此的了解。许多于 2009 年抵达的外国艺术家通过短暂的实践，通常急于批判现有的由西方为主导的艺术机构，而许多海地同行的艺术家则寻求进入这些机构，找到出售面向对象作品的市场。2010 年 1 月爆发的地震给第二届"贫民区现代艺术节"增添了压力，同时也对该艺术节与相关慈善机构的合并增添了困难。如何建立共识、避免主办地区的外国化处理以及增大参与的担忧一直存在。寻求策略来学习一个共同的主题以及解决一个共同的主题带来的挑战，免于偏颇视角的干扰，将下结论的学术会议改变成为包容性更强、参与度更大的社交活动。

艺术节与贫民区之间的冲突可以带来更多原生态的东西，随着全球化的日益加剧，更加需要像"贫民区现代艺术节"这样敏感、边界有争议但坚定不移地兑现承诺的活动。

艺术的对立面
吴昉

艺术与贫困是不是对立的？艺术与经济是成正比的对应关系吗？艺术是不是一定阶层独有的特权？海地贫民区现代艺术节让人思考这些带有思维定势的问题。远古社会、蛮荒之地，有人的地方就有艺术的萌芽，艺术的审美特征，最初是作为一种本能的需求而出现。当艺术成为温饱有余的点缀、文化繁荣的附属、经济富裕的量化标准，就已经掺杂了太多本无涉于艺术的成分。反观海地"贫民区现代艺术节"，"海地制造"、"垃圾教堂"、"废墟场所"，这些生机盎然的项目与创作，令人惊喜地显示出艺术圣地才有的活力。海地贫民区艺术节举办已有数届，一时兴起的热情，已发展成为可持续的文化项目，受当地居民的喜爱与欢迎。创意、审美、生活结合得恰到好处，才有所谓艺术，艺术的对立面从来不是贫困，而是观念意识上的狭隘与偏见。

Co-Founders: André Eugène & Leah Gordon
Location: Grand Rue, Port-au-Prince, Haiti
Media/Type: Various interventions
Date: 2009 (additional iterations 2011, 2013)
Commissioner: Primarily funded by participants with
some sponsorship support
Researcher: Jessica Fiala

"What happens when first world art rubs up against third world art? Does it bleed?" This provocative question introduced the first Ghetto Biennale in 2009, hosted by the artist group Atis Rezistans and based in the Grand Rue neighborhood of Port-au-Prince, Haiti. Drawing its original tagline from Gloria Anzaldúa's comparison of the Mexico-U.S. border to a raw wound, the Ghetto Biennale exposes disparities between first and third worlds and creates a "third space" beyond these binaries, where artists from different backgrounds can collaborate, share their work, and enter into debates and discussions.

Developed by artist André Eugène and curator Leah Gordon, the Ghetto Biennale responds to a contemporary art market that is increasingly global in scope, but frequently fails to grapple with inequalities faced by artists. Refused travel visas, artists from Atis Rezistans found that while their work could travel, often they could not. Already marginalized, they lost opportunities to network and a voice in exhibitions. The Ghetto Biennale responds to this lack of mobility by reversing the flow, bringing the international art community to Haiti, and in doing so, undermining obstacles to participation, providing access to fresh ideas, fostering dialogue across barriers, bringing greater visibility to Grand Rue artists, and creating an alternative to negative portrayals of Haiti in the media.

Visiting artists for the first Ghetto Biennale began arriving in late November 2009 and had three weeks to create projects on-site, collaborating with local artists and communities and using materials available in the Grand Rue neighborhood. The 35 resulting projects ranged in form and scope. Carole Lung Francis and tailor Jonas Labaze, established "Made in Haiti" —fashions created from second-hand clothing donated by charities. Jesse Darling and neighborhood children and teenagers built "Trash Church,"

a venue constructed from discarded materials that hosted music concerts, exhibits, and film screenings. Partnering with a healthcare organization, Seitu Jones created and dispersed over 1,000 clay balls filled with floral and grass seeds to "green" the city. Local teenagers took on the role of foreign reporters for the performance piece "Tele Geto," turning a critical, magnified eye on the biennale and the "outsider lens" often focused on Haiti.

Each iteration of the Ghetto Biennale (2009, 2011, and 2013) has brought over $60,000 into the community—particularly significant as the neighborhood is a designated "red zone," avoided by foreign workers and NGOs. The events have sparked a number of ongoing collaborations, projects, and opportunities, including overseas exhibitions and residencies for over 15 members of Atis Rezistans. In addition to visiting artists and academics, the events have been popularly attended by local audiences and have drawn participation from artists across Haiti.

In fitting with Atis Rezistans' aesthetic of repurposed cultural detritus, the Ghetto Biennale is an act of appropriation. Taking on the cultural institution of the biennale, it bridges normally distinct worlds—the biennale and the ghetto—and in doing so, holds the potential to problematize and develop a greater understanding of both. Many foreign artists arrived in 2009 eager to critique established, Western-dominated art institutions, often through ephemeral practices, while many of their Haitian counterparts sought to gain access to these very institutions and find markets for selling their object-based work. The devastating earthquake of January 2010 added pressure to the second Ghetto Biennale as well as the conflation of the art event with needed charity. Concerns linger over how to create common ground, avoid an exoticizing treatment of host communities, and expand participation. Each iteration seeks strategies to learn from and address challenges from a shared theme, to a "lens free" setting, to the replacement of a concluding academic conference with a more inclusive, participatory congress.

The friction between worlds can create rawness and as globalization expands, this makes projects like the Ghetto Biennale that take on these sensitive and contentious boundaries with unflinching and continued commitment all the more needed.

呼叫的游行
Call Parade

艺术家：多个艺术家
地点：巴西圣保罗的多个站点
形式和材料：装置
时间：2012 年
委托人：Vivo 电信公司
推荐人：里奥·谭

2012 年，巴西电信公司 Vivo 赞助了一个公共艺术项目 "呼叫的游行"，动用了 100 名艺术家和 100 个公共电话亭，在圣保罗全市进行展览，促进保护并爱护城市公共电话，使城市风光更具生气。作为一种概念艺术，这直接受到 Cow Parade 的启发。该项目包括创造基于单一牛体模型的 "设计师" 艺术家们以及牛雕塑在公共场所的展示（截至目前为止，该雕塑已在超过 75 个城市中展示）。然而，"呼叫的游行" 不同于 Cow Parade，它关注现有公共设施的转变，而不是随机预先介绍雕塑形式，比如，牛。从这一方面来说，这是一个更加本土化的项目。

"呼叫的游行" 的建设始于一个公共呼吁建议。然后建议提交到评判委员会并入围到 100 候选名单。其中，100 个参与艺术家被要求以原始的方式，转变 "8 个艺术电路" 沿线的电话亭。纪实照片的调查显示了多重系列形象和抽象设计，覆盖多重主题。其中，一个电话亭图形外观以人类的大脑形式呈现，另一个看起来像变形、装有镜子的迪斯科球，却展示给观众一种抽象的花卉图案。许多其他的亭子表现了人类肖像、城市以及乡村景观。Call Parade 的各种主题可能被认作一个整体，来构成圣保罗市公共或社会 "想象" 的多重面。

对于场所营造来说，展览结果在短期内很成功，改变了圣保罗市公众的视觉空间，而且众多引人注目的电话亭，赢得了居民和游客的注意。对于赞助商、专员振兴废弃公共电话亭，增加公共使用率的目标来说，该项目完全证实了它的价值。通过 Vivo 的测量，公共电话亭的损坏率下降了 18%，而且公用电话增加了 15%。我们可能会解释说，电话亭的审美转换，导致了当地的公共场所和设施的参与成员有了情感投入。这些情感投入或附属品对于场所营造至关重要。因为他们激励个人和社区积极使用和爱护共享场所。

我们可能还要注意一些更加令人不安的事情，即电信公司赞助的社会想象表达方式表明，想象中资本和消费主义结合的不确定性。可以推测，公共电话亭的"振兴"促进了当代消费特点，消费信息和沟通的强烈性和显著性（通过手机、平板电脑和笔记本电脑）。公共付费电话的使用并没有减少移动电话的使用，而仅仅增强了移动电话的使用量，从而创建一个更具媒体渗透的城市组合。当然，这也符合项目赞助商的利益，而是否符合公共利益则又是另一回事了。

公共情感的维护
吴昉

鸟兽草木皆有情，环境与人类的"观照"源自情感的交流，在现代化的生活环境里，公共环境与公共设施同样具备情感交流的潜质。由 VIVO 电信公司赞助的"呼叫的游行"公共项目，精彩之处在于通过对常设公共设施进行二次构建，创意带来的意外感冲击到人们的惯有思维，从而使普通的、习以为常的、大家的东西，拥有了神奇而个性的审美效果，引导使用者和行人对其产生全新的情感投入。一个特立独行的电话亭，勾起人们使用与保护的热情，创意的天马行空，使关注它、使用它的人也显得风格独特。因为不同而引人注目，因为注目而产生情感投入，又或者，因为共同的个性选择，让使用同一个创意电话亭的人们产生交集。人与外界的情感关联，就在彼此信号的感应下维系一处。

Artist: Multiple Artists
Location: Multiple sites throughout São Paolo, Brazil
Media/Type: Installation
Date: 2012
Commissioner: Vivo
Researcher: Leon Tan

In 2012, Brazilian telecommunications company Vivo sponsored a public art project entitled Call Parade pairing 100 artists with 100 public phone booths to create a citywide exhibition in São Paolo, motivate preservation and care for the city's public phones and enliven the urban cityscape. As a concept, this was directly inspired by Cow Parade, a project that involved artists creating "designer" cow sculptures based on a single cow body template, and staging the cows in public sites (to date the cows have appeared in more than 75 cities). Call Parade differed from Cow Parade, however, in that it focused on the transformation of preexisting public amenities, and not on the random introduction of pre-determined sculptural forms, such as the cow. In this way, it was a much more localized project.

Call Parade began with a public call for proposals. Submissions were then shortlisted to 100 by a jury. The 100 participating artists were asked to transform phone booths along "eight artistic circuits" in an original way. A survey of documentary photographs reveals a diverse range of figurative and abstract designs covering a multiplicity of themes or subject matter. One phone booth took on the graphic appearance of an exposed human brain, another looked like a malformed, mirrored disco ball and yet another presented audiences with abstract floral designs. Many of the other booths dwelt on portraits of people and urban as well as rural landscapes. The various themes in Call Parade may be considered as a whole to constitute various facets of the São Paolo public or social "imaginary."

In placemaking terms, the resulting exhibition was successful in temporarily altering the visual spaces of São Paolo's publics, with numerous eye-catching phone booths capturing the attention of residents and visitors. In terms of the sponsor/commissioner's goals of revitalizing disused public phone booths and increasing public use, the project certainly proved its worth. By Vivo's own measurements, vandalizing of public phone booths dropped by 18% and payphone calls increased by 15%. We might interpret these measures to say that the aesthetic transformation of the phone booths generated emotional investment by members of the participating neighborhoods in local public spaces and amenities. These emotional investments, or attachments, are critical in placemaking because they motivate individuals and communities to actively use and care for shared spaces.

We might also note something more unsettling, namely, that the sponsorship of this expression of the social imaginary by a telecommunications corporation suggests a problematic coupling of capital and consumerism with this imaginary. One might speculate that the "revitalization" of public phones contributed to the kind of consumerism characteristic of the contemporary moment, the intense and conspicuous consumption of information and communication (through cell phones, tablets and laptops). The use of public payphones did not reduce the use of mobile phones, but simply augmented mobile phone use to create an even more media saturated urban assemblage. This is, of course, in the interests of the project sponsor. Whether it is ultimately in the public interest is another matter altogether.

欧亚大陆
Eurasia

欧亚大陆

欧亚大陆横跨欧洲和亚洲两大板块，形成经度跨域最广的一片大陆。在数千年的文化发展中，以尼罗河流域底格里斯－幼发拉底河流域、印度河流域和黄河流域为中心的古代文明，为城市形态在人类社会的变迁发展奠定了坚实的文化基础。

城市的存在归根结底是人的存在。人的行为活动在昼夜交替的时光里创造着不同的文明。人与人之间、人与城市之间的互动最有趣也最为生动，它勾勒了历史的轮廓，也呈现了生活的真谛。

生活在欧亚大陆的艺术家，看到的都市不仅是关联着人类的生活方式、社会活动，还有时间、空间、运动、城市构建等重要因素。他们执着于重绘城市空间的尺幅，以信任为系带改变人的思维方式，呼吁人与人之间那份珍贵的交流。在这一过程中，艺术家并不看重艺术本身最终的视觉效果，而是更重视这一过程中，人与人之间的那份默契。

他们常说：

"非常有幸，在这里遇见了你……"

"那之后，我最大的收获是认识了很多朋友……"

"因为在这里，我会得到更多……"

而当你看到他们为你所呈现的，你一定会明白他们的话。

（张尚志）

IMVG

Itinerario Muralistico de Vitoria-Gasteiz

艺术家：克里斯蒂娜·魏克麦斯特、维拉尼卡·魏克麦斯特
地点：西班牙阿拉瓦省，维多利亚—加斯特兹
形式和材料：壁画
时间：2012 年
委托人：壁画公众报
推荐人：里奥·谭

"IMVG"是一个公共艺术项目，于 2007 年由克里斯蒂娜·魏克麦斯特、维拉尼卡·魏克麦斯特推出，作为大型特定场所壁画合作项目的最后一件作品。这个项目在 2007—2012 年的夏季推出，共同努力来复兴西班牙阿拉瓦省的维多利亚 - 加斯特兹中世纪区。每年创建了一到三个纪念性和永久性壁画（大小为 150—200 平方米），参与者来自各行各业和各个年龄层次。在此期间一共创建了 10 幅壁画。

对魏克麦斯特的壁画来说，一个关键部分就是其参与模式。用他们的话来讲，就是"我们设计了一个新方法，使市民能够通过艺术和场所营造与他们的城市互动。"每一幅壁画是在六周内完成的，由艺术小组协调参与者以及联合设计的壁画的最后实施。其中一个艺术协调员观察到"最大的挑战就是协调。"这是因为"IMVG"具有相对开放的框架，并邀请了具有不同美感和社会背景的公众参与。通过该框架，迫使参与者寻找求同存异，并采取积极措施通过壁画取得共识。

艺术家把该项目定性为草根或由下而上的社会干预项目。但是，跟一些干预项目（如涂鸦）不一样的是，"IMVG"壁画获得了建筑业主、政策制定者和城市规划者的认可和同意。值得注意的是，这些合作伙伴致力于壁画项目，但是事先不知道壁画会是什么样的，"IMVG"在社区获得了极大的信任。正如魏克麦斯特指出"政策制定者和艺术家必须准备好放弃全面控制，这样市民能体验到在他们的都市风景中留下永久性标志的力量和责任。"

作为一个旨在建立社会凝聚力以及培养美化和复兴维多利亚 - 加斯特兹中世纪区的共同意识的项目，个体参与者能有在他们社区留下永久性标志的体验，这点很重要。对艺术家和参与者的采访视频表明，在一系列壁画项目中取得了这个目标。这些公共艺术和艺术创造是不同的人居住在一个地方、进行协同努力以及让社区变得更像一个家的方式。这向我们提供了有关创意场所营造的非常成功的例子。

 "IMVG" 另外一个关键的成分就在有关"画笔之队" (Brush Brigades) ——一个政府资助的年轻人就业项目的创意，这个项目使得主要来自职业学校和美术学校的学生能通过壁画创造获得兼职收入。尤其是，在这个项目中，一群艺术家在创意领域提供了一个有关就业和体验的创意框架。通过在六年内成功完成七幅壁画，在 2013 年把项目拓展到另外一个社区，很明显，"IMVG" 被证明是一个可持续的艺术项目，能取得艺术家有关社会转型的多个目标。

壁画中的文化共创

张尚志

 "维多利亚的壁画之路"是以壁画的创作形式来塑造城市的纪念性和标志性的永久作品，它的特殊在于其基于城市公众间信任的创作形式，以创作家之间的协调为纽带，共同完成不同界面的壁画创作。整个项目最大化的特点是让人参与体验并同时提供了一个相关就业的形式，而项目本身的可持续性，也推动了该项目的发展。现代城市中的发展渐渐脱离了人与人之间直接的面对面交流，而该项目使不同背景、不同年龄段的人以艺术的形式，共同参与到同一个创作项目中，以一种现场交流、协调的方式，增强公众的参与性，是该项目作品最大的成功之处。对于文化的再塑，该项目打破了对根源的追溯，而是珍惜现有的人文，以不同的文化背景的作品，相互交织展现创作家对同一城市不同的文化见解。就像文化的海洋，激荡着朵朵浪花。

Artist: Christina Werckmeister and Veronica Werckmeister
Location: Vitoria-Gasteiz, Alava, Spain
Media/Type: Mural
Date: 2012
Commissioner: Muralismo Publico
Researcher: Leon Tan

Itinerario Muralistico de Vitoria-Gasteiz (IMVG) is a public art project launched in 2007 by Christina and Veronica Werckmeister, culminating in the annual collaborative creation of large-scale, site-specific murals. The project took place over the summers between 2007-2012 as a communal effort to revitalize the Medieval Quarter of Vitoria-Gasteiz in Alava, Spain. It produced between one and three monumental and permanent murals (150-200 square meters in size) each year, and it engaged participants from varied ages and walks of life. A total of ten murals were created during the period in question.

For the Werckmeisters, a vital component of IMVG is its participatory model. In their words, "We are designing a new way for citizens to interact with their city through art and placemaking." Each mural is produced over six weeks, with an artistic team facilitating and coordinating negotiations between participants and the final execution of the collectively designed mural. One of the artist-facilitators observes, "The biggest challenge is the coordination." This is because IMVG has a relatively open framework that invites members of the public with very different aesthetic sensibilities and social backgrounds to participate. Within this framework, participants are compelled to invent ways of co-existing with their differences while actively achieving a shared vision through a mural.

The artists characterize the project as a grassroots or bottom-up social intervention. However, unlike some interventions such as

graffiti, IMVG murals are produced with the knowledge and consent of building owners, policy makers, and urban planners.

What is remarkable is that these partners commit to the murals without knowing in advance what designs will be, a significant measure of the trust that IMVG has gained in the community. As the Werckmeisters remark, "policy makers and artists must be prepared to forego total control so that citizens can experience the power and responsibility of leaving a permanent mark on their own urban landscape."

As a project that aims to build social cohesion and nurture a sense of greater collective investment in the beautification and revitalization of Vitoria-Gasteiz's Medieval Quarter, it is critical that individual participants have the experience of leaving a permanent mark in their neighborhood. Video interviews with artists and participants provide evidence that this was achieved in a number of the mural projects. These public acts of art-making are a means for groups of people to inhabit a place, to stake collective claims to various publics, and to make a neighborhood a home. They provide us with an especially successful example of creative placemaking.

Another compelling component of IMVG is its creation of so-called "Brush Brigades," a government-funded youth employment program that enables individuals, mainly from trade schools and fine arts programs, to earn a part-time income while engaged in the mural creation. Notably, it is a group of artists in this case who provide an innovative framework for employment and experience in the creative sector. With the successful completion of several murals over six years, and an extension of the program into another neighborhood in 2013, it's apparent that IMVG is proving to be a sustainable artistic program achieving several of the artists' goals of social transformation.

自由屋社区研讨会
Freehouse Neighborhood Workshop
Floating School

艺术家：珍妮·范·海思薇基
地点：荷兰鹿特丹南部
形式和材料：壁画
时间：2008 年至今
委托人：Freehouse 社区研讨会
推荐人：里奥·谭

在 1900 年，当荷兰南部建立码头时，Afrikaanderwijk 作为工人房屋的开发项目也同时展开。自 20 世纪 70 年代，这个地区涌入了大量来自土耳其、摩洛哥、列斯群岛等地的移民。Afrikaanderwijk 成为荷兰第一个多种族和多文化社区。正如欧洲许多其他地区一样，移民、全球化和限制工业化所产生的迅速社会转型导致了社会局势紧张和新的挑战。根据艺术家珍妮范 海思薇基，这个社区一个关键的问题就是 "很少有人感觉到和公共空间及公共领域紧密相连。" "自由屋社区研讨会" 是一个珍妮·范·海思薇基构思的项目，从 2009 年到现在，艺术家、设计师以及 Afrikaanderwijk 居民通力合作，通过一系列研讨会来 "挑战人们在室外空间方面扮演更加积极的角色。"

这些研讨会是多样化的，但是分享一定的共同特征。例如，它们是基于对社区已经存在的技能或人才的识别和提升，并且它们向具有不同技能的人才提供就业机会。研讨会还紧随 "保持当地" 的宗旨，尽可能支持社区供应商和商业。举例来讲，最成功的研讨会是基于居民和缝纫技巧以及有关材料、专业时尚设计师等知识之间的代理合作关系以及时装秀和零售制造服装系列。这些研讨会在 Freehouse 中举行，这之前是一个商店，具有独特的开放式店面。

这个项目一个突出的特点就是高水平专业技能和社区所谓 "业余爱好者" 之间的合作以及一系列此类企业的成功。到目前为止，研讨会与 50 多名设计师进行合作，并对巴黎时装周的时装秀和沙博博物馆的展览作出了贡献。"自由屋社区研讨会" 不仅刺激了当地优质文化制造，还使社区及其居民在经济上受益。这样一来，就能抵挡全球化的潮流，这股潮流造成了限制工业化以及将制造业从欧洲外包其他地区，如亚洲的发展中国家。

作为一个场所营造项目，毫无疑问这个项目成功转变了 Afrikaanderwijk。它还基于本地文化产业开发了针对都市复兴的可缩放模型，面对类似挑战的不同社区可以引进和采用这个模型。通过把社区发展成为优质时尚和纺织品生产以及目标设计基地，这向居民提供了一种他们可以引以为豪的共

享身份，而非被社会和种族矛盾或冲突打破的社会身份。就艺术目的而言，该项目帮助提高了对公共领域和公共活动的投资或投入。其成功至少部分归因于艺术家融入了本地社区、为识别社区技能和资源而进行的广泛研究以及作为相关干预措施的研讨会的构思和实施所需专业知识。

重返公共空间

张尚志

现代科技的迅速发展，不仅为我们的生活带来了便利，更缩短了人与人交流的距离。全球化加速扩散了人类的移民，同一地区也容纳了更多的文化居民，而究竟以怎样的方式能让来自不同地域的人去交流，去共同生活与发展，或许公共空间的出现是个很好的解决办法。

公共空间绝不是简单的一块场地，也不是简单的一个景观，它不能脱离人的参与而独立存在，所以 Freehouse 出现在公众面前，它以一种参与性空间，并强化了空间的随机性，让更多的人可以以不同的需求参与到空间中，去交流，去体验。本项目作为场所营造项目，成功的将公共空间改变为针对都市复兴的转换模型，面对"文化多义"的社区，"空间多义"应对而生。人与公共空间及公共领域不在因为文化不同而相互独立，而是紧密相连。

Artist: Jeanne van Heeswijk
Location: Afrikaanderwijk district, South Rotterdam, the Netherlands
Media/Type: Mural
Date: 2008-ongoing
Commissioner: Freehouse Neighborhood Workshop
Researcher: Leon Tan

When docks were built in South Holland in 1900, Afrikaanderwijk came into existence as a housing development for workers. Since the 1970s the area witnessed an influx of immigrants from Turkey, Morocco, and the Dutch Antilles, among other places. Afrikaanderwijk became one of the first multiethnic and multicultural neighborhoods in the Netherlands. As is the case in many other parts of Europe, the rapid transformation of society through immigration, globalization and deindustrialization created social tensions and new challenges. A critical issue in this neighborhood, according to the artist Jeanne van Heeswijk, is that "less people feel connected to the public space and the public domain." Freehouse Neighborhood Workshop is a project conceived by Heeswijk, and realized between 2009 to the present day with a community of artists, designers, and residents of Afrikaanderwijk, as a series of workshops "that challenge people to play a more active part with respect to the space outside."

The workshops are diverse but share certain commonalities. For instance, they are based on the identification and enhancement of skills or talents that already exist in the neighborhood, and they attempt to connect individuals with different skills to employment opportunities. Workshops also stick close to a "keep it local" ethos, supporting neighborhood suppliers and businesses whenever possible. The most successful workshops were based on brokering partnerships between residents with, for example, sewing skills

and knowledge of materials, and professional fashion designers, to produce clothing collections for the catwalk as well as for retail. These workshops were accommodated in the Freehouse, a former store with a distinctive open storefront.

An outstanding feature of this project is the high level of professional expertise brought into collaboration with so-called "amateurs" from the neighborhood, and the success of a number of these enterprises. To date, the workshops have worked with over 50 designers, including Jean Paul Gaultier, and contributed to catwalk and other presentations at Paris Fashion Week and the Chabot Museum. Freehouse Neighborhood Workshop not only stimulated local and high quality cultural production, it also economically benefitted the neighborhood and its residents. In this way, it works against strong currents of globalization responsible for deindustrialization and the outsourcing of manufacturing from Europe to, among other places, developing countries in Asia.

As a placemaking initiative, there is little doubt that this project is successfully transforming Afrikaanderwijk. It is also developing a scalable model for urban revitalization based on local cultural industries, a model that may well be exported and adapted to different neighborhoods facing similar challenges. By placing the neighborhood on the map as a location for high quality fashion and textile production as well as object design, it provides residents with a shared identity that they can be proud of, in place of a communal identity fractured by social and racial tensions or conflicts. In terms of the artistic intention, the project has helped to develop a greater investment in, or commitment to, the public domain and public activities. Its success may be attributable, at least in part, to the artists' immersion in the local community, the extensive research conducted to identify skills and resources in the neighborhood, and the expertise with which the workshops, as relational interventions, were conceived and executed.

女报童
Papergirl

艺术家：Aisha Ronniger
地点：德国柏林
形式和材料：互动行为
时间：2006—2010 年
委托人：艺术家
推荐人：里奥·谭

"女报童"始于 2006 年，是一项基于传播艺术的自委托项目，其形式是在公共场所向陌生人随意发放卷起来的包裹。这个项目的灵感来自于柏林对涂鸦所施加的严格法律规定。基于美国大萧条期间的报童形象，最初的想法只是收集艺术作品，把它们卷起来，在步调很快的公共活动中发放，在这个期间，参与者骑自行车穿过城市不同的地点，朝行人扔卷包。虽然一开始这是一个单个事件，但是在柏林每年重复，一直到 2010 年。

"女报童"发展成为基于一小组简单规则的开源项目。艺术家简介地说明了项目的精神："参与、对话、非商业化和冲动"。游戏规则如下：
1. 通过提交或散播艺术，任何人都可以参加；
2. 跟报纸不一样的是，"女报童"卷包括通过邮件或个人亲自提交的艺术作品，没有进行编辑或印刷；
3. 这个项目是完全非商业的；不能购买卷包，也不会送至订阅者；
4. 其散播尽可能做到随机，因此嵌入了骑自行车的快步骤活动。

自从把卷包以礼品的形式发放给公共区域的接受者之后，而非画在或裱糊在建筑上，"女报童"规避了对公共领域艺术表达的越来越严格的法律限制。它展示了可等同于法律体系的一种社会活动，有效地批判了法律，并同时保证了在公共场合艺术表达的延续。就项目对社区的影响而言，它提供了令人惊喜的际遇，能让普通人在日常生活中接触艺术。正如 Ronniger 指出"我想给他们惊喜，在日常生活中突然收到一份礼物。"在 2006 年至 2010 年期间，一共发放了 1 500 个卷包。这个项目证明非常受欢迎，拓展到包括纽约、旧金山、温哥华、多伦多、巴塞罗那、开普敦、伊斯坦布尔、布里斯托尔、加的夫和贝尔格莱德的姐妹项目。

这个项目的一个最显著的特征就是坚持完全非商业化。Ronniger 解释道"它需要保持独立，这样才不会被利用。指定的艺术作品不应该被出售，行动应是一个惊喜。给予的精神不应该被金钱的动机影响或污染，也不该被误用来为公司做广告。""女报童"复兴了 Marcel Mauss 所研究的礼品经济，这是一个排除金钱的礼品交换礼仪系统。这个概念和项目的极高受欢迎程

度表明由很多艺术家和公众希望看到对生活几乎所有方面无情的商业化之外的替代性选择，包括艺术和文化。通过向接收者发送艺术礼品，"女报童"提供了一个模型，目标明确的小型组织可以使用这个模型来参与积极的文化获得，而不需要依靠高预算或商业买卖，而是基于慷慨和分享。

赠予的惊喜

张尚志

语言汇聚了最丰富的情感表达，传递着人的感悟，而艺术作品则将这种精神上的感悟具象化，凝聚着作者对生活，对这个世界的一切热情。"女报童"作为一项基于传播艺术的自委托项目，将个人的作品收集起来，并以随机重返的形式传递给不同的人，对作品的作者来说，突破传统意义上的批量制作，而是以一种独一无二的存在性将自己的精神注入艺术作品中，这种艺术形式未经更多的编辑与修改，显示出更多的纯粹，传达着他自己独有的思维，而对收到作品的人来说，他会去感受作品的作者所传达的情感。这种未曾谋面的交流以一件作品的彼此交错，相互影响，将公众的距离在精神上拉拢在一起，对于作品的理解与猜测，给人更多的惊喜与遐想，从中感悟更多作者与作品之间的思想融合。

Artist: Aisha Ronniger
Location: Berlin, Germany
Media/Type: Interaction
Date: 2006-2012
Commissioner: Artist initiated
Researcher: Leon Tan

Papergirl began in 2006 as a self-commissioned project based on distributing art in the form of rolled up packages to random strangers in public. It was inspired by the constraints imposed by tightening laws on graffiti in Berlin. Based on the image of American paperboys of the Great Depression, the initial idea was simply to collect artworks, package them into rolls, and distribute them in a fast-paced public event during which participants would ride through different parts of the city throwing out rolls to passers-by. While it started off as a single event, it was repeated annually until 2010 in Berlin.

Papergirl developed as an open source project based on a small set of straightforward rules. The spirit of the project is characterized succinctly in the words of the artists, "participatory, analogue, non-commercial, and impulsive." The rules of the game are as follows:

1. Anyone may participate by submitting and/or distributing art,
2. Unlike newspapers, Papergirl rolls consist of artistic contributions submitted by mail or in person, and are neither edited nor printed,
3. The project is strictly non-commercial; the rolls cannot be bought, nor are they distributed to subscribers,
4. Distribution is, as far as possible, spontaneous, and hence embedded in the fast paced activity of bicycle riding.

Since the rolls are delivered as gifts to recipients in public, and not

painted or pasted onto buildings, Papergirl sidesteps intensifying legal restrictions on artistic expression in the public sphere. It demonstrates a kind of social creativity that might be equated with jurisprudence, effectively critiquing the law while ensuring the continuity of artistic expression in public. In terms of the impact of the project on the community, it provided surprising encounters that bring ordinary people into contact with art in their everyday lives. As Ronniger says, "I want to surprise them, going about their everyday lives and suddenly getting a present." Approximately 1,500 rolls were distributed between 2006 and 2010. The project proved so popular that it expanded to include sister projects in New York, San Francisco, Vancouver, Toronto, Barcelona, Cape Town, Istanbul, Bristol, Cardiff, and Belgrade.

One of the most significant features of this project is its insistence on remaining completely noncommercial. Ronniger explains, "It needs to stay independent in order not to be manipulated. The designated artworks should not be sold and the action should be a surprise. The spirit of giving should not be influenced or contaminated by monetary motifs or misused by advertising for any companies." Papergirl resurrects aspects of the gift economy studied by Marcel Mauss, a system of ritual exchanges of gifts excluding money. The massive popularity of the idea and the project suggests that perhaps there are many artists, as well as members of the public, who would like to see alternatives to the unrelenting commodification of nearly all aspects of life, including arts and culture. By delighting recipients with gifts of art, Papergirl provides a model for how small but motivated groups can engage in life-affirming artistic gestures that do not depend on huge budgets, nor on commercial transactions, but on generosity and sharing.

游击队公共邮箱
Partizaning Public Mailboxes

艺术家：游击队集体
地点：俄罗斯莫斯科多个地点
形式和材料：互动行为
时间：2012 年
委托人：艺术家
推荐人：里奥·谭

"游击队公共邮箱（PPM）"是一个自代理艺术项目，旨在通过公共邮箱获得促进对话以及集体思维。在 2012 年，集体在莫斯科郊区设立了 15 个邮箱，每个邮箱提出了有关城市挑战的本地体验以及对将来期许的问题。邮箱刷上了明亮的颜色，使用扎线带系在社区的柱子上。邮箱上会有这样的提示："写下你觉得你所在社区缺少的东西，把信投到这个邮箱里。我们会制作相关问题和希望的地图，我们会尝试满足您的愿望或者找到能满足您愿望的人。"通过征集匿名信，"游击队公共邮箱"收到了许多有关当地生活和社区困难的手写故事和手绘图片。

自我定义为新兴 DIY（自己动手做）文化和战略都市运动的一部分，"游击队公共邮箱"旨在刺激社会行动、唤起人们"从底层重组他们的城市"并分析激励和维持游击队都市再规划或"黑客行动"的过程。艺术家强调使用邮箱的类似物而非数字模式，以影响各种数字文化活跃分子之外的不同参与者。他们发现邮箱通常能吸引小孩和老人。这点很重要，因为"如老人小孩或移民等弱势群体很少能发声，但是却是最主要和最脆弱的群体。"

邮箱本身构成类似物的方法，PPM 的网站 http://eng.partizaning.org 记录了相关发现。扫描了来自邮箱的信件，并且上传了展示邮箱变迁的照片，包括涂鸦、贴纸以及在一些情况下被摧毁的。该网站可广泛普及艺术研究。对艺术家而言，最主要的就是通过刺激不同社区有关共享问题的对话，PPM 从上而下地激活了社会变革。例如，在 Mitino 区，"游击队公共邮箱"获得了市政府部门的支持，他们解决了在邮箱中发现的公共路灯失修或非法食品供应商等问题。

作为一个场所营造项目，"游击队公共邮箱"成功地使多个莫斯科郊区参与其中，促进识别特定场所的问题和期望，并且人性化邮箱附近的公共空间。人们感觉到可以随意在邮箱上涂写、贴上贴纸，以各种方式回应表达人类的需求。在越来越数字化的世界，PPM 延伸到数字鸿沟贫穷端的参与

者，这点值得赞赏。尤其引人注目的是，在它企图解决本地问题的草根方案的愿景中，考虑到项目位于前苏联，基于从上而下中央集权的现代指令性经济的诞生地。尤其明显的是，在一些例子中，通过市政府部门的参与，它导致了社区中问题的直接改善。

把你的愿望寄出去

张尚志

数字时代的便捷无人能否，划过指尖的风带着无尽的信息。轻敲键盘，就可以让网络另一端的人知道自己的存在。但是，即使是在高度便捷的信息时代，网络并没有像空气一样，弥漫世界的所有角落。"游击队公共邮箱（PPM）"从现实中，延展了网络信息所不能触及的领域，它给更多群体，如老人、小孩等数字鸿沟贫穷端的公众更多的参与机会。公共邮箱虽然是个很小的装置物，但作为收集信息的容纳箱，不受任何限制，扩大了公众的参与性，而该项目的这种延展性，不仅成功地召集了公众，更依据当地政治的特殊性，通过市政府，直接进行信息传递而解决社区中的问题。城市的发展已经与数字信息相依不离，但有时，数字永远无法取代传统。敏锐的艺术家发现了问题并付出了努力。

Artists: Partizaning Collective
Location: Various locations, Moscow, Russia
Media/Type: Interaction
Date: 2012
Commissioner: Artist initiated
Researcher: Leon Tan

Partizaning Public Mailboxes (PPM) is a self-commissioned artistic project focused on stimulating dialogue, collective thought, and action through the interface of publicly sited mailboxes. In 2012, the collective set up 15 mailboxes in outlying areas of Moscow, and each mailbox posed questions about local experiences of urban challenges and wishes for the future. The mailboxes were brightly painted and attached to posts in the community with zip-ties. The mailboxes posed prompts such as: "Write what is missing in your area and drop a message in the box. We will make a map of the problems and wishes, we will try to fulfill them or find someone who could do it better than we could." By soliciting anonymous mail, PPM received many handwritten stories and hand-drawn images of local life and neighborhood difficulties.

Self-identified as a part of emerging DIY culture and the tactical urbanism movement, PPM was interested in stimulating social action, inspiring people to "reorganize their city from the bottom up" and analyze the processes involved in motivating and sustaining guerilla urban re-planning or "hacking." The artists placed special emphasis on working with analogue rather than digital means with the mailboxes in order to reach different participants than those active in various digital cultures. They discovered that the mailboxes often attracted children and the elderly. This was seen as significant because "minorities like the elderly and kids or immigrants rarely

get a voice but are the most important and vulnerable groups."

While the mailboxes themselves constituted an analogue approach, PPM's website, http://eng.partizaning.org, documented the findings. Letters from the mailboxes were scanned, and photographs of the mailboxes showing their transformation through graffiti, stickers, and in some cases, damage, were uploaded. The website made the artistic research widely available. Most importantly for the artists, PPM activated social change from the bottom up by stimulating dialogues within different communities about shared problems. For example, in the district of Mitino, PPM was supported and promoted by municipal authorities who resolved local street problems—public lights and streets in disrepair or issues with illegal food vendors—as identified in the mailboxes.

As a placemaking initiative, PPM successfully engaged a variety of suburban communities in Moscow, facilitating the identification of site-specific problems and wishes, as well as humanizing the public spaces in the vicinity of the mailboxes. People felt free to write on the mailboxes, place stickers, and in various ways, to respond to the human need for expression. In an increasingly digitized world, PPM is commendable for reaching out to participants on the poor side of the so-called digital divide. It is particularly compelling in its vision of grassroots solutions to local problems, given the location of the project in the former Soviet Union, the birthplace of the modern command economy based on centralized top-down planning. Notably, in at least some instances, it led to direct improvements in the community through municipal authority engagement.

凯旋门
Kattenburger

艺术家：克里斯蒂娜·勒库、大卫·史密森
地点：荷兰阿姆斯特丹 Kattenburg 岛
形式和材料：互动行为，建筑
时间：2012 年
委托人：艺术家
推荐人：里奥·谭

有关"凯旋门"的故事比较特别。在 Wiseguys（表演者）都市艺术项目的管理支持下，这个项目是阿姆斯特丹市选择在 Kattenburg 岛建立的四个项目之一。这个项目旨在结合本地居民的故事，创建一个特定场所的、永久的、独立的拱门。项目准备包括由岛上老年人组成的社区研讨会以及在现场为拱门准备地基的建筑公司，直到最后获得市议会有关在公共空间设立纪念碑的批准。但是，在获得批准后数天内，由于政客的抱怨，被取消了。这个挫折引起了岛上居民、市政府部门、Wiseguys 管理团队和艺术家之间的一系列争论。这些意料之外的对话可以说是这个项目最有趣的一方面。

对艺术家 Kristina Leko 和 David Smithson 而言，深入参与到本地社区是他们实践的基础。他们的工作可以被定性为以关系或对话为导向，这意味着研讨会和对话是他们的艺术媒介。原项目的研讨会有 25 名"凯旋门"老年居民参加，他们合作书写小岛的历史。该历史包括港口工人家庭的故事，他们居住在这个岛上超过两个世纪，尤其是在 20 世纪 60 年代，这些家庭需要抗争 20 多年以继续住在岛上。所提议的拱门项目旨在庆祝工人阶级以及积极分子为社区所作的贡献。

项目的突然取消不仅让"凯旋门"居民很意外，也让专员和 Wiseguys 管理团队大吃一惊。项目参与者组织到市政府进行游说，坚持以一定的形式实现项目。在一年多的艰难协商后，提出了一个新计划把故事整合到已经存在的社区综合体框架或"中枢"。管理团队获得了城市住宅公司以及阿姆斯特丹艺术基金会的支持。尤其是，市政府资助了新项目。

"凯旋门"是集体创意的一个出色的例子，参与者的持续参与是值得赞赏的。它显示了社区如何通过与艺术家的紧密合作来应对意料之外的挑战，并与政客和市政府部门进行艰难的协商。社区对项目的坚持为选民取得了真正的成功，在这个项目中，在特定场所建立纪念碑。项目同时也具有诗意的

一方面。作为对小岛社会运动的几年，它本身也是通过社会运动实现的。

艺术家们觉得这个项目很成功，因为它不仅在逆境中实现了公共项目，还因为它在参与者和艺术家之间产生了持久的友谊。Leko 解释她是如何评估项目的成功的："如果我交了几个好朋友，那么艺术的质量就是良好的。就"凯旋门"而言，可以说我们通过这个项目交了很多朋友。这个友谊是基于艺术工作体验（制作和感知）的相互尊重。"从艺术理论角度来讲，这表明所发展的质量关系是对话干预的艺术作品质量的可靠指示，这是一个非常合理的说法。

岂止于艺术？

张尚志

"凯旋门"的成功已经远远超出了艺术项目本身最后的成果，在项目过程中，参与者之间的交流以及对他们所生活的这个岛屿文化的探索与塑造，从各个角度将公众的身影纳入项目中，也正是基于这种大范围公众的参与，即使是项目在完成后的短时间内遭遇终止，依然引起了一系列的争论。这些争论正是该项目成功的标志。项目参与者在项目完成过程中，因为彼此间的交流而使陌生的人逐渐熟悉并成为好友，公众因为艺术而走到一起，基于同一个平台而相互尊重，这对于艺术品最终的样子来说，超出了艺术的表象，从更高的视角诠释了公共艺术的真谛，生活在这座城市的人，每一个人都与这座城市有着密切的联系，共同构成了一个整体，不仅是你所看到的，更是精神上的融合。

Artists: Kristina Leko and David Smithson
Location: Kattenburg Island, Amsterdam, the Netherlands
Media/Type: Interaction 、Building
Date: 2012
Commissioner: City of Amstedam-Center and De Key Housing
Corporation, with support from Amsterdams Fonds voor de Kunst
Researcher: Leon Tan

The story of the Kattenburger Triumphal Arch is an unusual one.
The project was originally one of four commissions selected by the
Amsterdam City authorities with curatorial assistance from Wiseguys
Urban Art Projects, to be realized on the island of Kattenburg in
Amsterdam. The proposal was for the creation of a site-specific,
permanent, freestanding arch, incorporating the stories of local
residents. Project preparation, including community workshops with
a group of elderly residents of the island and an architectural firm
prepping the site for the arch's foundation, proceeded all the way to
the final stage of receiving approval from the city board for public
spaces and monuments. However, days after being approved, it was
cancelled because of a politician's complaint. This setback provoked
a series of contentious conversations between island residents, city
authorities, the Wiseguys curatorial team, and the artists. These
unexpected conversations are arguably the most interesting aspect
of the project.

For the artists, Kristina Leko and David Smithson, deep engagement
with the local community is fundamental to their practice. Their
work can be characterized as relationally or dialogically oriented,
meaning that workshops and conversations are their artistic
medium. Workshops for the original commission involved the
participation of 25 elderly Kattenburg residents who collaboratively
wrote texts narrating the island's history. This history included
stories of the families of harbor workers, who had inhabited the
island for over two centuries, particularly accounts of how these
families had to fight for over two decades to remain on the island,
after its redevelopment in the 1960s. The proposed arch was

intended as a celebration of the working-class roots and activist struggles of the community.

The sudden cancellation of the project was shocking not only to the Kattenburg inhabitants, but also to the commissioners and the Wiseguys curatorial team. The project's participants organized themselves to lobby the city administration, insisting on the realization of the project in some form. As the result of difficult negotiations lasting over a year, a new plan was proposed to integrate the narratives into the framework of an already-existing neighborhood complex or "centrum." The curatorial team secured financial support from the city housing corporation as well as from Amsterdam Art Funds. Notably, the city administration itself contributed funds towards the new project.

Kattenburger Triumphal Arch is an outstanding example of collective creativity; the level of sustained engagement of the participants is commendable. It shows how a community, in close partnership with artists, is able to respond to unexpected challenges and engage in difficult negotiations with politicians and city agencies. The community's claim on the project produced real outcomes for the constituency in question, in this case, the realization of a site-specific monument. The project also has a poetic dimension. As a monument to the island' s history of social activism, it realized itself through social activism.

The artists consider the project highly successful since it not only realized a public work despite adversity, but also because it resulted in enduring friendships between participants and the artists. Leko explains how she evaluates a project's success: "If I made several good friends, then the quality of the art is good. In the case of Kattenburg, I would say that we made quite a few friends. This friendship is a mutual respect based on the experience of the artwork (production and perception)." In art theoretical terms, this asserts that the quality of relationships developed is a reliable indicator of the artistic excellence of a dialogical intervention, a most reasonable claim.

工作室 /TRANS305
Atelier / TRANS305

艺术家：斯蒂芬·尚克兰
地点：法国巴黎塞纳河畔伊夫里
形式和材料：互动行为，建筑
时间：2010—2012 年
委托人：塞纳河畔伊夫里市，文化和城市规划部门
推荐人：茉希·凯科拉

在 2007 年，当斯蒂芬·尚克兰在巴黎的一个郊区的塞纳河畔伊夫里启动项目 "TRANS305" 时，前总统尼古拉·萨科齐启动了一个大范围的都市计划叫作 "大巴黎"，包括巴黎郊区在内，重塑其作为 21 世纪国际都市的形象。除了建筑项目外，该计划还包括改进基础设施、公共交通计划、开发经济中心改造郊区的社会和空间，并且重新定义其管理。这个计划在国家中央政府、当地和地区政府部门之间创造了紧张局势。

作为回应，斯蒂芬·尚克兰说服塞纳河畔伊夫里（被称为 "红色郊区"，因为它是法国共产党的历史要塞）的当选议员考虑这些大规模都市变革，以制定可以并入新城市的创新艺术和社会计划。由于郊区是巴黎地区的一部分，但是不是由自治机关管理的，斯蒂芬·尚克兰构思了 HQAC——"Haute Qualité Artistique et Culturelle"（高艺术和文化质量），作为一个原型整合到都市规划开发商的合同中。

在 2010 年，通过创立 "工作室 / TRANS305"，开始了这个项目，根据其声明，这是 "一个理想的工作场所，以雕塑的形式结合到现实中去。" 这是一个不朽的移动结构，可以举办辩论、展览、研究项目、研讨会、作为艺术家居所和放映电影。受周围建筑的建筑学和美感的影响，工作室匹配它的环境，并同时体现了决策过程的透明性和渗透性，因为工作室是一个建造场地。

"工作室 / TRANS305" 是一个通用的和可调整的项目。它先被拆除，然后在 2012 年以不同的形式，并通过不同的编程重新安装在社区。根据斯蒂芬·尚克兰，首先这是一个 "在建筑场所中间的小岛"。在第二次安装时，工作室和城市居民严格相关，因为它靠近即将成为城市最重要公共空间的地点，在由 2000 间公寓组成的大型社会房屋开发项目旁边。

城市的转型成为非常规艺术干预的来源和舞台：艺术家同拆迁公司合作，在现代遗迹中建立特定场地的作品；来自各个学校和大学的学生就新空间和机构的执行进行辩论；CNRS（Centre national de la recherche

scientifique，一个国家研究机构）的研究人员调查了"TRANS305"的环境和社会影响。

"TRANS305"可以被视为艺术"按地区分配"的一个模型，颠覆原来模式并阐释"将地区作为"控制和绅士化模式的连续性，通常包括三个主要时间步骤：一个地区的知识分子、象征性教派、物质欲望和结构"catalogation"（程序）。通过开始创造 HQAC，然后创造工作室，斯蒂芬尚克兰打断了"按地区分配"的程序，通过自己的程序创建了艺术社区，并注入已经存在的系统。颠覆是通过工作室平台的政治和社会功能实现的，这是一个公共空间，允许参与者和居民能全面地观察周围的环境，转变都市环境。在一个生成新的都市和社会愿景的展示中，斯蒂芬·尚克兰转变了全景，这是一个可以显示全貌的出色工具。

城市之舟
张尚志

城市空间根生于一方文脉，延续历史发展，最终形成生活形态。历史的车轮不会脱离未来的轨迹重回昔日的印象。"工作室／TRANS305"的出现打破了城市重返往昔的空间转换，而是以一种随机改变的功能空间，在城市中以点聚焦公众对空间的体验。这个空间并非是一成不变的、不可移动，而是像城市之舟一样，穿梭于城市的任意场所空间，以不同的功能、形态出现在公众面前。它的出现并非偶然，而是根据地区重新诠释地域精神、物质与结构，以自己的独有的程序构建艺术空间，并将整座城市形成一个公共空间的艺术系统，让公众参与进来并观察周围的环境，在新的空间体验与展示中，展示了城市特有的全貌。城市也因此继续随着历史的车轮，在拂晓中继续前进。

Artist: Stefan Shankland
Location: Ivry-sur-Seine, Paris, France
Media/Type: Interaction, Building
Date: 2010-2012
Commissioner: The City of Ivry-sur-Seine, Department of Culture and of Urban Planning
Researcher: Giusy Checola

In 2007, when Stefan Shankland initiated the project TRANS305 at the Ivry-sur-Seine commune in a banlieue (suburb) of Paris, the former president Nicolas Sarkozy launched an extensive urban plan called Le Grand Paris (The Greater Paris) in order to include the banlieue in the Paris area and remake its image as a global metropolis of the 21st century. In addition to architectural projects, the plan included revised infrastructure, public transportation plans, and the development of economic centers that would transform the suburb socially and spatially, and redefine its governance. This plan created tension between the central state, local, and regional authorities.

In response Shankland convinced the elected members of Ivry-sur-Seine (which is called red banlieue because it was a historical garrison of the French Communist Party) to consider these large-scale urban changes as an opportunity to create innovative artistic and social plans that would merge in the new city. Since the banlieue is part of the Paris region but is managed by autonomous authorities, Shankland conceived HQAC, Haute Qualité Artistique et Culturelle (High Artistic and Cultural Quality), as a prototype that he managed to integrate in the contracts of the urban plan's developers.

In 2010 Shankland began the project by creating Atelier / TRANS305, which, according to its manifesto, is "an ideal workspace integrated in reality as sculpture." It is a monumental, mobile structure that hosts debates, exhibitions, research programs, workshops, artist residencies, and screenings of films. Inspired by the architecture and aesthetics of the surrounding buildings the Atelier suits its

surroundings, while simultaneously embodying the transparency and the permeability of the process of decision-making because the Atelier is a construction yard.

The Atelier was a versatile and adaptable project. It was dismantled, and then re-installed in the neighborhood in 2012 in a different form and with different programming. According to Shankland, in the first phase it was "an island in the middle of a building site." In its second installation the Atelier was strictly related to the city's inhabitants, since it was sited near what would soon be the most important public space of the city, next to a huge social housing development consisting of 2,000 flats.

The city's transformation became both resource and stage for unusual artistic interventions: artists worked with demolition companies on the installation of site-specific works among the modern ruins; students of the schools and universities debated the execution of new spaces and institutions; the researchers of CNRS (Centre national de la recherche scientifique, a national agency for research) investigated the environmental and social impact of TRANS305.

TRANS305 can be considered a model of artistic "territorialization" that overturns the schema and interrupts the continuity of "making territory" as a form of control and gentrification, often composed by three main chronological steps: the intellectual/symbolic denomination of a district, the material occupation, and the structural "catalogation" (procedures). By starting with the creation of the HQAC and then creating the Atelier, Shankland interrupted the "territorialization" process and created an artistic community model with its own procedures injected in the pre-existing system. The subversion of terms emerges by way of the political and social function of the Atelier's terrace, a communal space that allows participants and inhabitants to have a complete view of the surrounding, shifting urban environment. Shankland transformed the panoramic vision, which is the panoptical tool par excellence, in a display that generates new urban and social visions.

在中心的身份
Identità al Centro

艺术家：思考的艺术
地点：意大利阿雷佐蒙泰瓦尔基历史中心
形式和材料：互动行为
时间：2010—2011 年
委托人：80% 来自托斯卡纳大区，20% 来自蒙泰瓦尔基市
推荐人：茱希·凯科拉

"在中心的身份"是一个于 2010 年在蒙泰瓦尔基开发的长期艺术项目，这是一个位于意大利托斯卡纳大区的、拥有约 25 000 居民的一个小镇。这个项目的开发通过当地政府和市民以及不同学科和实践之间的复杂互动，包括设计、空间符号学和地图学，因为对于基于公共艺术集体的社区而言，Artway of Thinking 指的是"如果没有思考，就不会有艺术。"

在 20 世纪 90 年代晚期，很多市民搬离蒙泰瓦尔基的历史中心，生活和工作在新的现代郊区。这导致了非领土化过程以及老房子情况的恶化，还有很多弱势群体和少数民族的再领土化，这导致了古老市中心失去了身份认同、社会分化以及感知的个人安全的缺乏。

Artway of Thinking 在新居民、郊区居民和当地政府之间建立了一个互信的平台，通过组织各种活动，如非正式的晚宴、观察、安装临时办事处收集住房问题以及举办研讨会分析在总体物理性能中所收集的数据。通过制图对历史中心空间的分离和重新设计，艺术家模拟了各代和种族分化，并重新确定了新多文化社区的审美范式。社区制作了三个集体地图：爱的地图，有关分享空间；公共服务地图，包括 66 个色卡形式的请求；审美地图，在参与者所感知的丑陋的地方收集美感。

除了提高对艺术作为社会变革基本价值的公共意识外，"在中心的身份"还在文化、社会、都市和金融领域取得了创新成果。在社会领域，它提高了市民的自我组织，他们选择了一些"外国人"作为市议会的代表。在都市领域，它允许在"2011 年战略城市规划"中包括市民的请求。在金融领域，文化成为当地市政战略的一个关键因素，在这个方面，蒙泰瓦尔基的公共行政部门获得了 3 000 万欧元的公共基金用于项目开发。

在文化层面，"在中心的身份"对当前有关意大利历史城市中心的辩论作出了巨大的贡献，因为大多数意大利的城市都是古老的。在二战前，这些城市被认为"保存了传统"，应予以保护，以建立 Bel Paese（美丽国家）的统一形象。目前，它们大多数出现在明信片上，并根据它们吸引游客和

观众的能力进行评价。关于城市中心的第二个框架获得了更加直接的金融成功，但是同时也导致了社会结构和空间的弱化，最终改变了空间身份和社区。

通过重新确定由社区互动所导致的历史中心感知，Artway of Thinking 揭示了多文化成分所增加的人类遗产的财富，将"公共"空间转化为"共同"空间。

通过人们所表达的新的象征和文化价值，"在中心的身份"的艺术家成功地从政治和视觉／情感表达方面阐释项目成果。最后，通过使用统计共识方法衡量社会参与的质量，艺术家不仅提高了当地的政治，还使多文化社区对市中心产生一种归属感，其特征是呈现不同的历史。

重拾美好的遗失

张尚志

城市发展加速了人口流动，本土文化的缺失，使市民的身份认同与社会感知出现了前所未有的缺乏。Artway of Thinking 的出现为当地居民与政府之间建立了互信平台，公众通过各种活动走到一起来，在活动中交流、研讨，并通过艺术的制图，对城市历史、文化、空间进行设计，重新诠释了城市在公众心目中应有的美感。

"在中心的身份"的成功在于对文化的重新塑造，将曾经拥有的、流失的身份认知在艺术中重新找回，对当下城市文化的建设再现精神信心。项目完成过程中，通过公共艺术的手法，以文化重塑崛起社会经济结构转型，在维护城市形象的基础上，重拾公众对自有文化的认知。

城市历史遗留的遗产，在艺术中将遗失的心灵重新拾起，在地图中再构归属情感，重现往昔美好。

Artist: Artway of Thinking
Location: Historical Center of Montevarchi, Arezzo, Italy
Media/Type: Interaction
Date: 2010-2011
Commissioner: 80% Tuscany Region and 20% City of Montevarchi
Researcher: Giusy Checola

IDENTITÀ AL CENTRO is a long-term artistic project developed in 2010 in Montevarchi, a town of about 25,000 inhabitants located in the Tuscany Region of Italy. The project was developed through a complex interaction between local administrators and citizens, as well as between different disciplines and practices, including design, semiotics of space, and cartography, since, for the community-based public art collective, Artway of Thinking, "there is no art without a way of thinking."

In the late 1990s, many citizens moved away from Montevarchi's historical center to live and work in the new, modern suburbs. This caused a process of deterritorialization and deterioration of old houses, as well as a corresponding process of reterritorialization by many disadvantaged people and ethnic groups, which brought a loss of identity of the ancient city center, social divisions, and the perceived lack of individual security.

Artway of Thinking built a platform of trust between the new residents, the suburban residents, and the local government, by organizing several activities such as informal dinners, practices of observation, the installation of a temporary office for the collection of housing problems, and workshops that resulted in the translation of the data collected in a collective physical performance. Through the cartographic practice of classification and redesign of the historical center's spaces, the artists stimulated the blur of generational and ethnic divisions, and redefined together the aesthetical paradigms of the new multicultural community. The community produced three collective maps: the Love Map, about shared spaces; a public services map, composed of more than 66 requests in form of colored cards; and an aesthetics map, that collected the beauty and the ugly perceived by the participants.

Besides increasing collective awareness of art as a fundamental value for social change, IDENTITÀ AL CENTRO achieved innovative results in cultural, social, urban and financial terms. In social terms, it increased the self-organization of citizens, who elected some "foreigners" as representatives to the City Council. In urban terms, it allowed the inclusion of the citizens' requests in the Strategic City Plan 2011. In financial terms, culture became a key factor in local civic strategies, for which the Public Administration of Montevarchi obtained 30 million euros of public funds to develop ideas that grew out of the project.

In cultural terms, IDENTITÀ AL CENTRO has contributed enormously to the current debate on Italian historical city centers, since almost all Italian cities are ancient. Before WWII they were mostly considered as "containers of heritage" to be protected for the creation of the unified image of the Bel Paese (Beautiful Country); currently they are perceived mostly as postcards, and judged in relation to their capacity to attract visitors and spectators. This second framework for a city center achieves more immediate financial results, but simultaneously contributes to the weakening of social structures and spatiality, ultimately changing the space's identity and communities.

By redefining the perception of the historical centre as result of the community interaction, Artway of Thinking reveals the treasure of human heritage increased by multicultural components, for the transformation of "public" spaces into "common" spaces.

The artists behind IDENTITÀ AL CENTRO succeeded in translating the results of the project both in political language and in visual/emotional expressions through the new symbolic and cultural values expressed by the people. Finally, by introducing the need for quality of social participation to be weighed against the statistical consensus-based approach, the artists enabled both better local politics and the creation of a sense of belonging for the multicultural community to the city center, characterized by the presence of a different history.

媒体学会
Ælia Media

艺术家：巴勃罗·埃尔格拉
地点：意大利博洛尼亚艾米利亚罗马涅区
形式和材料：互动行为，建筑
时间：2010—2011 年
委托人：艾米利亚罗马涅区地方政府立法议会
推荐人：茱希·凯科拉

"Ælia Media" 是一个临时的文化新闻机构和移动广播中心，由巴勃罗·埃尔格拉在博洛尼亚创建。这个项目赢得了首个"国际参与式艺术奖"。在 2011 年 10 月，它被安装在城市的公共广场和 Villa delle Rose—MAMbo 内的一个画廊、博洛尼亚的当代艺术博物馆。这个项目是在首相贝卢斯科尼领导的第四届政府的最后几个星期安装的，他是意大利媒体帝国的所有人，而这个帝国被右翼利益主导。

当 "Ælia Media" 在 Puntone 广场揭幕时，15 000 名愤怒的抗议者进行抗议（在罗马组织，针对银行和金融政治的抗议），被 Black Blocs 转变为都市游击抗议。这很快使国家警察以公共安全和社会管理的名义到达现场。首个针对抗议的特殊法律于 20 世纪 70 年代中期——所谓的"沉重年代"通过，这是武装斗争和警察恐怖主义时期，这在 1980 年终止，那一年，新法西斯主义者在一个火车站进行了博洛尼亚大屠杀（许多市民仍然认为这是一个未解决的悬案）。在那个时期，博洛尼亚具有最古老的西方大学，数百名国外留学生在这里就读，在这所大学诞生了另类电台以及地区讽刺、批判和草根思维的重要传统。

首先，Helguera 通过会见大量独立艺术家，对当前的社会和文化问题进行了深入调查，以理解如何促进有关公共问题以及艺术和媒体的当前社会—政治角色。"Ælia Media" 作为替代性的多媒体渠道，旨在复兴过去的创造力以及 20 世纪 70 年代外国人和新市民之间的知识共享。与此同时，它还支持新兴和实验艺术观点，向媒体制作提供技能，并在艺术社区和社会积极分子之间建立对话。

"Ælia Media" 现场直播是在一个透明的移动装置中进行的，被构思为"第三空间"，通过与当地主要独立电台的合作，如 Radio Cittá Fujiko、Radio Cittá del Capo 和 Radio Kairós，可以鼓励共同成长，在 1977 年拆除了最重要的电台——游记电台 Radio Alice 后，这个电台被保存了下来。根据 Helguera，它对大众媒体的讽刺批判可以跟境遇主义者和未来主义者的观点相媲美，从中衍生了著名的意大利谚语："幻想会破坏力量，大笑会将你埋葬。"

　　"Ælia Media"指的是"Aelia Laelia Crispis"，16世纪有名的神秘铭文，意思是"博洛尼亚之石"，可能是献给死去的爱人的，是解码研究的对象；也指的是"Radio Alice"，作为它现在的换位含义，在未解开的神秘事件、具有创造力的过去和当前的文化参与之间创造祖先联系。

　　通过参考意大利媒体的矛盾情形，"Ælia Media"既是对意大利民主现状的回应，也是对当前本地社会和文化需要的回应，根据Helguera，在这个空间，我们可以假装玩耍和重新思考。

比媒体更媒体
张尚志
　　"Ælia Media"作为一个移动的广播中心，它更具有其他媒体无法比及的对话权。由Pablo Helguera在博洛尼亚创建的这个项目最大限度的拉近媒体与随机事件的距离。电台在外观上拆除了遮挡性的表皮，而以裸露的骨架呈现在公众面前。

　　媒体的制作以一种艺术的形式和社会积极分子之间建立对话。而媒体自身，在空间中并非真实中的自我。在这个空间，你所见所想并非逻辑中的推理，也并非现实或是假设的命题，一切存在的可能均由表象覆盖真实，亦或是潜在的涌动扭转因果。思考，这一人类特有的行为，在这里开始显得变幻莫测，它与媒体的对话、交流开始变得苍白无力却不显卑微，因为它值得人人拥有。它的庸常转换为非凡，梦想成为现实，而当你转身回望，过去却不曾拥有。

Artist: Pablo Helguera
Location: Emilia Romagna, Bologna, Italy
Media/Type: Interaction, Building
Date: 2010-2011
Commissioner: Legislative Assembly of the Emilia-
Romagna Regional Government
Researcher: Giusy Checola

Ælia Media is a temporary cultural journalism institute and mobile broadcast center created in Bologna by Pablo Helguera. The project won the first International Award for Participatory Art. It was installed in October 2011 in a public square of the city and in Villa delle Rose, a gallery within MAMbo, Museum of Modern and Contemporary Art of Bologna. This installation took place during the last weeks of the fourth government lead by Prime Minister Berlusconi, the owner of the Italian media empire, which was dominated by right wing interests.

On the day of Ælia Media's opening in Piazza del Puntone (Puntone Square), a protest by 15,000 indignados (organized in Rome against the power of banks and financial politics) had been transformed by Black Blocs into an urban guerilla protest. This quickly led to the National Police's arrival in the name of public security and social control. The first special laws against protests were approved in the mid-1970s during the so-called "Years of Lead," years of armed struggle and political terrorism, that ended in 1980 with the Bologna Massacre at a railway station by neo-fascists (which many citizens still consider to be an unsolved case). During that time Bologna, which hosts the oldest Western university attended by hundreds of foreign students, saw the birth of an important tradition of alternative radio as well as satirical, critical, and grassroots thinking in the region.

First Helguera did a deep investigation of the current local social and cultural issues by meeting with a remarkable number of

independent artists in order to understand how to facilitate dialogue about common issues and the current social-political role of art and media. Ælia Media was organized as an alternative multimedia channel that aimed to revitalize past creativity and share knowledge of the 1970s among foreigners and new citizens. At the same time it supported emergent and experimental artistic ideas, provided technical skills for media production, and created an archive of dialogues between the arts community and social activists.

The Ælia Media live transmissions were broadcast from a transparent mobile structure conceived as a "third place," which could encourage collective growth, in collaboration with the main local independent radio stations Radio Cittá Fujiko, Radio Cittá del Capo, and Radio Kairós, which remained after the dismantling in 1977 of the most important station, the guerrilla station Radio Alice. Its critique of mass media through satirical approach is, according to Helguera, comparable to Situationists and Futurists, from which the famous Italian expression "fantasy will destroy power and a laughter will bury you" was derived.

The name "Ælia Media" refers both to Aelia Laelia Crispis, a famous cryptic epigraph of the 16th century that has been known as "the stone of Bologna," likely devoted to a dead loved one and the object of decodification studies; and to the name "Radio Alice" as its contemporary transposition, by creating an ancestral connection between the unsolved mystery, the creative past, and the art engagement of the present.

By referring to the Italian media's paradoxical situation, Ælia Media was both a response to the state of Italian democracy and to the local current social and cultural need for the creation of art as independent space, where, according with Helguera, we can pretend, play, and rethink things.

迷宫
Labyrinth

艺术家：何塞普·帕蒂拉·尹·维拉尔
地点：西班牙吉罗那阿尔赫拉格尔 A-26 附近
形式和材料：建筑
时间：2011 年至今
推荐人：Jo Farb Hernandez
推荐人：茱希·凯科拉

创作大规模公共艺术的艺术家一般都经过专业训练，了解主流文化的趋势，但何塞普·帕蒂拉·尹·维拉尔是个例外。他对艺术兴趣浓厚，自学成才，创造了一件不朽的作品，充分代表了他所在村庄社会特征。除了丢在临时捐赠箱里的硬币，他没有接收过任何经济担保、资助或资金支持。然而，在三项独立的巨型工程上工作了超过 45 年后，他完全激发了村民对艺术的热情，他们正积极地投身于对他的作品的维护，并称之为当地的象征。

帕蒂拉是个没有地产的退休工匠，他在别人的私有土地上建造他自己的作品。他的第一件建筑作品样式简单，坐落在加泰罗尼亚比利牛斯山的山麓。第二件作品变得更具艺术特色，奢华和富有层次感。由现场发现的树枝建造而成，到 2002 年他的第二件艺术作品包括七个 100 英尺高的塔、无数的桥梁、避难所、走道、楼梯间以及超过一英里长的"迷宫"。

帕蒂拉的艺术作品最初只是挑战传统社会的审美标准，但观众对他作品的喜爱最终促使他继续努力，他的作品成为他所在地方的象征。尽管独自一人工作，但凭借着技术实力，他用表面脆弱的材料建造了强健的、耸入云霄的尖塔以及优雅的通道和避难所。他严格自律，安心地调整适应变化、机遇、不完美和材料缺乏或充足，即兴创作并聚合各种偶然因素。他的复杂建筑作品不隶属于任何成文的策划书，这一点可能会使艺术和建筑历史学家大吃一惊。

他几十年如一日的劳作证明了该过程与他的艺术作品一样重要，他对组件布置再布置，直至他的眼里和脑里出现了现场建筑的实物形态。他喜爱创新、优雅和强烈的审美观，面对周围的山麓触景生情，他也渴望当地人和路人能参与到他的工作中来。每年有成千上万的游客经过他的工作现场，可以和帕蒂拉的建筑作品进行近距离接触使他们精神焕发，情绪高昂："感谢使我们像变回了孩子一样，"一张匿名纸条上这样写道。

2002 年，尽管来自全世界的抗议支持者请愿如潮，迫于当地政府和州政府的压力，帕蒂拉被迫毁掉了他的第一件壮观的公共项目作品。但是，他并不屈服，于是搬到附近的一个地方，重头再来。他的第三件不朽作品 (2007

—2011 年）外形优雅，结构极其复杂，体现了他益发成熟的审美观和提升的技术。为增强耐久性，他又开始研究石材、混凝土和钢筋，很快他的建筑作品就成为了世界上最大的艺术作品，包括八座塔，一条新的迷宫和很多动态雕塑和瀑布喷泉。

然而，2012 年，帕蒂拉再次被迫拆除和焚烧所有的木制组件。今天，帕蒂拉与群情激奋的村民及成千上万的来自全世界的支持者正开展斗争，试图拯救和保护剩下的作品。得益于创新性的设计，作品的宏大以及他工作几十年强大的韧性，尽管一再遭遇挑战，他的作品已经成为阿尔赫拉格尔的象征，同时也日益获得全世界的认可。

帕蒂拉在 77 岁时仍然坚持创作。即使一再遭遇当局的阻拦，他的作品仍然成为代表当地特性和增强地域感的源泉。如果他的工作能因其贡献而获得认可和尊重，也许这将扩大公共艺术作品的涵义。

舞动建筑的人
张尚志

何塞普·帕蒂拉·尹·维拉尔对艺术的执着使他摒弃了传统意义上建筑学的法则，而是以一个工匠的视角重新诠释了建筑的形式与审美。把帕蒂拉说成是舞动建筑的人并不为过，因为他的作品最初只存在于他的脑海，凭借自己的技术与执着，容纳变化、机遇、不完美和材料的缺乏或充足，使建筑盘踞大地之上，以层叠的视觉挑战吸引成千上万的人经游现场。尽管迫于政府的拆除与焚烧，依然不能抹灭帕蒂拉的激情，甚至，他的热情感染了来自全世界的支持者与他共同奋斗，不仅拯救剩下的作品，更参与到作品的扩大化中。帕蒂拉的作品成为当地地域感的标志性源泉，他对艺术的执着如同海洋的宽广感着每一个公众。如今，这一作品得到更多人的认可与尊重，在公共艺术的意义上，他的成功不言而喻。

Artist: Josep Pujiula I Vila
Location: Near A-26, Argelaguer, Girona, Spain
Media/Type: Building
Date: 2011-ongoing
Commissioner: None
Researcher: Jo Farb Hernandez

While most large-scale public art are created by academically trained artists cognizant of mainstream cultural trends, Josep Pujiula i Vila does not fit this category. He is a self-taught artist who, driven by personal passion, created a monumental artwork that has become central to the shared public identity of his village. He has received no underwriting, sponsorships, or funding besides coins dropped into an improvised donation box. Yet, working for over forty-five years on three separate, massive constructions, he has ignited the passion of the villagers, who are now working actively to preserve his artwork and reclaim it as emblematic of their locale.

Pujiula, a retired factory worker without his own land, built his structures on someone else's private property. Nestled among the foothills of the Catalan Pyrenees, his first, straightforward architectural installation evolved into a second, more artful and expansive multilevel construction project. Created from branches found on-site, by 2002 this second work of art included seven 100-foot towers, innumerable bridges, shelters, walkways, stairwells, and a labyrinth over a mile long.

Although Pujiula's artwork initially challenged traditional community aesthetic norms, visitors' enjoyment of his project ultimately encouraged his continued efforts, and his work became emblematic of their own sense of place. Working alone, with impressive technical prowess, he constructed powerful, soaring spires as well as graceful passageways and shelters with superficially fragile materials. Rigorously disciplined, he confidently adjusted to changes, opportunities, imperfections, and a lack or abundance of materials, improvising and integrating contingent elements. Art and architectural historians might be startled to learn that no formalized written plans ever existed for his elaborate constructions.

His decades of labor confirmed that the process was as important as the evolving artwork, and he arranged and rearranged components as his eye and intent constantly responded to the site's evolving physical form. Driven by an innovative, graceful, and powerful aesthetic, evocative of the surrounding foothills, he was also inspired by the desire to involve community members and passersby in his work. Tens of thousands of visitors passed through the site annually. Their ability to physically interact with Pujiula's constructions energized and emotionally impacted them: "Thanks for making us feel like children," read an anonymous note.

In 2002, Pujiula was forced to destroy the first iteration of his spectacular public project by the village government and state authorities, despite petitions from protesting supporters worldwide. Yet, undaunted by the demolition, Pujiula moved to a nearby site, and began again. The third, monumental version (2007-2011) was elegant and extremely complex, as evidenced by his maturing aesthetic sense and his sharpened technical skills. To enhance durability, he also began working in stone, concrete, and steel, and soon his constructions again had become one of the world's largest art environments, comprising eight towers, a new labyrinth, and numerous kinetic sculptures and cascading fountains.

Nevertheless, in 2012 Pujiula was forced yet again to dismantle and burn all of the wooden components. Today Puijula—along with the energized villagers and thousands of international supporters—is fighting to save and conserve what remains. His work has become the very symbol of Argelaguer, while at the same time it has gained increasing global importance, thanks to his innovative design, the monumentality of the construction, and his formidable tenacity, working for decades despite repeated challenges.

At age 77, Pujiula continues to build. Despite challenges from the authorities, his work has been a positive resource that serves to support community identity and enhance its sense of place. If his work were recognized and honored for its contributions, perhaps it would expand the very definition of what public artwork can be.

研讨会

De Werkplaatsen

艺术家：哈罗德·斯考滕
地点：荷兰 Leidschendam Prinsenhof
形式和材料：互动行为，建筑
时间：2012 年
委托人：住房协会委托
推荐人：里奥·谭

"研讨会"是 Wiseguys 城市艺术项目策划的社会介入项目。其中包括艺术家哈罗德·斯考滕、彼得·范·德·希伊登、里斯贝特·帕里森、马塞尔·伊登、沃尔特·范·布鲁克胡伊岑、斯里·范·德·尔德、卡洛琳·德·罗伊、萨尼亚·梅迪茨、马尔科·库布和罗丽格·维塞克斯·布尔，他们与当地居民合作，通过艺术，在 Leidschendam(南荷兰) 转变 Prinsenhof 社会住房综合措施。Prinsenhof 是一个战后住房发展的地方，由 16 处建筑和大约两千一百套公寓组成。2006 年至 2012 年之间，受到当地住房协会的委托，10 个艺术家与公寓居民一起重新设计每个建筑物的主入口。

从字面意义上来说，De Werkplaatsen 可以翻译为英语"研讨会"。事实上，这个项目是以研讨会为基础的。作为一个平台，研讨会使居民能够选择他们想要一起工作的艺术家，以便对他们的公寓大门进行新的设计。此外，它还确保团队艺术家和居民参与谈判、作决定并更新这些地方的实际任务。另外，哈罗德·斯考滕和 Mira Kho 促进一系列文化研讨会，使新居民之间相互接触，从而促进邻里社会的凝聚力。其中，有一个烹饪研讨会，即不同年龄、不同文化背景的美食们可以在这里一起准备和分享食物。另一个是基于新食谱共享和食谱创建的研讨会。另外，还有针对儿童及剧院的数字研讨会。

这个项目的记录影像显示，不同居民和艺术家群体完成设计、改造、映像以及审美等方面的多元化。仅这一点就表明，从场所营造首创性来说，本项目很成功，将以前统一的场所转换为视觉上由集体努力塑造的不同居住地方。总之，这个项目前景广阔，需要获得建设部门的同意以及居民们的密切合作才能实现。与研讨会结合入口的协作改造，也独具首创性，这涉及感官而非视觉。例如，品味。传言表明此项目的影响包括小规模街头犯罪率的减少，Prinsenhof 的当地居民会有更大的安全感。

伴随社会改善目标的浪潮，"研讨会"属于不断增长的参与式艺术项目，并且以改造专业性、转变居民和艺术家新关系而广获赞誉。事实上，它确保艰苦时期的住房机构融资 (Wooninvest in Leidschendam 和 later Vidomes)，对于项目的价值感知来说，这方面很引人注目。可以想象，"研讨会"为公民居民展示了艺术的相关性，表明艺术是不需要成为超级富豪的奢侈商品，而可以成为改善集体生活的工具。如果放大社会效用以及艺术的日常相关性的说明，在荷兰艺术资助方面，这将可能会在一定程度上扭转这一预算削减趋势。

艺术的背后
张尚志

艺术作品源自艺术家独有的视角，而当一件作品不仅是与艺术家有关系时，与之相关的事事物物都显得极为重要。本项目是基于同一个目标，艺术家与公众共同努力实现最终成果的一个成功例子。由于项目的特殊性，战后住房发展的地方，有时公众的体验会比艺术家更有发言权。在项目完成过程中，公众参与到其中，与艺术家共同研讨，以自身的生活经验与见解，在艺术家的帮助下共同创造属于自己的艺术项目。以这种模式进行的住房发展避免了不必要的浪费，因为建筑本身凝聚着公众的意愿，也得益于最初的研讨所达成的共识。艺术的成功，有时需要放下某一方的"独特"，参与与交流更能促成对成果的共识。我们的艺术是融合于公众普遍性的感官而存在的自然关系，顺应共通意识，我们的环境才会真正改善，而不会让人担忧。

Artist: Harold Schouten
Location: Prinsenhof, Leidschendam, the Netherlands
Media/Type: Interaction, Building
Date: 2012
Commissioner: Commissioned by Wooninvest and Vidomes
Housing Associations
Researcher: Leon Tan

De Werkplaatsen is a social intervention curated by Wiseguys Urban Art Projects, involving artists Harold Schouten, Peter van der Heijden, Liesbeth Pallesen, Marcel van Eeden, Walter van Broekhuizen, Siree van der Velde, Caroline de Roy, Sanja Medic, Marco Cops, and Marjet Wessels Boer working with residents to transform the Prinsenhof social housing complex in Leidschendam (South Netherlands) through art. The Prinsenhof is a post-war housing development consisting of 16 buildings and approximately 2,100 apartments. Commissioned by the local housing associations, the 10 artists worked with apartment residents to involve them in redesigning each buildings' main entrance, between the years 2006-2012.

De Werkplaatsen literally translates to 'workshops' in English, and the project is, in fact, workshop based. As a platform, the workshop enabled residents to choose the artists they wanted to work with to create a new design for the entranceways of their apartment block. It also allowed teams of artists and residents to engage in negotiations, decisions and the actual task of renovating these sites. Harold Schouten and Mira Kho additionally facilitated a series of cultural workshops, making possible new encounters between residents and thus contributing to social cohesion in the neighborhood. One of these was a cooking workshop where gastronomes of different ages and cultural backgrounds came together to prepare and share meals. Another was a workshop based on the sharing of new recipes and the creation of a cookbook. There was also a digital workshop for children as well as a theatre.

A survey of the documentary images for this project reveals an astonishing aesthetic diversity in the completed redesign and renovation, reflecting, perhaps, the diversity of aesthetic sensitivities in different resident and artist groups. This alone suggests that the project was successful as a placemaking initiative, transforming previously uniform spaces into visually distinct living places with identities shaped by collective efforts. The project was ambitious, requiring agreements with building authorities and the close cooperation of many residents for its realization. It was also innovative in combining the collaborative renovation of the entranceways with workshops involving senses other than the visual, for example, taste. Anecdotal accounts suggest that the after-effects of the project included a reduction in small-scale street crime, with residents reporting a greater sense of security in the Prinsenhof complex.

De Werkplaatsen belongs to a growing tide of participatory art projects with socially ameliorative goals, and is commendable for the professional execution of the renovations, and the brokering of new relationships between residents and artists, and between residents themselves. The fact that it secured financing from housing authorities (Wooninvest in Leidschendam and later Vidomes) in a time of arts funding austerity is also notable, speaking to the perceived value of the project. Conceivably, De Werkplaatsen demonstrated the relevance of art to citizens-residents, showing that art need not be a luxury commodity for the ultra-wealthy, but can be a tool to improve collective life. This demonstration of the social usefulness and everyday relevance of art, if amplified, might go some way towards reversing the trend of budget cuts to arts funding in the Netherlands.

露天电影院
Open Cinema

艺术家：玛利亚·赖瓦多斯卡
地点：葡萄牙 Guimarães
形式和材料：互动行为，建筑
时间：2012 年 10 月 20 日—2013 年 7 月 5 日，
 2013 年 9 月 12 日—10 月 20 日
委托人：Gabriela Vaz-Pinheiro, Reakt, Views and Processes
推荐人：里奥·谭

"露天电影院"是一个位于葡萄牙，Guimarães 市（欧洲文化之都，2012）内的艺术和建筑项目。该项目涉及到将两个不同的流程——建筑和艺术融合，从而创建一个场地定制的公共电影活动。具有 16 个电影院"吊舱"来容纳观众的可移动建筑是福涅尔建造的，而吊舱中放映的影片内容的设计由玛利亚·赖瓦多斯卡通过广泛的研究以及与参与者召开的专题会开发而成。

"露天电影院"的建筑形式为：一只具有 10 根黄色管子的钢箱，钢箱可升高 1.2 米。这些管子通向吸音软木塞内部，在听觉上与环境相隔绝，且装有一个屏幕和环绕声系统。每个吊舱每次最多可容纳三名观众，使观众能单独或与他人共同观看该项目。该建筑物，当完全被观众占据时，呈现出多脚型半人机器外观。该建筑物，坐落于 Largo Condessa do Juncal—Guimaraes 古镇的一个公共空间中，其每个管状开口的明亮颜色以及整体形式的陌生感，可供路人进行有趣的探索。

设计则建立在对当地工人阶级研究的基础上。玛利亚·赖瓦多斯卡对"Guimarães，作为文化之都，对工人在其日常生活中的意义"有着浓厚的兴趣。因此，她邀请了当地制造业（纺织和鞋）的工人参与放映的策展，也就是选择值得个人及其家庭纪念的一系列电影预告片。工人在午餐时间，在放映专题会中，通过投票选择预告片。因此，电影（诸如甜蜜生活、公民凯恩、外星人、现代启示录以及其他影片）的时长约为三分钟的预告片成为露天电影院的内容。

该项目，作为基于对现已存在的文化活动（如 Guimarães 的电影俱乐部）的理解而创建的临时设施和活动，成功地顺应了当前对电影和政治的狂热这一潮流，从而吸引了 18 000 名居民和游客（根据组织者的信息）。考虑到 Guimarães 自身的人口只有 50 000 多，18 000 名游客是一个相当

惊人的数量。毋庸置疑，对历史以及已有社会关系和动态的关注，促进了"露天电影院"作为场所营造举措以及与因全球化而理想破灭的社区合作这一决策（指的是将制造业外包给劳动力成本低的国家）作为深思熟虑的慷慨行为的成功。

对玛利亚·赖瓦多斯卡和福涅尔来说，"在当代商业和消费文化中，社会参与、开放和慷慨为制定艺术创作和参与的替代方法，提供了一个对立立场"。考虑到晚期资本主义的到来意味着新的人民阶层的出现—不稳定性无产者（其共同经验是经济不稳定，工作没有安全感以及生活环境不稳定），当前不可持续的晚期资本主义模式的社会组织和生活的替代方法的搜寻，与当前时事密切相关。"露天电影院"是来自截然不同的世界和环境中的小群体如何在不受产权规范约束，而是在慷慨和互惠这一逻辑上运作的临时领域中，在肯定艺术活动的生活中，聚集到一起的典范。

驻足回首

张尚志

"露天电影院"成功地将城市装置与多媒体结合到一起，带给人意想不到的趣味效果。在喧嚣嘈杂的现代都市，露天电影院以独特的造型，带给人外观上的神秘感，调动了公众的参与性。舱体的形态造型满载过往时空，将公众聚集在一起，在多媒体设计上，更是以艺术的形式基于当地地域文化倾向，以公众的意愿作为展放内容，它所包含的潜在信息，足以让我们去了解一个社会的风俗、习惯、历史。当你因为一丝好奇而俯首进入，在你未看到作品之前首先把你带入意境，先在空间氛围中与你沟通，再与你靠近，这是神秘、惊喜和发现的三部曲。"露天电影院"不必具有都市标志意义，却蕴含着某种探索精神的传递，有那么一刹那，我们驻足在同一个空间，有那么一刹那，我们共同回首不曾忘却的过去。

Artist: Marysia Lewandowska
Location: Guimarães, Portugal
Media/Type: Interaction , Building
Date: 2013.09.12-10.20 , 2012.10.20-2013.07.05
Commissioner: Gabriela Vaz-Pinheiro, Reakt, Views and Processes
Researcher: Leon Tan

Open Cinema was a collaborative project by Marysia Lewandowska and Colin Fournier, commissioned by Gabriela Vaz-Pinheiro for ReaKt, Views and Processes, a program of art and architecture in the city of Guimarães, Portugal (European Capital of Culture, 2012). The project involved the merging of two distinct processes, architectural and artistic, resulting in a site-specific, public cinematic event. While the portable architecture housing 16 cinema 'pods' for spectators was constructed by Fournier, the programming of the film content screened in the pods was developed by Lewandowska through extensive research as well as workshops with participants.

The architecture of Open Cinema took the form of a steel box elevated by 1.2m on four legs, housing 16 yellow tubes. The tubes opened into sound-absorbent cork interiors, acoustically isolated from the environment, and equipped with a screen and a surround sound system. Each pod accommodated up to three spectators at a time, allowing audiences to view the program individually or in intimate groups. Visually, the structure, when fully occupied by viewers, took on the appearance of a many-legged semi-human machine. Sited in Largo Condessa do Juncal, a popular public space in the old town of Guimarães, the bright coloration of each of the tubular openings, as well as the strangeness of the whole form, lent themselves to playful exploration by passersby.

The programming was based on research into local working classes. Lewandowska was interested in "what Guimarães, as Capital of Culture, meant to the workers in their daily lives." Consequently, she invited workers in local manufacturing industries (textiles and shoes)

to participate in the curation of the screenings, which turned out to be a series of trailers of films memorable to individuals and their families. Workers selected the trailers in screening workshops during lunch hours by voting. Trailers, of around three-minutes in length, for films such as La Dolce Vita, Citizen Kane, ET, and Apocalypse Now, among others, thus became the content for Open Cinema.

As a temporary installation and event based on understanding the meanings of pre-existing cultural initiatives (such as the Cine Club of Guimarães), the project successfully aligned itself with contemporary currents of enthusiasm for film and politics, thereby engaging up to 18,000 residents and visitors (according to information from the organizers) over its 75-day duration. 18,000 visitors is quite a remarkable number given that the population of Guimarães itself stands at little over 50,000. This attention to history and the importance of established social relations and dynamics no doubt contributed to the success of Open Cinema as a placemaking initiative, together with the decision made to work with communities disillusioned by globalization (meaning the outsourcing of manufacturing to countries with lower labor costs as a deliberate act of generosity.

For Lewandowska and Fournier, "Social participation, openness, and generosity offer a counter position in setting up alternative approaches to art production and participation, in a contemporary culture of commerce and consumption." The search for alternatives to unsustainable late capitalist modes of social organization and life is extremely topical, given that the advent of late capitalism has meant the emergence of a new class of peoples, a 'precariat' whose common experience is one of financial instability, job insecurity, and generally precarious existential circumstances. Open Cinema is an exemplary model of how small groups of people, from very different worlds and circumstances, may come together in a life affirming artistic event within a temporal territory that is not subject to the norms of property ownership, but functions rather on the logic of generosity and reciprocity.

彩虹公园
Rainbow Park

艺术家：亚当·卡利诺夫斯基
地点：英国伦敦南岸中心
形式和材料：互动行为，彩色沙石
时间：2012年7月16日—9月9日
委托人：凯茜·马杰与艺术活动供应商南岸中心
推荐人：里奥·谭

凯茜·马杰委任波兰雕塑家亚当·卡利诺夫斯基来设计"彩虹公园"，为2012年6月16日至9月9日于伦敦南岸中心举行的世界节作准备。该工程由150吨等级不同的彩色沙与纯色沙和约25个形似巨石的多彩雕塑像组成，它给混凝土构造的皇后步道（从伦敦眼到泰特现代美术馆之间的大道）带来了别样的艺术效果，顿时增添了几分活力，吸引路边行人驻足停留。

在为期三个月的世界节期间，该作品随着观众的参与而变色，卡利诺夫斯基将彩色沙沉积在混凝土人行道上，让人不禁回想起彩虹带。当观众走在沙上，经过有色区时，各色区连接的颜色混合起来，产生新的颜色。例如，红色区和蓝色区连接的地方混合产生紫色。

混合沙粒产生新的颜色，使人自然地回想起19世纪新印象派的实验。这些实验包括将颜色分离成的单个点，这些点又可以合成光，观察者通过认知、感性行为能够达到这种效果。然而，卡利诺夫斯基与19世纪的艺术家不同，他搬上舞台的是颜色混合与感知的协作实验。在这里，观众自然而然、不约而同地将艺术家个人当成了颜色点的组成部分。

"彩虹公园"是个小憩的好去处，富于想象力的所在。散布在沙砾上的巨石可以发挥座位区的功能，观众走过来休息片刻，可以暂时从他们的日常生活及感知习惯与思维方式中释放出来。这一对习惯的短暂中止放飞了想象力，并通过色彩联想产生了公共环境下的不同情感体验。卡利诺夫斯基在采访中提到，他对于"把色彩所强调的社会意义融入到真实世界中去"的兴趣由来已久。从这个意义上说，"彩虹公园"像他的其他作品如"自由实践Ⅰ、Ⅱ"一样，探讨了色彩在公共领域的社会意义和在情感价值。

卡利诺夫斯基也对人类认知及其与改变的关系兴趣浓厚。关于"彩虹公园"，他是这么说的："对这些作品的感知融合了作品元素的物理、触觉、视觉和肖像方面的意义，他们可能具有一些治疗功能。"他的观点是，彩色沙带来的环境变化唤起了特别（非习惯）感知，这些感知可能会导致观众社

会实体的积极改变。尽管特别感知及其后效应在持续期间难以追踪，许多纪实图片提供了充足的证据证明了当地人士通过在该项目现场休息、祈祷和玩耍参与了作品本身。

据南岸中心估计，大约有 780 万人参加了世界节并观赏了"彩虹公园"。卡利诺夫斯基估计参与该项目的积极参与者至少有 150 万。如果游客数量可以作为衡量项目成功与否的定量标准，那显然"彩虹公园"是很成功的。它体现了场所营造的创新力，毫无疑问，该项目在世界节期间成功地改变了现场的面貌，促进了观众积极参与社交祈祷活动。

"彩虹公园"是一个色彩在公共领域参与社会活动的大规模实验，吸引了超过一百万人参与到特别感知的体验，可以说是艺术家意图表达的成功。它提供了大量的可供分析的研究资料，即使用于此类分析的严谨的研究方法仍然还不健全。

都市遐想
张尚志

"彩虹公园"的出现无疑给冷静思索的都市带来一丝活跃。作品本身将不同的彩色沙拼接成线，放眼望去，就像穿梭在城市中的彩虹。在彩色沙中点缀的 25 个形似巨石的多彩雕塑更像是提示公众从忙碌的都市中跨越到梦幻般的彩虹中。如果说参与其中的公众人数是评判该项目成功与否的标准，那么"彩虹公园"是成功的。人们从不同的方向进入彩色沙，他们的身影模糊了不同色彩间的边界，更让自己融入缤纷的气氛中。在强烈的、活跃的气氛中，即使是匆匆而过的行人，也会驻足稍许让自己的心情在这纯净的色彩氛围中放纵一下，这是属于每个人内心的小小世界。"彩虹公园"的意义正如它所带来的感官震撼，当我们被它吸引并走近它的一刹那，我们会意识到，我们的生活就像其中的一粒彩砂，只有当我们在一起时，才会如此灿烂。

Artist: Adam Kalinowski
Location: Southbank Centre, London, United Kingdom
Media/Type: Interaction, Colored Sand
Date: 2012.07.16-09.09
Commissioner: Cathy Mager Curator and Arts Events Producer
Southbank Centre
Researcher: Leon Tan

Rainbow Park by Polish sculptor Adam Kalinowski was commissioned by Cathy Mager for the London based Southbank Centre's Festival of the World between June 16 and September 9, 2012. Consisting of a monumental installation made of 150 tons of colored and plain sand of different gradation, and 25 or so multi-colored sculptures resembling boulders, the project transformed the concrete atmosphere of Queen's Walk (a stretch that leads from the London Eye to Tate Modern), making the space at once more lively and attractive to passers-by.

Conceived as an installation that changes as a result of audience participation over the 3-month duration of the festival, the artist deposited colored sand over the concrete walkway in sections reminiscent of the bands of a rainbow. As the audience walked over the sand and played within and across the colored sections, they effectively mixed the colors across each section's edges, producing new colors. The mixing of red and blue, for example, produced violet along the edges.

The mixing of grains of sand to produce new colors naturally recalls nineteenth century experiments in pointillism and divisionism. These experiments involved separating colors into individual dots, which would be synthesized optically, i.e. through a cognitive/perceptual act by the viewer. Unlike the nineteenth century artists, however, Kalinowski staged a collaborative experiment in color mixing and perception. Here, the collective audience, albeit unintentionally, displaces the individual artist in the composition of the dots of color.

Rainbow Park was created as a place of respite and imagination. The boulders scattered across the sand thus serve as seating areas,

allowing participants moments of respite from their everyday routines as well as their habits of thinking and feeling. This temporary suspension of the habitual facilitates the imagination, and, through color association, the experience of different emotions within a public setting. In an interview, Kalinowski mentions his long-standing interest in color "put outside into the real world where the role of social meaning of a color is emphasized." In this sense, Rainbow Park, like the artist's other projects including The Practice of Freedom I & II, is an investigation into the social meanings and affective possibilities of color in the public sphere.

Kalinowski is also interested in human perception and its relation to change. Regarding Rainbow Park, he says, "The perception of these works is a fusion of physical, tactile, visual and iconological meanings of its elements, and they might have some therapeutic feature also." The idea is that the changing environment of colored sand provokes unordinary (not habitual) perceptions, and that these perceptions may lead to positive changes in the social reality of participants. While the duration of unusual perceptions and their after-effects are difficult to track, the many documentary images provide ample evidence of members of the community engaging with the artwork through rest, contemplation and play in the project site.

According to Southbank Centre estimates, approximately 7.8 million visited the Festival of the World and viewed Rainbow Park. The artist himself estimates the number of active participants in the project to be at least 1.5 million. If visitor numbers can be taken as a quantitative measure of a project's success, Rainbow Park was clearly a hit. As a placemaking initiative, there is little doubt that the project successfully transformed the site for the period of the festival, stimulating positive audience participation in social and contemplative activities.

As a large-scale experiment in social engagement with color in the public sphere, with over a million participants engaging in unordinary perceptual experiences, it would seem to be a success in terms of the artist's intentions. It provides a great deal of research material for analysis, even if appropriately rigorous methodologies for such analysis remain to be developed.

善举
Acts of Kindness

艺术家：迈克尔·兰迪
地点：英国伦敦地铁系统各站台
形式和材料：互动行为，印刷
时间：2011—2012 年
委托人：地铁艺术塔姆辛·狄龙，路易丝·柯欧西和凯茜·海恩斯
推荐人：里奥·谭

地铁艺术将迈克尔·兰迪在 2011 年的"善举"作为伦敦地铁乘客的公共艺术品。顾名思义，该项目包含分享伦敦地铁里不同地点的日常善举的故事。迈克尔·兰迪要求公众通过该项目网站提供个人经历的有关善举的故事。然后，他从收到的大约一千个故事里筛选出一部分，作为 2011 年至 2012 年沿中央线的展品。有几个站点被选中，一些中央线列车上也会展示展品。

迈克尔·兰迪的项目完全是出于个人兴趣而不是商业性的考虑，在这种情况下"善举"的价值是随机的，陌生人可以在匿名的公共场所如地铁列车车厢里自由付出他们的时间、努力和关心。"善举"的想法来自于 2001 年，那一次他用尽了个人财产。迈克尔·兰迪解释说："我感兴趣的是，在基本意义上，除了我们的经济特性之外，人之所以为人的东西；特别是当人们有经济上的烦恼，世界似乎充满困扰，我们都不知道该如何继续前行的时候。我想找出人之所以为人东西，在物质生活之外，如何进行联系。对我来，说答案就是同情心和善意。这个项目就是一种集中探索这一想法的途径。"

在浏览了项目网站上归档的众多故事后，他看到了一条贯穿许多故事的共同主线。陌生人对身边乘客的同情和关心感到惊喜，许多人在这些不求回报或非商业化的交流中找到了引自于其中一则故事的"团结"意识。这个项目使参与的观众意识到非商业价值的重要性，即基本的人类价值，如移情、团结和善良，更有当我们自己在只处理经济特性时往往会掩盖的那部分东西。

"善举"具有成功的场所营造创新力，它将人性和团结带入到在其他情况下疏远的公共环境里；它使伦敦的公共交通系统的站台和列车变得人性化起来，因为每年有大约 40 亿人次的乘客乘地铁穿过那条隧道。然后该项目的进一步举措更加成功，伦敦电影协会委任兰迪选取其中的一些故事制成动画微电影，叙述部分由提交故事的乘客和作者他们自己来完成。

该项目用一种富于同情心的方式对当代社会的商品化和普通人的私人化进行了批判。它强调了他们的生活特质，冒着被猖獗的物质主义所损害的风险，却对相随的个人主义和竞争意识不管不顾。反对不和谐与冷漠，该项目探讨了原始的善意，用迈克尔·兰迪的话来说，意思就是："我们相亲相近，我们是一家。"

传递无尽
张尚志

川流不息的城市交通，每天运载着成千上万的乘客，迈克尔·兰迪正是以伦敦地铁作为传媒的场所空间，通过视觉传达传递善举信息。作为一项公共艺术，迈克尔·兰迪在作品的素材收集形式上让更多的人将自己的或身边的故事集中起来，经过筛选重现于地铁公共空间。不仅是作品的形成过程让公众参与其中，更在后期，以视觉刺激的方式引起公众的注意。就信息传递而已，"善举"是可以讨论的。对于它所传达的信息，以不同的感动感染着每一个看到他的人，而城市中的人也被迫随着这份感动深思着这座城市存在的真正意义。以至于后来受其影响，从"善举"传媒中的故事改编微电影，增强了它的影响力度，形成无限传递的公共义举。

Artist: Michael Landy
Location: Various locations, London Underground Network, London, United Kingdom
Media/Type: Interaction, Print
Date: 2011-2012
Commissioner: Tamsin Dillon for Art on the Underground, curated by Louise Coysh and Cathy Haynes
Researcher: Leon Tan

Art on the Underground commissioned Michael Landy's Acts of Kindness in 2011 as a public artwork for commuters on the London Underground. As the title suggests, the project involved sharing stories of everyday acts of kindness in various locations in London's subway. The artist invited the public to submit stories of personal experiences of kindness through the project website. Subsequently, Landy made a selection from the approximately 1,000 stories received, for installation along the Central Line between 2011 and 2012. Several stations were involved, and stories were also included in a number of Central Line trains.

Landy's project grows out of a personal interest in non-commercial values, in this case, the value of random acts of kindness, in which complete strangers give freely of their time, effort, and caring to each other in anonymous public spaces like the carriage on an Underground train. The idea for Acts of Kindness came after Break Down (2001) where the artist destroyed all of his personal belongings. Landy explains,

I'm interested in what makes us human in a basic sense apart from our economic identities, especially at this time when people have lots of financial worries and the world seems to be a very troubled place and we don't really know how to move forward. I want to find out what makes us human, and what connects us, beyond material things. For me the answer is compassion and kindness. This project is a way of collectively exploring that idea.

When reviewing the numerous stories archived on the project website, one is struck by a common thread of surprise that runs through many of the narratives. Strangers were pleasantly surprised by acts of compassion and caring from fellow passengers, and many discovered in these free or non-commercial exchanges a sense of "solidarity" to quote from just one story. Audiences in this project were effectively reminded of the importance of non-commercial values, basic human values such as empathy, solidarity, and kindness, aspects of ourselves that tend to be overshadowed whenever we operate solely on the basis of our economic identities.

As a successful placemaking initiative, Acts of Kindness brings a sense of humanity and solidarity to an otherwise alienating public environment; it humanizes the public platforms and trains in London's mass transit system, for some of the 4 billion passengers that pass through the Tube annually. A further measure of this project's success is the subsequent commission Landy received from Film London to develop some of the stories collected into animated short-films, narrated by the authors themselves, those passengers who had submitted their stories.

The project is, in a compassionate way, a critique of the commodification of contemporary society and the privatization of the commons. It criticizes by highlighting those human dimensions of life that run the risk of being eroded by rampant materialism, with its attendant ideology of individualism and competition. Against divisive and alienating forces, this project explores the original sense of kindness, which in Landy's terms, means "we're kin, we're of one kind."

标志性的公共艺术项目
Landmark Public Art Program

策展人：盖纳·塞维
地点：爱尔兰朗纳湖
形式和材料：互动行为，装置
时间：2012 年
委托人：梅奥郡议会，Gaynor Seville
推荐人：里奥·谭

位于卡斯尔巴内朗纳湖一个未充分利用的自然地点中的"标志性的公共艺术项目"，由梅奥郡议会（爱尔兰）委派。"标志性的公共艺术项目"由盖纳·塞维开发和策展，包括两个场地定制的永久设施、临时项目，艺术家住处以及向新兴艺术家提供的补助。2010 年发布了项目简介，公共艺术项目的官方启动则开始于 2012 年 4 月。"LPAP"由百分比公共艺术计划室（Percent for Art）资助。

关于这两个永久设施的创建，On Sight（视野）被委派给克利里·康纳利，而 Landmarks 则被委派给伊莱恩·格里芬。On Sight 是一个位于湖周围，具有四根双筒观察柱的雕刻视频装置。尽管观察柱令人回忆起世界各地受欢迎的旅游景点所看到的那些装置，但它们为"LPAP"观众带来了一次截然不同的视觉体验。观众们，并没有看到预料之中的景观，而是看到了不同时间里他人参与不同活动的景观。

Landmarks 是一组小型雕塑，也位于湖泊周围。Griffin 进行了与该区域相关的广泛历史研究，除其他物体外，还为早前人们居住的朗纳湖周围的房子创建了一个微缩模型以及为这些人穿的传统鞋子创造了青铜复制品。和 Connolly 的作品相似，Griffin 的作品唤醒了这一地点的深厚历史，使观众在探索、重振朗纳湖的同时，经历非比寻常的体验。

"朗纳湖涟漪"和"Bridging Sounds"是委派的两个当代作品，分别由 Rob 和 Matt Vale 以及 Fionnuala Hanahoe 创作。"朗纳湖涟漪"是湖边的一个交互式光效应，而"Bridging Sounds"包括观众在穿过朗纳湖的新桥时所触发的声音装置。这两者都有效地改变了观众对湖泊的体验，激发了现场的现象和反思。

"LPAP"也包括湖泊自身内的"亲密的戏剧体验"：表演公司（艺术家），按汤姆·索芙特编写、路易斯·罗维执导的作品，通过船夫操控，用木船渡运三位乘客。该表演称为"穿过湖泊"，包括故事和歌曲，唤起人们对

穿越湖泊、通向彼岸这一古老传说的想象。

奖励给作曲家兰·威尔逊的住处，是与当地社区召开会议，记录在当前经济形势中他们挣扎的故事的所在之处。然后这些体验被翻译成音乐作品，向该社区演奏。

　　"LAPA"由于其设计的融贯性以及其与当地社区和场地的明确相关性，是值得称赞的。"LAPA"作为一个涉及到议会其他项目的场所营造举措，如建设一座新的桥梁，它对朗纳湖产生了兴趣，与之接触，从而激发人们重新思考该自然资源，将其作为提供前所未有的休闲、娱乐，享受和应用机遇的空间。"LPAP"为艺术参与城市和议会对振兴和重建项目的规划提供了一个有用的模式，说明了在从心理上和社会上将当地社区和被忽视的、未充分利用的空间联系起来中，该公共领域的艺术家如何发挥了关键作用。
　　"LPAP"作为行内的标杆，在2012年Allian Business to Arts奖项的两个分类中获得了好评——最佳试运行实践以及该社区最佳创意实践。

距 离
张尚志

景点设施总会显得乏味无力，而克利里·康纳利的视频装置则是重合了不同时空的人参与不同活动的景观，虽然外貌普通，却在参与中释放不同的体验兴奋感。"朗纳湖涟漪"则依靠现代科技的力量，创造了一个交互式光效应，在声与光的交错中，散开无尽的想象。"亲密的戏剧体验"也通过对文化的发掘，揭开传说中对神秘憧憬面纱，让人们的想象力进一步在自然体验中释放无限。"LPAP"为艺术参与城市和议会对振兴和重建项目的规划提供了一个有用的模式，空间的散落是公众在心理上和社会上常常忽略的元素，而"LPAP"的艺术，正是以艺术家敏锐的见解，通过不同的手法使之成为空间纽带，拉近自然与人的距离，最终发挥公共艺术在资源中的发酵作用，以弥补缺失的空间体验，还原我们与自然间最真实的触感。

Curator: Gaynor Seville
Location: LoughLannagh, Ireland
Media/Type: Interaction/ Installation
Date:2012
Commissioner: Mayo County Council, Developed and Curated by
Gaynor Seville
Researcher: Leon Tan

The Landmark Public Art Program (LPAP) was commissioned by
Mayo County Council (Ireland) for Lough (lake) Lannagh in Castlebar,
an underutilized nature site. Developed and curated by Gaynor
Seville, LPAP consisted of two permanent site-specific artworks,
temporary projects, an artist residency and bursaries for emerging
artists. Project briefs were released in 2010, and the official launch
of the public art program took place in April of 2012. LPAP was
financed by the Percent for Art scheme.

The two permanent commissions were awarded to Cleary Connolly
for On Sight, and Elaine Griffin for Landmarks. On sight is a
sculptural video-installation of four binocular viewing posts around
the lake. While the viewing posts are reminiscent of those found
in popular tourist attractions around the world, they presented
a very different visual experience to LPAP audiences. Instead of
seeing the landscape as would be expected, viewers instead saw the
landscape at a different time, populated with other people engaged
in different activities.

Landmarks is a set of small-scale sculptures, also sited around the
lake. Griffin conducted extensive historical research relating to the
area, and created, among other objects, a miniature model of the
kind of houses people actually inhabited by Lough Lannagh in
previous times, and a bronze replica of traditional shoes worn by
these people. Like Connolly's work, Griffin's evoked the deep history
of the site, enabling audiences to experience something out-of-the-
ordinary while exploring, and revitalizing, Lough Lannagh.

Lough Lannagh Ripples and Bridging Sounds were two of the

temporary commissions, created by Rob and Matt Vale and Fionnuala Hanahoe respectively. Lough Lannagh Ripples was an interactive light performance by the lakeside, while Bridging Sounds comprised a sound installation triggered by audiences when they crossed the new bridge at Lough Lannagh. Both effectively transformed the experiences of the lake by audiences, stimulating imagination and reflection on the site.

LPAP also included "an intimate theatrical experience" in the lake itself, with The Performance Corporation (artist) ferrying three passengers at a time in a wooden boat steered by a boatman performing a work written by Tom Swift and directed by Louise Lowe. Called Across the Lough, the performance included stories and songs, and evoked the ancient notion of crossing water as a passage to "the other side."

The residency, awarded to the composer Ian Wilson, involved meetings with local community members and documenting their stories of the struggles of the current economic situation. These experiences were then translated into a musical piece performed for the community.

Ambitious in scope and scale, LPAP is commendable for the coherence of the programming as well as its clear relevance to the local community and site. As a placemaking initiative that also involved other Council projects such as the building of a new bridge, it functioned to generate interest in, and engagement with, Lough Lannagh, facilitating a re-imagination of the natural resource as a space that offered previously unthought-of opportunities for leisure, recreation, enjoyment, and use. LPAP offers a useful model of artistic engagement for cities and councils planning revitalization and redevelopment programs, demonstrating how artists working in the public sphere can play a crucial role in connecting local communities psychologically and socially to neglected or underused spaces. As a measure of its excellence, it was highly commended in two categories in the 2012 Allianz Business to Arts Awards—Best Commissioning Practice and Best Use of Creativity in the Community.

低地
Lowlands

艺术家：苏珊·菲利浦斯
地点：英国格拉斯哥格拉斯哥大桥、加里东铁路大桥和乔治五世大桥
形式和材料：互动行为、装置
时间：2012 年 4 月 16 日—2012 年 5 月 3 日
委托人：格拉斯哥国际视觉艺术节
推荐人：里奥·谭

　　"低地"是 2010 年特纳奖获得者苏珊·菲利浦斯创作的大型声乐展品，在 2010 年的格拉斯哥国际视觉艺术节上展出。她第一次到现场参观时，注意到大桥栏杆上放置的花朵，那是对一名自杀者的匿名纪念。因此受到启发，她决定她的作品将取材于 16 世纪苏格兰民谣"远方的低地"，这首民谣有三个版本。苏珊·菲利浦斯解释说："三个版本都讲述一个溺死的女人变成幽灵后回来，对她再也不能与情人相守感到哀痛的故事。" 苏珊·菲利浦斯对她自己演唱的三个版本都录了音，然后在三座大桥展出，每个版本的扬声器都不同，这三座大桥是格拉斯哥桥 (1772 年建成)、加里东铁路大桥 (1905 年建成) 和乔治五世大桥 (1928 年建成)。

　　三个版本的录音开始部分相似，但慢慢地每个版本都有所变化，到后来不同版本的歌唱时不时地出现了重叠。然而，它们又只以一个声音收尾。观众可以看见大桥上的扬声器，但他们却很难直接识别声音的来源。这增强了单独歌唱的分离感，达到与歌曲主题共振的效果：溺水女人的灵魂脱离肉体，哀痛失去的爱。结合当代那些生活在大桥底下的人，苏珊·菲利浦斯是这么说道："不少人在这里从事毒品交易、吸毒和酗酒。"该作品的美感有些怪异，又使人难忘。

　　歌声未经训练却很悦耳不吵杂，经过大桥的桥拱后声音被放大，与水声和观众的声音一起，在现场错落有致，相得益彰，大桥使人不禁想起格拉斯哥 18、19 和 20 世纪那段总被文化记忆遗忘的历史。对观众来说，该作品勾起了他们个人有关失去或过去的爱的回忆，也使他们时常注意到阅历与生活本身的无常和变幻莫测。

　　说到场景营造的项目，低地充分展示了现场、其带有悲剧色彩的历史以及充满艰辛的当代生活环境的美学敏感性。在公共领域对声音介质的熟练应用，以及通过演绎 16 世纪的老歌把苏格兰的过去和现在捏合起来，这是很

值得称赞的。"低地"促使人们去反思和思考，它将观众的个人记忆、想象和情感以及三个历史现场联系了起来。这些联系在场景营造的过程中是很重要的，因为在此基础上个人和社会才能在同一个空间里获得更多，并对未来更加保持信心。

声与波的影子

张尚志

苏珊·菲利浦斯创作的大型声乐展品，是一件利用听觉设备所创作的广播传媒系统。三个版本的"远方的低地"在桥下的光影中，伴着泛泛的金属光波纹，混响着歌声。都市的繁华充满诱惑与险恶，生活在其中的人们想融入其中，却发现不得不面对的现实。

作品巧妙地利用环境所形成的光影，并以声音刺激感官，在音阶的鳞次栉比中，渲染历史的悲伤气氛，唤醒听众内心深处的感悟。这一切，都是站在现代社会都市的阴冷的角落，对人性的冷峻反思，它给人带来的空间体验是暂时的平静，在凝想中，将对历史的思绪在水纹中轻轻地展开，在歌声的波动下，让历史的轮廓重现眼前。这些联系在场景的营造中交相呼应，带给每一个参与者不可磨灭的映象。

Artist: Susan Philipsz
Location: Glasgow Bridge, the Caledonian Railway bridge and the George V Bridge, Glasgow, United Kingdom
Media/Type: Interaction , Installation
Date: 2010.04.16-05.03
Commissioner: Glasgow International Festival of Visual Art
Researcher: Leon Tan

Lowlands is a large-scale sound installation by 2010 Turner Prize winner Susan Philipsz, commissioned for the Glasgow International Festival of Visual Art in 2010 (April 16-May 3). On her first site visit, the artist noticed flowers on the rails of the bridge, an anonymous memorial to a suicide. Taking this as a cue, Philipsz decided to base her work on a sixteenth century Scottish ballad "Lowlands Away," of which there are three versions. According to the artist, "Each version tells the story of a drowned woman returning as a ghost to mourn the fact that she will never be with her lover again." Philipsz created recordings of herself singing each version, and installed the recordings, each coming from a different speaker, at three bridges, the Glasgow Bridge (completed in 1772), the Caledonian Railway Bridge (completed in 1905) and the George V Bridge (completed in 1928).

The recordings began similarly, but each changed over the duration such that the different versions of the song overlapped at times. However, they always ended with a single voice. Despite the visibility of the speakers at the bridges, it was difficult for audiences to directly identify the source of the sound. This would have had the effect of amplifying the sense of disembodiment of the singing voice(s), an effect resonant with the song's theme: the disembodied spirit of a drowned woman mourning the loss of her love. Combined with the contemporary conditions of those living under the bridges,

of which the artist says, "There's lots of drug dealing and glue sniffing and Buckfast drinking here," the aesthetic sense of the work might well be characterized as eerie or haunting.

The untrained but tuneful voice without a body, magnified by the acoustic qualities of the bridges with their arches, not to mention the sound of water and the sight of its passage, was well aligned with the site, insofar as the bridges reflect aspects of Glasgow's history over the eighteenth, nineteenth and twentieth centuries, a history prone to the forgetfulness of cultural memory. For audiences, the work likely evoked personal memories as well, of lost or past loves, perhaps also serving as a reminder of the transience and unpredictability of experience and of life itself.

As far as placemaking projects are concerned, Lowlands is a work that demonstrates aesthetic sensitivity to the sites, to their sometimes-tragic histories as well as to their contemporary circumstances as a kind of seedy underbelly of everyday life. It is commendable for its skillful use of the medium of the voice in the public sphere, and for bringing together the Scottish past and present through the resurrection of a sixteenth century song. Stimulating quiet reflection and contemplation, it forged connections between the personal memories, imaginations and emotions of audiences and three historical sites. These connections are important in any placemaking endeavor, since they are the foundation upon which individuals and communities can invest themselves more in a space and commit themselves more to its future.

他处
Elsewhere

艺术家：丹宁·鲁伊兹·海特莱斯
地点：瑞典马尔默中央车站
形式和材料：互动行为，装置
时间：2010 年
委托人：瑞典国家公共艺术委员会和瑞典交通署
推荐人：杰西卡·菲亚拉

"他处"具有两个交叉点，一个针对临近沉浸式环境的内部关注，从一点朝向更大网络和地区的外部视角。地方是一个含有独特结构、质量和社区的容器，同时设定了关系——跟所有不存在的"他处"相比，这里是存在的。丹宁·鲁伊兹·海特莱斯的"他处"把"这里"和"他处"之间的关系带到镜头前，要求观众在广泛的世界中加入个人的位置。

"他处"把瑞典地下马尔默中央车站的混凝土墙转变为火车的车窗。覆盖两边 180 米长的车站，46 投影"车窗"展示了 1—12 分钟驶过的城市风景、乡村风景、工业风景和城市郊区，在三年制作期间在全球取景。一共 1 500 序列、90 小时的连续镜头，将这些小片断连接到负责的软件程序，并根据特别设计的语法进行连续录制。因此，观众可以观看到不断变化的序列组合，与车站另外一端的切换部分并列放置。

这些影片没有展示主要城市的经典明信片风景，而是通常被忽视的匿名地方。这是一个秘密，一个游戏，也是问题的开放式结尾的起点。受 19 世纪晚期全景图像、火车旅行和新生电影的影响，"他处"展示了旅行本身这一行为，尤其是哥本哈根飞机场和 87 000 日常通勤者提供服务的车站。一方面，画面跨越空间联合在一起，展示了一种"普遍的存在"在各个大洲同时展开。但是，这个项目同时也显得遥远——强调一个人在广大的世界中微小的位置。轻松流动的图像通过新技术反映了对空间的同时挤压，但是这个看上去的轻松程度也提出了批判——暗示看不见的障碍以及旅行和特权之间的联系。随着软件在车站墙壁的任意一端展现新的并列，它通过不同和相似的频谱循环，从个人、哲学和地理政治学角度做出阐释。

"他处"跟地点紧密相关，同时也与时间相关。平均每四分钟，会向等待的行人展开拍摄的序列，暗指行人自己的旅行和车站本身——各个地点之间的地点。随着新技术发展，所有时间转化为"有用的"时间和"无用的"等待时间，它们自相矛盾地扮演一个独特的角色。这是自由设置的时间，人们在这个时间无法做什么，从这个角度来看，它提供了一个潜在思考的

机遇。这在马尔默的中央车站更加有说服力，在这里，建筑项目提供了一个没有广告的空间，包括"他处"的镜头。不是被广告轰炸，向乘客展现了"他处"具有诗意的、唤起回忆的和偶尔有趣的画面。广告可能提供一个结尾——对人生问题的答案，而"他处"展示了一个开始，一个对话的开始，一个没有结束的探询。

作为一个场所营造项目，"他处"既提高也反映了它的场所。作为车站设计和安装的一部分，它被盛赞为"欧洲最大的永久视频艺术作品"，并把冰冷的水泥土转变为开放的风景。它暗示技术变迁，从全景装置到汽车、因特网和高速列车，邀请观众享受、思考、质疑和考虑历史和现在无数"他处"之间的关系。它没有把"世界"放在"其他地方"，但是同时既远又近、相互联系的和当前的。在这么做的过程中，"他处"提供了有关"地点"这一概念的新体验，既是独特的又是广阔的。

光影旅行
张尚志

也许，没有人会在意各种图像对生活的意义，对图像的感知模式与状态总在不经意间被忽略。事实上，图像是一种潜在的流动的力量，他可以给人的内心留下最深刻的情感记忆，也可以构成思维与意识的本源。作者将不同地方的影像汇聚起来，以一种旅行的编排方式展现在观众面前，而作者所选取的极为平凡的影像也构成了"他处"的另类计划。光影记录的影像同时在时间中交叉相错，将富有哲学气息的"有用的"时间和"无用的"等待时间转换为同一个矛盾体。当你站在这闪动的影像前，身体和心灵同时开始游离，在天空与大地间不断地进行虚虚实实的切换、变化，当灵魂像挣脱了线的束缚在空中飘扬的风筝时，那种轻盈的飘零仿佛珍重回归释然。这也是艺术家有关"地点"的一种全新诠释。

Artist: Tania Ruiz Gutiérrez
Location: Malmö Central Station, Sweden
Media/Type: Interaction , Installation
Date: 2010
Commissioner: National Public Art Council of Sweden and
Trafikverket
Researcher: Jessica Fiala

"Place" entails a dual-pronged focus—both an inward attention toward an immediate, immersive environment, and outward view from one node toward larger networks and regions. Place is a vessel containing unique structures, qualities, and communities and also a set of relationships—this place as it exists in comparison and contrast to all the "elsewheres" it is not. Tania Ruiz Gutiérrez's Elsewhere (2010) brings this dynamic relationship between here and out there to the fore, inviting viewers to engage with their own personal positions within an expansive world.

Elsewhere turns the stark concrete walls of Sweden's underground Malmö Central Station into the windows of a train. Covering both sides of a 180-meter stretch of the station, 46 projected "windows" present 1-12 minute films of passing cityscapes, countrysides, industrial landscapes and city outskirts, filmed internationally during three years of production. Totaling 1,500 sequences and 90 hours of footage, these vignettes are connected to a complex software program and are continuously reordered according to a specially designed grammar. Viewers are therefore met with an ever-changing combination of sequences flowing one after the other and juxtaposed with shifting counterparts on the other side of the station.

Rather than iconic postcard images of major cities, these films feature anonymous spaces often overlooked or passed through. There is a mystery here, and a game, as well as an open-ended starting point for questions. Inspired by late 19th century panoramas, train travel, and nascent cinema, Elsewhere showcases the act of travel itself, particularly relevant for a station serving Copenhagen airport and 87,000 daily commuters. On the one hand, the images unite across space, indicating a "universal present"

that unfolds concurrently across continents. But the work also distances—heightening one's small position within a vast world. The leisurely flow of images reflects the contemporary compression of space through new technology, but this seeming ease is also laid bare for critique—hinting at unseen barriers and the connection between travel and privilege. As the software curates new juxtapositions on either side of the station walls, it cycles through a spectrum of differences and similarities, offering interpretations ranging from personal and philosophical to geopolitical.

Elsewhere is significantly tied to place, but it is also importantly connected to time. Averaging four minutes in length, the filmed sequences unfold for waiting passengers, referencing the passengers' own travel as well as the station itself—the place between places. As new technology increasingly enables the transformation of all time into "productive" time, the "unproductive" time of waiting paradoxically serves a unique role. It is time set free, when one can do nothing and, in this respect, it potentially provides an opening for contemplation. This is even more potent in Malmö Central Station where a building initiative creates an advertising-free space, including Elsewhere's footage. Instead of being bombarded by ads, riders are presented with the poetic, evocative, and at times playful images of Elsewhere. Whereas ads might offer closure—an answer to life's problems—Elsewhere presents an opening, the start of a conversation, an unending inquiry.

As a placemaking project, Elsewhere both enhances and reflects upon its site. Designed and installed as part of the construction of the station, it has been heralded as the "largest permanent video artwork in Europe" and transforms cold concrete into open landscapes. It alludes to varied technological shifts from panoramas and locomotives to the internet and high speed trains, inviting viewers to enjoy, ponder, question, and consider relationships to myriad "elsewheres" historic and current. It positions "the world" not "somewhere else," but rather simultaneously near and distant, interconnected and present. In doing so, Elsewhere offers a renewed experience of the concept of "place" as both distinct and expansive, an apt installation for a place of transit.

反思人类的能量
Rethinking Human Energies

艺术家：亚历山大·卡尔博尼
地点：意大利 Guilmi 公共场所
形式和材料：互动行为
时间：2010 年
委托人：有关艺术家
推荐人：茱希·凯科拉

小村庄 Guilmi 坐落于意大利中南部阿布鲁佐乡村地区的一座小山上。这里人烟稀少，人们很难估计有多少人居住。2010 年，多学科艺术家和研究者亚历山大·卡尔博尼抵达 Guilmi 时，大约有 100 个居民，过去两年来没有孩子出生。Guilmi 似乎将要绝后了。代际非领土化导致许多建筑物衰落，当地农业经济也随之解体。亚历山大·卡尔博尼决定探索居民区 Guilmi，他通过潜在的身体运动感知，将人、村子和景观作为单一能量系统的一部分，组成同心圆，建立与他们的物理和概念化关系。

"反思人类的能量"是名为 Overlapping Discrete Boundaries 项目的一部分，于 2009 年在亚洲启动。在这个项目中，亚历山大·卡尔博尼探索了 12 个城市的紧张情况，它们远离 Guilmi 并且截然不同。然而，它们却有共同的文化和地质问题。由于自然和社会因素，如频繁的地震和就业机会的缺乏，这些问题包括整个社区的地理定位。

受到人类学家 Marc Augé 和人口统计学家 Hervé Le Bras 的启发，亚历山大·卡尔博尼反映大型城市中心和小乡村关系的提升。尽管农村城镇人口减少，但全球网络可以创造城市和乡村社区之间的联系。

该项目的开发分为三个阶段：展览照片、视频、地图和图纸，这些在与河源头的当地人互动之后产生，80 个家庭捐赠的 80 个灯运用于公共广场作为最终表现，将灯和摄影图像返还给捐赠者，作为人类和物质能量交换的证据。

亚洲大都市城市生活流动和开放性很大。但是在 Guilmi，夜晚很寂静，白天也只能看到老人坐着观望生活。城市人口流动唯一失去控制的时候是在停电期间，这是亚历山大·卡尔博尼在 Guilmi 期间使用的案例。然后他为社区提供光照 40 分钟。总之，他创造了一个全球城市化网络的暂时中断。在意大利，这可能会导致社会、经济、文化遗产的消失，所以这些构成了小领土和他们的文化多样性。

作为礼物形式的灯具集合使居民熟悉艺术研究，认识到它作为 xenia 当地
传统的形式。这种实践起源于古希腊，用于调节客人和主人之间的关系，
它在 Guilmi 和其他意大利小村庄依然存在。

这个项目，完全达到自给自足，创建了当代艺术实践和居民之间的信任感，
让文化运营商和当地管理员在今后能够进行良性文化介入。此外，艺术家
也经历了地缘文化映射的另一种方法，这主要基于意想不到和前所未有的
类比，考虑了开发现代城市的扩张、身份形式以及流行文化阻力的全球普
遍紧张性。

光，信息，交流
张尚志

地域文化滋生于人的行为，人的活动改变了自然形态，这一过程所留下的
痕迹在过往岁月的加乘累计下，逐渐成脉相传。而当社会变迁，文化开始
变得并不像想象中那样坚不可摧，随着人口的流动，逐渐消失。

亚历山大·卡尔博尼在人类学家 Marc Augé 和人口统计学家 Hervé Le
Bras 的启发下，以照片、视频、地图和图纸为缘起，通过光与网络的交换，
创建了居民与艺术实践间的信任感。

整个项目在公众的参与中，深度挖掘了埋藏在记忆深处的感受。艺术家以
独特的视角对事件发生本质进行探索，引发思考，不仅是照片、图像中所
传达的表象，更是在精神层面发人深省。光，是距离天堂最近的力量，它
照亮了人与人之间的距离，城市间，村落间，即便是天际的边缘，只要存
有信任，一切就会变得清晰可见。

Artist: Alessandro Carboni
Location: Guilmi, Italy
Media/Type: Interaction
Date: 2010
Commissioner: Artist initiated
Researcher: Giusy Checola

Guilmi is a small village situated on a hill in the countryside of the Abruzzo region in south-central Italy. It's so depopulated that no one can estimate how many people actually inhabit it. In 2010, when multidisciplinary artist and researcher Alessandro Carboni arrived in Guilmi, there were approximately 100 inhabitants and no children had been born in the last two years. Guilmi was going to be "turned off." The generational deterritorialization caused the decay of many buildings and the disintegration of the local agricultural economy. Alessandro Carboni decided to explore the human settlement in Guilmi through the perceptive potential of his body in motion, considering the people, the village and the landscape as part of a single energy system consisting of concentric rings by establishing physical and conceptual relationships with them.

Rethinking Human Energies is part of the larger project entitled Overlapping Discrete Boundaries, launched in 2009 in Asia. In Overlapping Discrete Boundaries Carboni explored the urban tensions in 12 cities that are very different and distant from Guilmi, but with whom they share cultural and geological problems. Such problems include the geographic repositioning of entire communities due to natural and social factors, like frequent earthquakes and the lack of job opportunities.

Inspired by anthropologist Marc Augé and demographer Hervé Le Bras, Alessandro Carboni reflected on the increased relations between big urban centers and small rural ones. Despite the dwindling populations of rural towns, there is a global network that creates connections between urban and rural communities.

The project was developed in three stages: an exhibition of photographs, videos, maps, and drawings, produced after the tracing of the river's course from its source and the interaction with local people; the final performance run in a public square made by 80 lamps donated by 80 families; the return of the lamps to the donors and the donation of photographic images to them, as evidence of the exchange of their human and material energy.

In the Asian metropolis city life is always moving and "turned on." In Guilmi the silence of the night is profound and during the day the only presence is the elderly who sit and watch life go by. The only time when cities lose control of their moving identities is during a blackout, which Carboni used in Guilmi during the performance. He then lit up the community for 40 minutes. He created a temporary interruption of the global network of urbanization, which in Italy could lead to the disappearance of its social, economic, and cultural heritage, consisting of small territories and their cultural diversity.

The collection of lamps in the form of a gift allowed residents to become familiar with the artistic research by recognizing it as form of the xenia's local tradition, a practice originally from ancient Greece that regulates the relationship between guest and host, which is still alive in Guilmi and in other small Italian villages.

The project, entirely self-sustained, created a sense of trust between contemporary art practices and residents that will allow cultural operators and local administrators to produce virtuous cultural interventions in the future. Moreover, the artist experienced an alternative method of geo-cultural mapping based on unexpected and unprecedented analogies developed in consideration of commonly held tensions worldwide between the expansion of modern urban and identity forms and the resistance of popular culture.

祝你平安
A'Salaam Alaykum: Peace Be Upon You

艺术家：泽娜·厄尔·哈利勒
地点：意大利都灵梅茨基金会
形式和材料：互动行为，装置艺术
时间：2009年7月28日—8月2日
委托人：梅茨基金会
推荐人：里奥·谭

"祝你平安"是意大利都灵的梅茨基金会于2009年委派的公共设施，其创作人为黎巴嫩艺术家泽娜·厄尔·哈利勒。该作品由一个3.8米高，旋转的阿拉伯文"阿拉"标志构成，该作品的材质为镜面玻璃砖。伴随着DJ音乐以及日常生活的图像投影，该标志反射出安装在盆地周边的聚光灯。创建的这种氛围能使人回忆起黎巴嫩的夜总会，而反射的标志从视觉上使人回忆起意大利先锋艺术家穆尼尔·瑟尔霍迪·法尔曼的作品。"祝你平安"坐落在基金大厦外部的"盆地"区域，即用来安装先前的加热设备槽的空间，这是Maria Centonze策展的四个设施中的第三个设施。

目前，EI Khalil的作品朝着关系艺术这一轨迹发展，它发生在观众能沉浸其中的社交场合中。它也可被认为是"体验式设计"的一种形式。毕竟，艺术家已经确定在不同领域加入了足够的熟悉线索，DJ音乐、闪光灯、镜面反射等等，从而使该设施易于访问（而非概念上不透明的），能带来"振奋人心的"体验。该项目欢迎观众把这个空间当成迪斯科来跳舞。

艺术家公开承认她相信跳舞具有治疗功能："跳舞，似乎已经成为一种治愈方式，跳舞似乎能帮助我们忘却彼此反目成仇的原因。在这个狭小的黑暗空间里，随着我们身体的移动，我们开始意识到肉身就是肉身，我们都是一样的。在兴奋控制我们的身体后，我们试图忘记我们利用宗教来杀戮的原因。我们为什么发现彼此是如此不同。"似乎通过邀请观众来跳舞，"祝你平安"能在艺术上促进冲突的解决。

可通过关注该项目对都灵的某一社会角落的影响来评估这一项目的意义。该城市，作为工业（汽车）制造中心，吸引了移民劳动力，许多来自摩洛哥。这些观众对该设施的反应最为热情，很多孩子自发地加入到了跳舞中。孩子们还与该艺术家合作，教土生土长的意大利人阿拉伯舞蹈动作，据这位艺术家说，这是附近的摩洛哥移民首次参加梅茨基金会活动。

"祝你平安"在不同的居住文化之间架起了一座桥梁，使彼此之间可能永远不会发生关系的不同人和社区联系在一起。它作为一项营造场所活动获得了成功，因为它通过促进跨文化交往增强了社区的团结性。它还有助于

增强混乱移民社区的文化认同过程。来自北非的许多观众告诉艺术家,在此次活动中"他们最为强烈地感受到了对其社区的归属感"。从某种意义上来说,该作品为北非移民营造了一个家的感觉,将这种家的感觉及其仪式带回到了意大利的东道社区中。

变成舞者

张尚志

艺术家以最普通的材料与形式,将"祝你平安"置于人更容易接近的场所,以声、光、影等常见元素唤起人们对舞池的记忆。这种氛围更容易让参与者推开陌生的屏障,走进艺术家所设想的舞池中,在音乐中,忘却自我,抛弃都市中的一切尘嚣,展现真实的自己。在光与声的律动中,似乎能听见人与人之间的那种连绵不绝的私语,这种交流让参与者模糊了自己位置的边界,对生活中所拥有的痛苦与伤口,消隐了疼痛。

有时都市的喧嚣并非来自真实,而是紊乱无序的思维。我们即使是在舒适、轻松的环境下,也无法找到一个能愈合梦想的安身之处。而一个舞池,更容易将记忆深处的自己唤醒,不管大家来自何方,在这里,一切变得更容易交往,更容易抚平自己的创伤。

Artist: Zena el Khalil
Location: Fondazione Merz, Turin, Italy
Media/Type: Interaction , Installation Art
Date: 2009.07.28-08.02
Commissioner: Fondazione Merz
Researcher: Leon Tan

A'Salaam Alaykum: Peace Be Upon You was a public installation by Lebanese artist Zena el Khalil, commissioned in 2009 for the Fondazione Merz in Turin, Italy. The work consisted of a rotating 3.8-meter-tall "Allah" sign in Arabic script, executed in mirrored glass tiles. The sign reflected spotlights installed on the periphery of the basin, and was accompanied by a DJ set by Ayla Hibri as well as by projections of images of everyday life. The atmosphere created was one reminiscent of a Lebanese nightclub, while the mirrored sign visually recalled works of pioneering Iranian artist Monir Shahroudy Farmanfarmaian. A'Salaam Alaykum was sited in the "basin" area just outside the Foundation building, a space that used to accommodate tanks for the former heating plant, and was the third of four installations curated by Maria Centonze.

El Khalil's work can be located in the trajectory of relational art, insofar as it staged a social situation within which audiences could immerse themselves. It might also be considered a form of "experience design." After all, the artist made sure to include enough familiar cues in different registers, the DJ music, flashing lights, mirrored reflections, and so on, to render the installation easily accessible (rather than conceptually opaque) and oriented towards "uplifting" experiences. Audiences were welcome to dance as though the space was a disco.

The artist has publicly stated her belief that dancing can be therapeutic: "Dancing, it seems, has become a form of healing.

Dancing, it seems, helps us to forget why we turned against each other. As our bodies move together in small dark spaces, we begin to realize that flesh is flesh and we are all the same. As euphoria takes over, we try to forget why we used religion to kill. Why we found each other so different." It appears that by inviting audiences to dance, A'Salaam Alaykum was intended as an artistic contribution to conflict resolution.

The significance of this project is best evaluated by attention to its impact on a certain sector of Turin society. As an industrial (automotive) manufacturing hub, the city has traditionally attracted migrant laborers, many from Morocco. These audiences responded the most enthusiastically to the installation, with many children spontaneously joining the dance. The children also worked with the artist to teach native Italian audiences Arabic dance moves, and according to the artist, it was the first time that the Moroccan immigrants from the neighborhood had ever visited a Fondazione Merz event.

A'Salaam Alaykum formed a bridge between different resident cultures, bringing diverse individuals and communities together, who might otherwise never have interacted with each other. It succeeded as a placemaking event insofar as it contributed to community resilience by promoting cross-cultural engagement. It also helped to strengthen cultural identification processes for a dislocated migrant community. Many members of the audience hailing from North Africa told the artist that the event "was the closest they felt to being back in their community" (of origin). In a sense, the work created a home away from home for North African migrants, introducing this home and its rituals to a host community in Italy.

北美洲
North America

北美

本届参选的 18 例北美地区公共艺术作品中，来自美国和加拿大的共有 16 例。这两个国家文化差异不大，有着相似的历史与宗教传统，经历过工业化的飞速发展后，城市在发展的同时，社会问题渐渐凸现。这使得后工业时代公共艺术作品的两方 —— 公众，艺术家都发生了变化。一方面，公众对公共艺术的需求从物质层面转化到精神层面，从追求自然生态到追求文化生态，从被动欣赏到主动参与，人们对公共艺术的审美日益成熟。另一方面，艺术家们则由最初传统的设计者、建筑师转为构思创意、协调沟通，组织者。最终，呈现出来的公共艺术作品，由最初的在公共空间展示艺术作品演化为不拘泥于传统的多元化、综合化表现形式。

于是我们看到了尝试拓展公共空间的"丹佛"、"巴斯特空间"、"艺术棚屋项目"；精心营造氛围的"哀悼公园"；传达理念，搭建对话平台，尝试发现并解决热点问题的"冲突餐厅"、"土壤厨房"；旨在重塑社区，城市复兴的"城市花园"、"在水边"、"大西洋艺术公园"；探索将现代科技融入公共艺术以增强交互性的"云之森林"、"音乐秋千"，直至探讨公共艺术作品在公共空间中地位转变问题的"摩尔街市场"……某种程度上，他们代表了公共艺术发展的新趋势。

而两例来自墨西哥的公共艺术案例："瞭望台"、"卡马拉 蓝道玛"则结合当地独特的文化传统和宗教习俗，在继承了传统公共艺术特点的基础上更加侧重探索自然与人们心灵的共鸣，试图使自然、人的心灵与艺术完美统一。而这正是后工业时代的艺术家们所追求的。（冯莉）

1.26 丹佛
1.26 Denver

艺术家：珍妮特·埃切勒曼
地点：美国科罗拉多州丹佛市
形式和材料：雕塑，纤维
时间：2010 年
委托人：丹佛市
推荐人：杰奎琳·怀特

丹佛文化局的公共艺术项目委托珍妮特·埃切勒曼创建一个临时公共艺术作品，以庆祝两年一度的美洲节日的开幕，这是一个在丹佛举办的有关艺术、思想和文化的国际节日。为了在西半球 35 个国家中探索互联性，珍妮特·埃切勒曼从 NASA（美国国家航空和宇宙航行局）喷气推进实验室的声明中获得灵感。该声明称 2010 年 2 月发生在智利的地震，使地球的质量分布发生轻微改变，导致地球的日长缩短了 1.26 微秒。国家海洋和大气治理署对地震之后发送到太平洋的海啸进行了计算机模拟，而这模拟图成了她雕塑结构的基础。

珍妮特·埃切勒曼是一个自学成才的雕塑家，她具有绘画和心理学的研究生学位，以多节网状天线形式的设计而闻名。最初她在印度做富布莱特高级讲师时，从看到的大渔网获得灵感。让人们关注在城市风景中通常被忽视的空间，如建筑之上和之间的区域。

但是，作品"1.26 丹佛"的临时性质以及海啸形状的复杂性让艺术家无法使用钢结构，她本来希望使用钢结构来支持巨大的网状，这个形状需要具有永久抵抗大风以及大量的雪和冰的能力。

了解到不能采用钢骨架，她考虑使用软性网格来重建海啸的几何结构，埃切勒曼研究了模型性质，了解到蜘蛛如何吐丝以铺展它们的网状结构，如何使用更细、更轻的丝合成多股线。纤维技术的新发展使埃切勒曼得以完成作品，通过使用 Spectra® 纤维来创建支撑矩阵，这个材料比相同重量的钢结构牢固 15 倍，并基于颜色、流动性和其他线的成本选择其他纤维材料。这一环保超轻设计，能够很方便地把雕塑直接连接到建筑的正面。

悬挂在七层高的丹佛美术馆屋顶，正好在市中心交通要道的上方，"1.26丹佛"起伏的形状赋予有棱角的都市风景以柔和性。其工程设计模仿手工制作蕾丝的错综复杂结构，在风的作用下，起伏的形状唤起人们的敬畏，

向忙碌的城市生活引入崇敬。晚上，黑暗遮盖了支撑电缆，人们看到的是彩色灯光照耀下，天空中的一个漂浮的发光形体。

"1.26 丹佛"还扮演了全球大使的角色，通过共享的艺术体验连接各大洲。在丹佛展出后，"1.26 丹佛"于 2011 年在澳大利亚展出，悬挂在悉尼市政厅。之后，从 2012 年 12 月至 2013 年 1 月 "1.26 丹佛" 在欧洲展出，它被安装在荷兰的阿姆斯特河畔，作为阿姆斯特丹灯光节的标志性项目。

风的形状
冯莉

夜色中，一张巨大的网漂浮在城市上空，仿佛在描绘风的形状。流动的曲面在天空映衬下和棱角分明的建筑物形成鲜明对比；色彩斑斓的网格结构与河中的倒影交相辉映，为这座城市带来一丝灵动。来自丹佛的这一公共艺术作品为人们营造了梦幻般的场景。令经历了一天繁忙工作，身心疲惫的人们摆脱了束缚，自由地感受风的形状。而这一复杂结构背后的数据来自于计算机模拟的地震海啸图，艺术家通过现代纤维技术在空中再现了海啸模拟图像，1.26 微秒的时间变化内涵，以彩色网格形式表现在城市天空，让每个驻足仰望的人在新奇之余，有了更多思考。单调乏味的日常生活之外，多关注我们的城市，地球，环境，还有生活在其他地区的人们，这正是艺术家想唤醒人们的。该作品结合了艺术和科技，借助被人们忽视的建筑物间空间，传达信息，关注当代问题。

Artist: Janet Echelman
Location: Denver, Colorado, USA
Forms and Materials : Sculpture, Fiber
Date : 2010
Budget Source/Commissioner: City of Denver
Researcher: Jacqueline White

The Denver Office of Cultural Affairs' Public Art Program commissioned Janet Echelman to create a temporary public art installation to celebrate the inaugural Biennial of the Americas, an international art, ideas, and culture festival hosted in Denver. To explore the theme of interconnectedness among the 35 nations that make up the Western Hemisphere, Echelman drew inspiration from an announcement by the NASA Jet Propulsion Laboratory that, by slightly redistributing the earth's mass, the February 2010 earthquake in Chile had shortened the earth's day by 1.26 microseconds. A computer-generated simulation by the National Oceanic and Atmospheric Administration of the earthquake's ensuing tsunami rippling across the Pacific became the basis for her sculptural form.

Echelman, a self-taught sculptor who has graduate degrees in painting and psychology, is known for monumental knotted mesh aerial forms, originally inspired by massive fishing nets she encountered in India on a Fulbright Lectureship. She seeks to bring attention to spaces that are typically unseen in cityscapes, like the areas above and between buildings.

However, the temporary nature of the Denver installation, as well as the complexity of the tsunami shape, precluded the artist's use of the steel armature she'd previously used to support her gigantic mesh forms, which need to permanently withstand high winds, as well as plenty of snow and ice.

Realizing that instead of a steel skeleton, she would need to use a soft grid to recreate the tsunami geometry, Echelman looked to nature for models and learned that spiders employ a stronger silk to lay out the structure of their webs and a thinner, lighter silk for the connecting strands. New developments in fiber technology allowed Echelman to fully realize her vision, using the new Spectra® fiber, a material 15 times stronger than steel by weight, to create the support matrix and choosing other fibers based on color, fluidity, and cost for the other strands. The low-impact, super-lightweight design also made it possible to temporarily attach the sculpture directly to the façade of buildings.

Suspended from the roof of the seven-story Denver Art Museum above downtown street traffic, the undulating forms of 1.26 Denver brought softness into a hard-edged urban landscape. Engineered to imitate the intricacy of handmade lace, the billowing shape, animated by the wind, evoked awe, introducing the possibility of reverence into the daily bustle of city life. At night, darkness concealed the support cables while colored lighting illuminated a floating luminous form.

The monumental aerial sculpture debuted on July 6, 2010 and was originally scheduled for removal on July 31. However, its exhibition was extended another week to provide the public additional viewing time.

1.26 has also served as a sort of global ambassador, connecting continents through a shared artistic experience. After its display in Denver, 1.26 traveled to Australia in 2011, where it was suspended from Sydney Town Hall. Then, from December 2012 through January 2013, 1.26 visited a third continent, where it was installed over the Amstel River in the Netherlands as the signature project of the Amsterdam Light Festival. The lighting program of undulating colors was reflected on the water below in the signature canals of the city.

生命之树和回忆之叶
Tree of Life & Leaves of Remembrance

艺术家：克拉克·威格曼
地点：美国华盛顿州西雅图市
形式和材料：青铜雕塑
时间：2012 年
委托人：WHEEL
推荐人：杰奎琳·怀特

位于华盛顿州西雅图的青铜雕塑"生命之树和回忆之叶"是全球首个永久安装在公共场所的无家可归者纪念碑，传达了公共纪念的新趋势。传统纪念碑通过纪念社区先锋、领导或战争英雄来赞颂市政骄傲，今天的纪念碑则以纪念受害者的方式，对具有挑战性的社会问题展开社区对话。

"生命之树和回忆之叶"最初的动机来自于为无家可归的人所举办的无声守夜行动。人们在西雅图的大街上或附近的国王县常常会发现死去的无家可归者。于是每周在西雅图司法中心台阶上，散发传单为死者举行无声守夜。但由于路人们通常会扔掉列有死者名字的传单，守夜活动的组织者和她的支持者们决定寻求一种能长期存在的纪念方式。申请过程历经九年，成为了一场具有争议的斗争，涉及五个市政部门，包括三次重新设计，最后产生了戏剧性的颠覆——对派克市场历史委员会对无家可归纪念碑项目的拒绝提出了上诉。

争论的主要问题集中在，在哪里纪念过世的西雅图无家可归者。无家可归者通常选择的地点是大萧条期间流动厨房所在地，但该地区邻近派克市场，是这座城市的主要旅游目的地。如今，雕塑伫立在维特史坦柏公园的一角，晚上，华丽的灯光照射在雕塑旁那条到达市中心滨水区的通道上。雕塑的建成体现了城市态度，充分承认了所有西雅图的居民。

"生命之树"雕塑包括两个青铜翅膀形状，其中，树叶轮廓暗示树的形状，从圆形玻璃广场中升起，夜晚，灯光照耀下，整个树形轮廓被绿光笼罩。

由于城市法令禁止公园中的纪念碑刻名字，雕塑家克拉克·威格曼、风景建筑师凯伦·柯斯特和建筑师吉姆·洛肯还创造了"回忆之叶"与"生命之树"的轮廓相呼应，每个树叶上面刻有死去的无家可归者的名字。好像是被风吹散的样子，树叶使用环氧基树脂材料粘附在人行道中，代表承认个体生命，并同时提醒人们无家可归者们的现实。树叶的位置以及有关死者的回忆记录可在网站 fallenleaves.org 上查到。

作为年度守夜活动的聚集场所，"生命之树和回忆之叶"帮助促进正在进行哀悼仪式，每年12月举办，纪念在前一年死于西雅图的无家可归者。随着越来越多无家可归者去世，更多的树叶被贴在了人行道上。

对于在生命中缺少固定住址的人而言，"生命之树和回忆之叶"可以作为他们生命旅程的公开见证。鉴于无家可归者还将不断增加，这个项目将会持续进行下去。

温暖
冯莉

无家可归者，由于种种原因主动或被动地脱离社会、家庭，进入社会边缘。令人感动的是无声守夜行动的支持者们与市政长达九年的不懈抗争，从作品中可以感受到艺术家强烈而复杂的情绪。树叶萌芽的形状正象征了人们不懈的努力如同生命力一样顽强；残缺的叶片暗示了死者飘摇的人生；似合拢手掌的双翅预示了人们对无家可归者的关怀和对未来的希望；阳光下，灯光里，雕塑呈现出温暖的金色，这迟到的关怀也许会给无家可归的观者带来些许温暖。

雕塑坐落在城市旅游地的一角，与"生命之树和回忆之叶"一起向人们传达着这座城市的温暖与公民的尊严。作品提醒人们关注被忽视的群体，呼唤爱、平等与关怀，通过公共艺术表达出城市态度。

Artist: Clark Wiegman
Zone: Seattle, Washington, USA
Media/Type: Bronze sculpture
Date: 2012
Budget Source/Commissioner: WHEEL
Researcher: Jacqueline White

The world's first permanent publicly sited homeless memorial, the bronze Tree of Life in Seattle, Washington, exemplifies a trend in public memorials. Whereas the traditional memorial extols civic pride through celebratory tributes to community pioneers, leaders, or war heroes, memorials today increasingly commemorate victims, thereby facilitating community conversations about challenging social issues.

The impetus for Tree of Life emerged from weekly silent vigils held on the steps of the Seattle Justice Center for homeless people who had died on the streets of Seattle or in surrounding King County. Observing that passersby often threw away the flyers that listed the names of the people being honored, the homeless women who organized the vigils, along with their supporters, sought to create something more lasting. Their quest became a nine-year—often-contentious—battle that involved five city departments, included three major redesigns, and culminated in the dramatic overturning, on appeal, of the Pike Place Market Historical Commission's rejection of the Homeless Memorial Project.

Where exactly to permanently acknowledge that some Seattle residents died without an address was a major source of the controversy. The location that was the overwhelming choice of the homeless community had been the site of a WPA soup kitchen during the Depression, but was also adjacent to Pike Place Market, the top tourist destination in the city. That the sculpture now

stands, dramatically lit at night, in the corner of Victor Steinbrueck Park that is slated to serve as a gateway to a renewed central waterfront, speaks to a process that helped usher in an enlarged civic commitment to fully recognize all of Seattle's residents.

The Tree of Life sculpture consists of two bronze, wing-like shapes, in which cutouts of leaves suggest a tree form. The tree rises out of a round glass plaza that showcases both the park's original landscape and a network of roots that glow green at night.

Because a city ordinance prohibits memorials with multiple names in a city park, sculptor Clark Wiegman, landscape architect Karen Kiest, and architect Kim Lokan also created Leaves of Remembrance. The bronze leaves, which echo the cutouts in the Tree of Life, each bear the name of homeless person who has died. Scattered as if by the wind and attached by epoxy to the sidewalks, the leaves acknowledge individual lives while also reminding pedestrians of the reality of homelessness. The location of the leaves and memories about the deceased are catalogued at fallenleaves.org.

Tree of Life helps facilitate ongoing rituals of grieving by serving as a public gathering place for an annual vigil, held each December, for homeless people who have died in Seattle over the preceding year. As more people die without adequate shelter, more leaves are affixed to Seattle sidewalks. Dedication ceremonies for the leaves occasion pilgrimages by family members.

For people who lacked a permanent address during their lifetime, Tree of Life and Leaves of Remembrance serve as an enduring public testimony to their inherent humanity. Because homelessness persists, the sculpture is an ongoing work in progress.

瞭望台
The Lookout Point

艺术家：HHF 建筑师
地点：墨西哥哈利斯科州
形式和材料：混凝土建筑
时间：2010 年
委托人：哈利斯科州政府和国家旅游局
推荐人：马洛里尼·则姆

鲁塔德尔·佩雷格里诺是一个自 17 世纪起就被采用的天主教朝圣路线。每年有二百多万人走这个横跨墨西哥哈利斯科州的 117 公里长的路线，以显示虔诚、信仰和净化。每年四月中旬的圣周，朝圣之旅人数最多，从阿梅卡到塔尔帕德·阿连德，其中罗萨里奥圣母教堂是最终目的地。由于路线的条件很艰苦，行程本身成为净化仪式的一部分，也是朝圣之旅最主要的因素。

在 2008 年，哈利斯科州政府设立项目，以改进旅行条件并提高鲁塔德尔.佩雷格里诺的体验。在哈利斯科州政府和国家旅游局的财政资助下，墨西哥建筑师塔提阿娜·毕尔堡和德里克·德勒坎普着手管理该项目，从世界各地选择建筑师和艺术家，在路线的关键地点创造一系列站点。

总体规划包括带有基础设施和标志性建筑的生态走廊，以提高宗教仪式的体验。好的结构可以提高路线的精神重要性，并通过提供基本服务，保持一种持久感，如庇护和瞭望台。它们能提供平静、灵感和自省，协助朝圣者的情感之旅。

"瞭望台"被设计用来作为朝圣者朝圣之旅的额外路线。它曲折的、环形的旅程是朝上的，类似于迷宫的设计，尾端是一个栖息在高平台的瞭望台，在这里可以看到哈利斯科的全景。之后，旅客可以从蜿蜒的路线下来，继续他们的旅程。"瞭望台"提供了一个平静和自省的场所，以完成项目精神层面的升华。结构的特殊用材以及重复的环形便于引出简单冥想的体验。这些情感因素和空间的实际功能交织在一起。"瞭望台"几乎完全由混凝土制成，下半部分凉爽、遮阴，这个让人放松的环境设计被用来模拟避难所。登上瞭望台，朝圣者可以俯瞰全景；在下面，他们能放松和休息。

这个场所还成为了社会聚会场地，并给地区旅游业创造了经济来源。伊万·巴恩在他的专题摄影中提到："在那里，人们立即知道要做什么——他们坐在阴凉处，交谈、喝茶，他们的孩子在旁边玩耍……这个场地成为一个聚会点……因为人们在附近开了小餐馆。"

从艺术角度来讲，该项目使朝圣者能够 "以旅程的形式体验建筑本身" （诺拉·施密特）。无压力的设计，为朝圣者提供了一个机会陷入沉思，而不会被打断。

材料的简单性跟旅程的精神重点以及哈利斯科的现代生态学产生共鸣，某些用户在社交媒体上认为该构造似乎 "没有意义和格格不入……" 而一些追随者怎认为这个项目 "令它左边的小屋相形失色。"

"瞭望台" 与朝圣者的需要产生共鸣，为他们提供休息场所、慰藉并提高自省。这一构造不仅仅在审美上或象征意义上具有重要性；更成为一个互动性项目，提升了沿着鲁塔德尔.佩雷格里诺路径行走的朝圣者体验。

朝圣路上的心灵庇护所

冯莉

"瞭望台" 是公共艺术与地域文化、宗教结合的案例。因其修建在朝圣之路上这一特殊的地理位置，被赋予了更深的意味。 上半部分螺旋形上升的结构，体现出自然与生命之美。旋转上升的过程仿佛进入鹦鹉螺壳中，神秘而又令人期待。类似的结构出现在 2010 年世博会的丹麦馆，优美的弧线，螺旋，结合现代的材料，焕发出新的生命力。古老的，重复的环形引出简单冥想的体验，与宗教的克己、自省不谋而合。而登顶后猛地从狭窄的空间到视野开阔的平台，从山顶俯瞰全景，回望来时路，恰如顿悟的过程。下半部分的圆形穹顶，四周由多个拱形门进入，营造出庇护所般的环境。混凝土材质与墨西哥的气候、朝圣之路上蜿蜒山路及地貌相契合，和环境融为一体，更符合朝圣之路上克制、简朴的生活要求。

想起电影 《The Way》（朝圣之路）中的一句话： " Choose a life, not live one."（选择你的生活而非仅仅活着）。在此停歇的人们若能悟到些什么，艺术家的目的就已达到。

Artist: HHF Architects
Location: Espinazo del Diablo, Jalisco, Mexico
Forms and Materials: Concrete building
Date: 2010
Budget Source/Commissioner: Government of Jalisco and the
National Tourism Agency
Researcher: Mallory Nezam

The Ruta del Peregrino is a Catholic pilgrimage route that has been traveled since the seventeenth century. Each year, more than two million pilgrims walk the route, which extends 117 kilometers across Jalisco, Mexico, as an act of devotion, faith, and purification. The pilgrimage is most heavily trafficked during the Holy Week in mid-April and spans from Ameca to Talpa de Allende where the Virgen del Rosario is the ultimate destination. Because the conditions of the route are harsh, the difficulty of the journey itself is a part of the ritualistic purification, and perhaps the most important element of the pilgrimage.

In 2008, the government of Jalisco set forth to improve travel conditions and the pilgrims' experience on Ruta del Peregrino. With financial support from the Government of Jalisco and the National Tourism Agency Government, Mexican architects Tatiana Bilbao and Derek Dellekamp managed the project and selected architects and artists from around the world to create a series of stations at key points along the route.

The master plan consists of an ecological corridor with infrastructure and iconic architectural pieces that add to the experience of the religious ritual. Successful structures enhance the spiritual importance of the route and also add a sense of permanence by providing basic services, such as shelter and look out points. They should also allow for serenity, inspiration, and introspection, assisting the pilgrims' emotional journey.

Lookout Point is one of those architectural additions to the route,

and it's designed to be an additional literal loop in the pilgrim's path. Its meandering, circular journey upwards, resembling the design of a labyrinth, ends at a lookout point perched on a high ledge, offering sweeping views of Jalisco. Visitors then descend back down the winding path and continue on their journey.

Lookout Point successfully integrates the spiritual emphasis of the project—a site of serenity and introspection—through its use of materials, its design and its enterprising placement as an elevated lookout point that spans for miles. The structure's singular material and repetitive circular design facilitate an experience of meditative simplicity. These emotive qualities are woven in with the practical functionality of the space. Made almost entirely out of concrete, the lower half of Lookout Point is cooling and shaded. This relieving environment is designed to feel like an intimate sanctuary and includes a brick opening in the shape of a cross. On top, pilgrims reach a cathartic panoramic view; below, they find solace and respite.

This site has also become a social gathering space and created economic opportunism around tourism. Photographer Iwan Baan reflects on these benefits in his photo essay on the Ruta del Peregrino. "There, people immediately know what to do—they sit in the shade, they chat and drink, the children run around...(the site) become(s a) meeting point."

Artistically, the structure leads the pilgrim to "experience the architecture as a journey," said Nora Schmidt, writer for Architonic. The unimposing design gives the pilgrims an opportunity for contemplation.

The simplicity of the material resonates with the spiritual focus of the journey, as well as the sparse vegetation of Jalisco. Ultimately,

Lookout Point addresses the pilgrims' needs by affording rest, solace, and an opportunity for introspection. The structure is more than an aesthetically or symbolically significant piece of design; it is an interactive work that enhances the pilgrim's experience along the Ruta del Peregrino.

在水边

Jiigew

艺术家：爱德华多·阿奎诺、凯伦·杉斯基、布鲁克·麦克罗伊
地点：加拿大安大略省桑德贝市
形式和材料：钢管与新数字媒体组件
时间：2011 年
委托人：桑德贝市
推荐人：莱恩·柏格森

于 2011 年竣工，永恒艺术装置"在水边"是加拿大安大略省桑德贝社区的代表作品。这是该社区首个包括听觉和新数字媒体组件的艺术作品，也是该社区首个由艺术家和建筑师合作创造的项目，到目前为止，也是这座城市最大型的公共艺术项目。

Jiigew 在布瓦族中的意思是"在水边"，由 20 米高的柯尔顿钢柱，内嵌级联 LED 灯和摩尔斯电码的录音叙述组成，坐落在这个海滨社区的海岸线上，起着灯塔的作用，连接了水、陆地和人。艺术家爱德华多·阿奎诺和凯伦·杉斯基通过与建筑师布鲁克·麦克罗伊合作，将该项目体现为三个不同的方面：实体存在（高度、灯光、声音），连接（当地成分、场所特殊性、精确阐释）和吸引人们（到访、感受、讨论）。

居民和城市管理者认为这个艺术家团队达到了这一目的。"这个艺术作品与其他作品截然不同，可以让你做出不同的阐释，"桑德贝市文化服务和活动协调员利亚·贝利说："它们看上去很神秘、不真实，就像在实体上不可能存在的一样。而且在晚上，它的级联灯光所传达的摩尔斯电码信息也很神秘。"

除了巨大尺寸和使用重工业材料外，雕塑——当地人称之为"灯塔"——保留着一种亲近和简单的感觉。从海滨公园的另一端可以看到级联灯光，这吸引了成千上万的观众来此。灯塔附近，人们会听到令人放松的声音。该作品鼓励听众靠近，并反思人们几千年来同苏必略湖互动的方式。利亚·贝利补充道："无法在其他地方或为其他社区创造'在水边'。因为它根植于当地的布瓦文化、语言、故事，历史和地理。它融合了工业设计和人类的声音以及居住在该地区的人所讲述的故事。"

桑德贝的这个新成员在社区激发了人们的情感并促进了对话，促使居民探讨项目的性质以及公共艺术在社区更广泛的作用。利亚·贝利说，它还帮助居民"重新唤醒"城市的天际线。习惯于看到自然地标、一成不变的海岸线和旧工业遗产，包括已倒闭的谷物升降机和历史铁矿石码头的社区居民，现在能在地平线上看到一个漂亮的新焦点。坐落在这些天际线下其他永久性建筑之中的"Jiigew"象征着城市的发展和复兴。

"包含了所有功能，"贝利说，"'在水边'作为一个独特的、互动的艺术作品脱颖而出。"

天际线——城市心电图

冯莉

水边，岸上，天边……高耸的灯塔雕塑的出现打破了桑德贝市民们印象中的城市天际线。夜晚神秘的信号灯，附近似人们低语般的音乐，仿佛在向人们讲述布瓦人的历史。而简洁的外型，加上现代多媒体技术，又令这个以旧工业社区为主的城市焕发出了新的生命力。远望城市天际，市民们与这个城市新焦点对话，回溯历史，展望未来。天际线上的突兀，仿佛人们心中的悸动，预示着城市未来发展的变革。该公共艺术作品成功地改变了城市面貌，重新塑造了城市文化生态，与当地居民产生共鸣。声光电与建筑结构的结合，继承传统又打破陈规，成为城市复兴的地标性建筑，是地方重塑的典型案例。天际线正如一个城市的心电图，是当地居民的心灵悸动，记录着城市心跳，在每个夜晚，和灯塔信号灯一起搏动。

Artists: Eduardo Aquino and Karen Shanski in collaboration with
Brook McIlroy
Location: Thunder Bay, Ontario, Canada
Forms and Materials: Steel pipe and New media components
Year Completed: 2011
Budget Source/Commissioner: City of Thunder Bay
Researcher: Laine Bergeson

Completed in 2011, the permanent art installation Jiigew is a first on
many fronts for the Canadian community of Thunder Bay, Ontario.
It is the first artwork in the community to include both an auditory
and a new, digital media component; it is the first project in the
community created by an artist/architect collaboration; and, to date,
it is the city's largest-scale public art project.

Comprised of two 20-meter-high Corten steel pillars embedded with
cascading LED lights and a recorded audio narration of Morse code,
Jiigew—which means "by the water" in Ojibwe—is situated along
the shoreline in this coastal community where it serves as a sort of
beacon, connecting water, land, and people. Artists Eduardo Aquino
and Karen Shanski, in collaboration with Brook McIlroy Architects,
sought to have the project embody three different aspects: physical
presence (height, lights, sound), connection (local content, site
specificity, subtle interpretation), and ability to move people (to
visit, to feel, to discuss).

Residents and city administrators believe the artistic team achieved
its goal. "It became the first work of art that, in the end, didn't look
like anything else which led to a host of other interpretations," says
Leah Bayly, Cultural Services and Events Coordinator for the city of
Thunder Bay. "They look mysterious, unreal, like they aren't even

physically possible, and there is mystery and wonder at the Morse code messages in its cascading lights at night."

Despite their large size and use of heavy, industrial materials, the sculptures—which are referred to by locals as "the beacons"　— retain a feeling of intimacy and simplicity. The cascading lights, which can be seen across the waterfront park, draw viewers toward the piece. Once nearby, people are drawn even closer by the soothing audio. The piece encourages listeners to get closer and reflect on the ways that people have interacted with Lake Superior for thousands of years. Bayly adds, "Jiigew could not have been created anywhere else, or for any other community. It draws upon local Ojibwe culture, language, stories, local history and local geography. It fuses industrial design with human voices, the stories of the people that live in the region, told in the languages of those people."

This new addition to Thunder Bay has stirred emotion and inspired dialogue in the community, prompting residents to talk about the nature of the project and the broader role of public art in the community, says Bayly. It has also helped residents "reawaken" to their skyline. Community members, who used to seeing the natural landmarks, unchanging shoreline, and old industrial relics— including a defunct grain elevator and historic iron ore dock—now have a beautiful, new focal point on the horizon. Situated among these other permanent pieces of the skyline, Jiigew symbolizes growth and revitalization in the city.

"Encompassing all that it does," says Bayly, "Jiigew stands as a unique, interactive, and ultimately transformational work of art."

时间之手
Hands of Time

艺术家：克里斯托尔·普日比勒
地点：加拿大不列颠哥伦比亚省维多利亚市
形式和材料：青铜雕塑
时间：2013 年
委托人：加拿大维多利亚市
推荐人：格里高利·多尔

"时间之手"是安装特定场所的作品，12 个小雕像被安装在不列颠哥伦比亚省的维多利亚市不同的地点。作品由不列颠哥伦比亚省的艺术家构思和执行，在纪念维多利亚成立 150 周年公共艺术竞标中胜出。

雕塑由青铜铸成，展现了一系列劳动的手，其中有雕刻艇桨、握住铁路道钉、拿着毯子等，展示出维多利亚市的城市历史厚度。

通常，纪念性质的公共艺术作品设计，出于安全的考虑，往往充斥着传统雕像和纪念碑。而维多利亚市政府的官员另辟蹊径，选择了这一交互式、多场所安装的系列雕塑作品。这些雕像展示了广泛的文化内涵。为确定放置地点，普日比勒咨询了许多社区居民，包括当地原住民，最终将 12 个青铜雕塑放置在城市特定地点。欣赏并寻找这些作品的过程成了一个交互式的寻宝游戏。

跟许多城市一样，维多利亚市也曾为实现成功的公共艺术项目历经波折，"这些作品不是被静静地欣赏，就是被热烈地争论。"公园娱乐和文化部门的城市社区协调员尼古拉·雷丁顿说。"时间之手"项目在社区民众中产生了巨大的影响。"这个作品激发出公众强烈的社区主人翁意识，"雷丁顿说。具体表现为很多形式：当地人自发形成的组织会定期检查项目，在出现问题时，会给雷丁顿的员工打电话。当地居民和访客也通过信件和因特网，向他们发来了数量众多的反馈信息。

特别的是，人们发现搜索 12 个雕像变成了有趣的互动元素，使得公众更加亲近作品，欣赏作品，不被艺术领域的阐释阻碍。

雷丁顿同时指出该作品并非只注重对历史的美化和渲染，也包括了描述维多利亚市历史中的困难时期，这无疑冒着一定的艺术风险，但最终获得了市民的认可："艺术家非常准确和诚实地，但是令人尊敬地，反思了我们的过去，包括整个社区。"

经过艺术家努力，作品获得了巧妙的平衡，这在公共艺术领域很罕见。青铜渲染的作品，具有一定的不锈钢成分，采用了所有观众在公共艺术作品中所熟悉的传统技巧。所描绘的形象没有回避传统题材，但在处理过程中考虑到了敏感性及普遍性。最后，通过在特定场所的交互模式防止更小、更传统的雕像，艺术家向更广泛的观众提供了欣赏公共艺术的全新体验，这超越了装饰物，获得更富挑战性和分享性的体验。

诚实地面对历史

冯莉

作为纪念性质的公共艺术案例"时间之手"能脱颖而出，最可贵的一点是能够诚实地面对历史。对维多利亚建市 150 年来值得纪念和值得反思的历史都真诚面对，这比通常的歌颂粉饰历史更加真实，所以才赢得了市民的肯定，并且得到了他们的积极参与和保护。维多利亚市建于 1862 年，因 19 世纪 50 年代发现金矿而迅速成为冒险家和淘金者的乐园，随后成为著名的深水不冻港和海军基地。150 年间富兰士河见证了城市的发展，也目睹了人性的贪婪与丑恶。艺术家选择在 12 个不同地点放置有代表性的青铜雕塑，正像是历史的 12 处见证，而居民或者游客找寻这 12 个雕塑的过程，如同亲身经历了城市的这些历史节点，这种引导性的交互过程——回顾并反思，正是该作品的过人之处。

Artist: Crystal Przybille
Location: 12 locations throughout Victoria, BC, Canada
Forms and Materials: Bronze Sculpture
Date: 2013
Budget Source/Commissioner: City of Victoria, BC, Canada
Researcher: Gregory Door

The Hands of Time is a site-specific installation of twelve statuettes throughout the city of Victoria, British Columbia. The piece, conceived of and executed by the BC-based artist, Crystal Przybille, was the winning bid in a public art process to commemorate the 150th anniversary of Victoria's founding.

The sculptures, cast in bronze, depict hands engaged in activities—carving a canoe paddle, holding a railway spike, and carrying blankets—that show the breadth of Victoria's history.

At first sight, the project concept (including the pun in its title) seems underwhelming. Victoria, after all, is the capital of British Columbia, and while small (under 80,000), it might have set its sights higher than this fairly predictable project. On the other hand, it is worth considering that commemorative public art designed for civic memorializing is frequently a safety zone filled with conventional statues and monuments.

What the project has going for it is that it is a moderately interactive, multi-sited permanent installation that illustrates a broad cultural swath and doesn't shy away from exploring controversial topics.

Like many cities, Victoria has a checkered history of success with public art projects, "They've either been quietly appreciated or hotly contested," says Nichola Reddington, the city's Community Recreation Coordinator in the Arts & Culture division of the Parks, Recreation, and Culture Department. The Hands of Time Project has had a very different impact.

"There is a great sense of community ownership with the work," says Reddington. This ownership takes many forms: self-formed groups of locals regularly check up on the works and telephone Reddington's staff when there's a problem. Feedback, via letters and internet, from both residents and visitors, has been vast.

Placed throughout the city, the works together form an interactive treasure hunt. According to Reddington, people find the "search" for all twelve sculptures to be an engaging interactive element that allows the public to approach the artwork through a common framework and unimpeded by art-world interpretations.

Reddington also points to the artistic risks the artist took in depicting the difficult passage of Victoria's history, rather than sticking to safe renderings of sanitized history. Przybille consulted with numerous communities, including local indigenous groups, in executing her commission. "The artist tells a very accurate and honest—but respectful—reflection of our past, and embraces the whole community," says Reddington. For example, two sculptures depict the canoe paddle carving and basket-making crafts of the First Nations tribes of the area. Another statue commemorates Chinese contributions to the region's history.

While admittedly not groundbreaking work, the artist struck a tricky balance, rare in public art. The works, rendered in bronze with some stainless steel elements, adopt a conventional vernacular familiar to virtually all viewers as a public art idiom. The imagery depicted doesn't shy away from controversial subjects, yet treats them with a sensitivity and universality calculated to not raise alarms. Finally, by placing smaller, more conventional sculptures in a site-specific, interactive pattern, the artist gives an expanded audience a new appreciation for public art that goes beyond ornamentation and toward a more challenging and shared experience.

哀悼公园
Shadow Lines

艺术家：阿尔瓦多·加尔
地点：美国印第安纳州印第安纳波利斯市
形式和材料：钢筋石笼，石块，半地下建筑
时间：2010 年
委托人：印第安纳波利斯艺术博物馆
推荐人：莱恩·柏格森

作为一个进行艺术创作的建筑师，阿尔瓦多·加尔创作了"哀悼公园"作为室外博物馆，作品的结构简单明了，却很难将它归类。它是个独立的房间，却没有墙壁；它是画廊空间，却没有陈列任何艺术作品；它位于城市，但是，一旦进入其中，却只能看到天空。安静是这个空间唯一的声音。

设计成园中园的样式，位于印第安纳州印第安纳波利斯美术馆旁边的"哀悼公园"，是个由几百万块石头在钢筋石笼墙中组成的方形区域，只能通过地下通道才能到达。一旦参考者穿过通道进入室外的房间，他们就会被高耸的石墙、植被和天空包围。这个空间旨在作为人类回归自然的隐居处（石头、植被和天空），帮助人们应对这世界上人为的恐慌。

加尔说钢筋石笼墙代表了 20 世纪和 21 世纪的暴行和受害者。"游客必须穿过这些笼子包裹的黑暗的通道才能到达公园，我们邀请他们使用这个平静的避难所反思、思考和冥想我们的生活和这个星球上的所有悲剧，"阿尔瓦多·加尔说，""哀悼公园"雕刻出一个空间用于隐居和治愈，在这里，生活的希望萌发自石堆。"

通过营造出让参观者逃离世界又同时有荣辱感的氛围，项目对社区产生了深刻的影响。"通过创建园中园，并采用自然的能呼吸和具有生命的边界，"哀悼公园"拉紧了自我和集体、远离和参与、思考和行动之间的模糊却紧密的联系。"阿尔瓦多·加尔说。

当今世界，能停留下来在公共空间进行思考的机会越来越少，该项目提供了一个安静和冥想的空间——在这里，人们可以退回到私人领域，来寻求平静与治愈。阿尔瓦多·加尔继续说："当代社会创造了一种危险的关联，私人空间意味着平静，公共空间代表着混沌与嘈杂。"

　　"哀悼公园"反其道而行之,指出相反的情况是可能的:即,我们可以在公共区域寻找平静和集体治愈。它还提醒参观者公共艺术有更加广泛的力量。经过精心设计的公共空间,如"哀悼公园",可以影响公众,对过去的悲剧进行弥补,并且促进和谐。

　　"我希望观众走的时候能够意识到,还有其他的思考方式,"阿尔瓦多·加尔说,"对我而言,艺术的空间就是自由的空间。这是非常特殊的空间。"

空间的力量

冯莉

寂静的通道中,四周的石笼墙令人感到压迫,像要将人困住,只有出口的亮光在指引方向。紧张,恐惧逼近,加快脚步,来到出口,闻到青草的气味,感受到热烈的阳光,眼前豁然开朗。拾级而上,天空、草木、石墙,这是隐居者的乐园,难以想象它就在美术馆的旁边。艺术家利用精妙的空间设计给人们带来了超乎寻常的体验,回忆身处钢筋石笼墙中的压抑和困顿,对比平台上的开阔与豁达。狭窄开放的空间变化像为人们模拟了一段人生经历,更提醒人们不要忘记人类曾经历的苦难和暴行。石笼墙代表的东西因人而异,而平台上的感受只有经历过通道的人才能体会,繁华过后的平淡,狂热之后的反思,暴行过后的忏悔……该作品充分展示了空间的力量,身处嘈杂的市区却可享受冥想的空间,是运用公共空间营造禅宗思考方式的典型。

Artist: Alfredo Jaar
Location: Indianapolis, Indiana, USA
Forms and Materials: Metal cages and stones
Year Completed: 2010
Budget Source/Commissioner: Indianapolis Museum of Art
Researcher: Laine Bergeson

An architect who makes art, Alfredo Jaar created Park of the Laments to serve as an outdoor museum—though the piece, while straightforward in structure, defies easy categorization. It is a self-contained room, yet it has no walls. It is a type of gallery space, yet it contains no additional art. It is located in the city, yet, once inside, visitors can only see the sky. Silence is the only intentional sound in the space.

Designed as a park within a park and situated adjacent to the Indianapolis Museum of Art in Indianapolis, Indiana, Park of the Laments is a square area enclosed by millions of rocks in metal cages and accessible only by an underground tunnel. Once visitors pass through the tunnel and enter the outdoor room, they are surrounded by the towering stone walls, vegetation, and sky. The space is intended as a retreat to nature—the rocks, vegetation, and sky—that helps people cope with the man-made horrors of the world.

Jaar has said that the stone walls represent the atrocities and victims of the 20th and 21st century. "Visitors, who must cross the darkness of a tunnel directly below these cages before reaching the park, are invited to use this haven of peace in order to reflect,

ponder, and meditate about all the tragedies that are now ingrained in our existence and in our planet," says Jaar. "Park of the Laments carves out a space for retreat and for healing, where hope and life is literally sprouting from the rubble."

The project has impacted the community by allowing visitors to feel removed from the world while simultaneously engaging with it. "By creating a park within a park and employing breathing, living borders made of nature, Park of the Laments tightened the ambiguous but essential link between the self and the collective, removal and involvement, contemplation and action," says Jaar.

The project is a space for silence and meditation in a world where opportunities to pause and reflect in the public sphere are becoming scarce—and where people often find themselves retreating into the private realm to seek peace and healing. This, says Jaar, "has created dangerous correlations in contemporary society, assimilating private with peace, and public with chaos."

Park of the Laments specifically defies this trend and proposes that the opposite is possible: that we can find peace and collective healing in the public sphere. It also serves to remind visitors of the power public art more generally. Thoughtfully designed public spaces like Park of the Laments can reach out to the public, make amends for past tragedies, and promote harmony.

"I hope the audience leaves with the notion that there are other ways of thinking about situations," says Jaar. "For me, the space of art is a space of freedom. It is a very privileged space."

大西洋城艺术公园

Artlantic

艺术家：多位艺术家合作（策展人：兰斯·冯）
地点：美国新泽西州大西洋城
形式和材料：雕塑，建筑，装置，当代艺术展览
时间：2012 年
委托人：大西洋城
推荐人：莱恩·柏格森

在飓风桑迪肆虐之前，大西洋城有过美好的日子。由于高失业率和非常少的洁净和安全公共空间，这个城市以过去几十年来一直采用的方式发展着——把资源投入购物、赌博和夜生活。

这个趋势在几年前发生了转变。随着对更多公共空间要求的增长，该市委托策展人兰斯·冯创造了"Artlantic"，这是一系列临时艺术展览，由多个艺术家利用闲置空间进行艺术安装，其中，几个是国际知名艺术家，有两位是在当时还不知名的本地艺术家。艺术家们通过不同的媒介进行工作，包括风景设计、园林设计、建筑、雕塑和古怪的装置，如在黑暗中发光的灯箱。目的是为公众——本地市民、游客和城市雇员——创造一个开放的、充满艺术的空间。

"公园在飓风桑迪摧毁后的那个礼拜开放，当地人从那时起就一致拥护 Artlantic 公园，"兰斯·冯说，"到目前为止，我们创造了三个独立的类似于公园的场所，从春天一直开发到秋天。"

除了向居民提供漂亮的公共空间，该项目还在每年夏天向三千多万游客介绍新的艺术作品，大多数游客对当代艺术不是非常理解。如伊利亚和伊米莉亚卡巴科夫有趣的海盗船，标题为"恶魔的愤怒"。"船"从地面上升起，让人想起新泽西海岸的沉船以及伴随大西洋城赌博业的"赃物"。参观者可以在半埋的船只内外走动。

艺术家罗伯特·巴里的装置艺术是一串被照亮的文本，由 23 个鼓舞人心的词汇组成，包括"思考"、"辉煌"和"可能"。这件作品被用来和沿着这座城市标志性木板路旁的粗体历史标识进行对话。琦琦史密斯的繁茂花园——纪念她死于艾滋病的妹妹——则完全由红色植物构成。深红色的花园环抱着史密斯的青铜雕塑"她"——一位妇女怀抱着一只雌鹿。

环境艺术家约翰洛夫的大型迷宫雕塑，部分是舞台，部分是迷宫，部分是

光影幻象。"洛夫的装置中展出的大量照片让人叹为观止,"兰斯·冯说,"当人们四处走动,并阅读每件艺术作品前的介绍时,我们感到很有成就感。"

把"Artlantic"从许多其他公共艺术项目区别开的一个关键因素,是对艺术作品的选择过程。冯策划他喜欢的展览—没有任何限制,这在公共艺术领域是前所未闻的,通常情况下总会有不同的利益相关者进行干预。

"从策展的角度来看,跟在MOMA(现代艺术博物馆)策划展览没什么区别,但是这与为公共空间选择艺术作品还是有很大差别的,"兰斯·冯说,"感觉很微妙,但其实是有很大差异,结果是产生了不同类型的公共艺术。"《纽约时报》和《华尔街日报》对"Artlantic"的评论"不必在公共艺术范畴内讨论装置艺术"。兰斯·冯指出:"应当仅从当代艺术的角度进行评论。"

飓风过后
冯莉

作为有多位艺术家参与的系列艺术展,最大的特点是作品的多样性,而如何挑选艺术家,如何挑选契合主题的作品是最考验策展人的。兰斯·冯成功的关键在于把握住了大西洋城公众的需求,并把这一需求和城市发展结合。以艺术作品展览的方式传达信息,重塑文化生态。

所谓不破不立,飓风"桑迪"的灾难过后,大西洋城面临一个转型,人们渴望重塑一个和过去不同的大西洋城,包括更多充满文化气息,艺术气质的公共空间。兰斯·冯正是利用这一契机,和热爱这座城市的艺术家们一起完成了这一系列的艺术展览,并使其成为展示当代艺术作品的窗口。虽然展览是临时的,但是它带给人们的思考和启示却是长久的。

飓风过后,在大西洋城,公共艺术参与区域建设,改变城市风貌,为当代艺术的交流、共享提供平台,并引发了公众对艺术的热情与思考。

Artists: Multiple Artists; Curator: Lance Fung,
Location: Atlantic City, New Jersey, USA
Forms and Materials : Sculpture, construction, installation art, and
Exhibition of contemporary art
Year Completed: 2012
Budget Source/Commissioner: Atlantic City
Researcher: Laine Bergeson

Even before Hurricane Sandy devastated the region, Artlantic City
had seen brighter days. With very high unemployment and very
little clean, safe public space, the city hummed along the way it had
for many decades—by pouring resources into what it was known
for: shopping, gambling, and nightlife.

The tides started to turn several years ago. With growing demand
for more public space, the city commissioned curator Lance Fung
to create Artlantic, a series of temporary art exhibitions that utilizes
vacant spaces for art installations by various artists, among them
several of international renown and two until-then-unknown
local artists. The artists worked in a variety of different mediums,
including landscape design, garden design, architecture, sculpture,
and quirky installations such as glow-in-the-dark light boxes.
The idea was to create open, art-filled space for everyone: locals,
tourists, city employees.

 "Since we opened the week after Hurricane Sandy devastated the
region, the locals have embraced Artlantic Park," says Fung. "To
date we have created three separate park-like locations, and they
are packed spring through fall."

Beyond giving residents beautiful open spaces to enjoy, the
project has introduced new artwork to over 30 million tourists
each summer, most of whom have very little knowledge of
contemporary art. Some of the marquee works included in the
project are: Ilya and Emilia Kabakov' s playful pirate ship titled
Devil's Rage. The "ship" rises out of the ground, evoking both the

sunken ships that line the ocean floor off the New Jersey coast and the "booty" associated with the Atlantic City gambling industry. Visitors are encouraged to walk in and around the half-buried vessel.

Artist Robert Barry installed an illuminated text piece, comprised of 23 inspiring words, including "Wonder," "Glorious," and "Possible." The piece is meant to stand in dialogue with the bold, historic signage that lines the city's iconic boardwalk. Kiki Smith's lush garden—in part an homage to her sister who died of AIDS—is entirely composed of red plants. This crimson-colored garden surrounds Smith's bronze sculpture, Her, in which a woman cradles a doe.

Environmental artist John Roloff's large-scale labyrinthine sculpture is part stage, part puzzle, and part optical illusion. "The abundance of photos taken at the Roloff installation is astounding," says Fung. "It's rewarding when I see people walking around and reading all of the didactic signs in front of each artwork."

One of the elements that separates Artlantic from many other public art projects was the actual artwork selection process. Fung was asked to curate an exhibition of his liking— and with no restrictions, something virtually unheard of in the public art realm where a variety of different stakeholders often weigh in.

"Curatorially, it was no different than curating a show at the MoMA, but very different than selecting a work of art for a public space, says Fung. "It seems subtle, but it made a huge difference and also resulted in a different kind of public art." Reviews of Artlantic, which have appeared in the New York Times and Wall Street Journal, among many other publications, "do not discuss [the installation] in the context of public art," notes Fung, " but just as great contemporary art."

Young Circle 艺术公园 / 千禧喷泉
ArtsPark at Young Circle/Millennium Springs

艺术家：田甫律子、马尔吉·诺特哈特
地点：美国佛罗里达州好莱坞
形式和材料：景观艺术，公园重建
时间：2009—2011 年
委托人：好莱坞布劳沃德城
推荐人：格里高利·多尔

位于 "Young Circle 的艺术公园" 是由 Glavovic 建筑事务所的建筑师马尔吉·诺特哈特所领导的重建项目。该公园的一个主要部分是由公共艺术家田甫律子设计的 "千禧喷泉" 特色雕塑。艺术公园分两阶段完成（2009 年和 2011 年），重建了位于佛罗里达州，好莱坞中的原 Young Circle 公园。新公园用多功能公共艺术和娱乐中心（提供了观看、参与艺术、游戏和娱乐的机遇）代替了先前 10 英亩的草坪 / 花园区域。

好莱坞佛罗里达州是于 20 世纪 20 年代早期在 "城市美化" 运动中，由约瑟夫·卫斯理·杨所开发的规划社区。该城市，具有近 15 万的人口，是拥有着风景宜人的木板人行道的旅游胜地。Young Circle 公园，离海滨有一英里多的路程，是该城市美术设计的一个关键的原始元素——是您在市区可驻留的一大片绿色空间。但是，到 21 世纪早期，这一设施匮乏的绿色开发空间对公众来说已经过时了。

21 世纪初期，公园重建成为重建周边社区这一大型计划中的首要任务。这一任务被委派给了马尔吉·诺特哈特（他最近赢得了设计局对佛罗里达州南部区域颁发的 "最佳设计师" 奖项）。诺特哈特构思了一个 "文化艺术胜地和被动的自然环境"，并非仅仅是一个简单的公园，而是将建筑、景观设计，与 "公园爵士乐" 的积极艺术设计方法相结合的产物。马尔吉·诺特哈特的方案将艺术设计放在其设计中心，融合了两大特征：一是表演，二是旨在为公共空间提供演示和授课的艺术工作室。

马尔吉·诺特哈特的建筑作品一般对环境比较敏感。在艺术公园中，出现了日本公共艺术家田甫律子在 "水 / 自然" 项目中提出的关键特征。该作品的重头戏，即 "千禧喷泉"，由 126 英尺长的 "水雕塑" 构成，水雕塑包含由计算机控制的水枪。

田甫律子受到了公园现有的猴面包树的启发——面包树也称为 "生命之树"，在其几千年的生命中，它吸收了数百加仑的水分。"猴面包树作为水和生

命的来源这一重要性，不可低估"，田甫律子在其建议书中写道，"因此，它自然而然的，应该在定义水雕塑的实际形式中发挥关键作用。"为了将树融入其最终设计中，田甫律子利用了树的电子信号，用它们来确定水枪的算法高度和喷射时间。田甫律子的其他设计，歌颂了好莱坞的创始人，他们的雕像位于圆形区域中。

公园的重建取得了显著成就，在整个社区赢得公众的称赞。定期组织的活动包括月光下的集体鼓乐、互动性家庭艺术实践活动、迪斯科舞厅以及定期的快餐车活动。尽管经济衰退阻断了社区周围活动的发展，但也有迹象表明复苏会带来新的投资。最近的一所特许高中为这一区域中艺术和科学高中的新建开启了大门。

公共综合体

冯莉

该公共艺术案例有两大特点：其一，通过改造使公共绿地变成公众活动中心，教育中心，表演中心，极大地提高了利用率和互动性。通过对公共空间的开发，发挥公共艺术民众教育的功能，影响和改变着人们的生活。其二，"千禧喷泉"的设计独特，以猴面包树的电信号驱动喷泉，体现了自然与科技的统一。

随着社会的发展，人们的对公共艺术的需求从物理空间层面扩展到精神层面，从追求自然生态到追求文化生态。作为公园重建项目，该作品代表了当下的趋势，即公共空间不仅仅提供一个公众活动的绿地，还是集表演、艺术工作室、艺术作品、建筑设计于一身的综合体，为社区重塑发挥作用，从而改变社区面貌，甚至影响到周边地区。

Featured Artist: Ritsuko Taho , Margi Nothard
Location: 1 Young Circle, Hollywood, FL USA
Forms and Materials: Landscape Art, Park reconstruction
Year Completed: 2009/2011
Budget Source/Commissioner: Broward Co/City of Hollywood
Researcher: Gregory Door

The ArtsPark at Young Circle is a redevelopment project led by architect Margi Nothard, Glavovic Studio. A key component of the park is the Millennium Springs water/sculpture feature, designed by public artist Ritsuko Taho. ArtsPark, completed in two phases (2009 and 2011), redeveloped the original Young Circle Park in Hollywood, Florida. The new park replaces the former 10-acre lawn/garden expanse with a multi-faceted public arts and entertainment complex that includes opportunities for viewing and participating in art, play, and entertainment.

Hollywood, Florida is a planned community developed in the early 1920s in the "City Beautiful" tradition by Joseph Wesley Young. With a population of nearly 150,000, the city remains a tourist destination with a popular, developed boardwalk. More than a mile from the waterfront, Young Circle Park was a key original element of the city's Beaux-Arts design—a large green space to anchor the downtown area. By the early twenty-first century, however, the open green space with few amenities was out-of-date for modern users.

In the early 2000s, the park redevelopment became a priority as part of a larger plan to redevelop the surrounding neighborhoods. The commission was awarded to an ambitious design from Nothard (who recently won a Design Bureau "best architect" award for the South Florida region). Rather than a simple park, Nothard conceived of a "cultural arts destination and passive landscaped environment" that combines architecture, landscape design, and an active arts programing approach that surpasses the usual "jazz-

in-the-park" offerings of such spaces. Placing arts programming at the heart of her design, Nothard's plan features two pavilions—one for performance and the other for art studios designed to provide public space for demonstrations and classes.

Nothard's architectural work is often environmentally sensitive. At the ArtsPark, a key feature emerged in the water/nature project proposed by Japanese public artist Ritsuko Taho. Called Millennium Springs, the centerpiece of the work consists of a 126-foot-long "water sculpture" with computer-manipulated water jets. (The city had specified a "spectacular fountain" in its RFP).

Taho was inspired by the park's existing baobab trees—which are known as the "tree of life," and during their thousand-years lifespan absorb hundreds of gallons of water. "The significance of the Baobab Tree as a source of water and of life itself cannot be underestimated," Taho wrote in her proposal, "and thus it is natural that it should play a key role in defining the actual form of the water sculpture." To incorporate the tree into her final design, Taho harnessed the electronic signals from the trees and used them to determine the algorithmic height and duration of the water jets. Other elements of Taho's design pay tribute to Hollywood's founder, whose statue stands in the circle.

The finished version of the park is a remarkable success, earning accolades throughout the region and winning popularity from the public. Regular events include a moonlit drum circle, interactive/hands-on family arts activities, a discothèque, and regular food-truck events. While the economic recession has interrupted development around the circle, there are signs that a recovery will bring new investment. A charter high school recently opened the doors of a newly constructed arts and sciences high school located on the circle.

土壤厨房
Soil Kitchen

艺术家：Futurefarmers 艺术家团体
地点：美国宾夕法尼亚州费城
形式和材料：互动行为，装置
时间：2011 年至今
委托人：费城艺术、文化和创意经济办公室
推荐人：莱恩·柏格森

在费城 2 号大街角落一个废弃的大楼里，你可用自家院子中的土样换取一碗汤。然而，这一交换只是这一独一无二的公共艺术装置的一个方面，它的目的在于提醒人们对日益严重的环境问题的关注。

"土壤厨房"是一个多层面的公共艺术项目。作品旨在呼吁重视城市土壤的退化状态，教育居民如何应对环境中可能的污染物。包括：一个临时厨房，一个用于检测土壤中的污染物的科学实验室，由风车提供能量的绿色能源示范项目，一个教育中心（可在研讨会中教授水土保持与烹饪知识）。为了向伟大的虚构人物堂吉诃德致敬（其中一座雕像骄傲地站在角落），艺术家们集体构思该项目,值得一提的是,居民通过将土壤样本带到"土壤厨房"，真正参与其中。

"土壤厨房"的理念与环境保护署的全国棕色地带会议主旨一致，这一想法是在与费城绿色协调合作 2015 年倡议时，构思出来的。风车意在呼吁人们将更多的注意力放到城市中心绿色能源的建设上。更希望能唤起堂吉诃德精神（他是 17 世纪虚构的人物，勇于反对巨头）。在堂吉诃德精神的感召下，"土壤厨房"立场坚定地反对某些机构和拥有立法权的人对能源选择的粗暴干预。

当创建"土壤厨房"时，Futurefarmers 同样也借鉴了堂吉诃德的想象力。艺术家设想了一个多层级项目，提醒人们关注，经历过上百年的工业化的地球，这个我们称之为家的地方，自然资源被渐进破坏的严重程度。并鼓励人们思考问题并提出解决方案。同时，他们也想和参与者创建一个对话空间，提供风力涡轮机建设、都市农业、土壤修复、土壤和堆制肥料方面的免费教育研讨会，并举办土壤科学家和烹饪课讲座。

Futurefarmers 由艺术家丹·阿连德、伊恩·考克斯、艾米·弗兰切尼斯基以及建筑师罗德·弗朗肯组成。费城艺术、文化和创意经济办公室运用

威廉姆·佩恩基金会的资助对此工作进行了授权。委员们和艺术家希望该项目能够汇聚公民、艺术家、设计师、科学家、开发人员和决策者的智慧，碰撞并交流思想。

公共艺术与环境问题

冯莉

"土壤厨房"以自己的方式就公共艺术如何解决环境问题作了初步尝试。一群有责任感的艺术家、科学家们展开了一系列活动，充分发挥了公共艺术教育，组织、激励的功能，吸引大众关注环境问题，并让公众亲自参与，尝试各种解决方法。堂吉诃德的反抗精神与天马行空的想象力正和"土壤厨房"的精神相契合。吸引着有共同追求的人们参与其中。这一公共艺术案例体现了他们共同的价值观与追求。另一方面，艺术家们成功地将现代实验室融入公共艺术作品中，用科技的力量解决环境问题。随着科技的进步，作品内容可以不断扩展，引入不同的实验设备，绿色科技项目，使其成长为可持续设计的、有长久生命力的作品。

Artists: Futurefarmers
Location: Philadelphia, Pennsylvania, USA
Forms and Materials: Interaction and installation art
Duration: 2011-present
Budget Source/Commissioner: Philadelphia's Office of Arts, Culture and the Creative Economy
Researcher: Laine Bergeson

In an abandoned building on the corner of 2nd Street and Girard in Philadelphia, you can bring in a soil sample from your yard in return for a bowl of soup. But this dirt-for-food exchange is just one aspect of a unique public art installation designed to raise awareness about pressing environmental issues.

Soil Kitchen is a multifaceted public art project. At once an installation, a temporary kitchen, a science lab (the soil is tested for contaminants), a demonstration of green energy (the project is powered by a windmill), an education center (workshops are being taught on everything from soil conservation to cooking) and an homage to the great fictional character Don Quixote (a statue of whom stands proudly on the same corner) the piece calls attention to the degraded state of urban soil and aims to educate residents on how to respond to possible contaminants in their environment. Futurefarmers, the artists' collaborative that conceived the project, notes that by bringing soil samples into the Soil Kitchen headquarters, residents are literally "taking matters into their own hands."

Soil Kitchen will coincide with the Environmental Protection Agency's National Brownfields Conference, and it was conceived in conjunction with Philadelphia's Green by 2015 initiative. The

windmill calls attention to the need for (and call to) incorporate more green energy in urban centers. It also evokes the spirit of Don Quixote, the seventeenth century fictional character who took a stand against giants. In the spirit of Quixote, Soil Kitchen takes its own stand against the complex institutional and legislative factors that govern many of today's energy choices.

Futurefarmers also looked to the imaginative power of Quixote's character when creating Soil Kitchen. The artists envisioned a multifaceted project that would raise awareness about centuries of industrial wear on the earth and the gradual destruction of the natural resources in the places we call home. They wanted to encourage people to think both about the causes of the problem and the solutions. They also wanted to empower participants by creating a space for dialogue and offering free educational workshops in wind turbine construction, urban agriculture, soil remediation, and composting, as well as lectures by soil scientists and cooking lessons.

Futurefarmers is comprised of the artists Dan Allende, Ian Cox, Amy Franceschini, and architect Lode Vranken. The work was commissioned by Philadelphia's Office of Arts, Culture and the Creative Economy, using a grant from the William Penn Foundation. In the end, the commissioners and the artists hope the project will bring together a mix of citizens, artists, designers, scientists, developers, and policy makers in a robust exchange ideas and resources.

城市花园
Citygarden

艺术家：纳尔逊·伯恩·沃茨
地点：美国密苏里州圣路易斯
形式和材料：雕塑，景观艺术
时间：2009 年
委托人：盖特威基金会
推荐人：莱恩·柏格森

15 年来，密苏里州圣路易斯市中心的一块两分区的土地一直被闲置着。这块地虽然就在 Gateway Mall（盖特威商场）的附近，却因缺乏视觉效果及人气而很少被关注到。

随着"城市花园"（一个三英亩的公园）在 2009 年的开放，这一切都发生了转变。囊括了由全球最著名的艺术家设计建造的众多大型当代雕塑以及结合该地区的生态学特点创造的特定景观艺术。受盖特威基金会的委托，这个雕塑公园和公共聚会场所全年面向公众免费开放，通过吸引各个年龄层及不同社会背景的游客，帮助复兴市中心区域。

项目的一个突出的特征是可接近性。孩子们可以触摸甚至爬上艺术作品，或在下面奔跑（其中一个雕塑是由 100 个垂直水柱构成）。20 种林荫树点缀着公园，向人们提供了躲避夏日骄阳的遮阴处。1 150 英尺长的座位墙迂回在多年生植物和灌木花园中。公园里的断崖上有一个玻璃咖啡馆。盖特威基金会的一名管理人员卡罗尔指出，项目的一个目标就是促使访客欣赏公共艺术和城市空间，并开发社区。"简言之，'城市花园'能吸引各个社会层面，"卡罗尔说，"这里是一个非常民主的空间。"

景观建筑师 Nelson Byrd Woltz 对花园的设计进行探索，试图模拟并展示该区域的自然历史与生态。雨水处理系统能够妥善的解决积水问题；公园内三个独特的生态区域——河岸峭壁、冲积平原和河流阶地——反映了它们在圣路易斯内和附近的天然现象；一个 550 英尺长的密苏里石灰岩墙令人想起附近的河岸峭壁。地区原生态植物自然地将公园及访客与当地地理条件联系起来。

公园环境的成功营造，促进了人们对附近社区的投资。"项目复兴了公民精神，人们骄傲并且乐观，"卡罗尔说。她还补充道，"城市花园"促进了该区域其他项目的开发，包括拱门和滨河地区的改造计划。在 2011 年，作为开发催化剂，"城市花园"获得了城市土地学会颁发的 ULI 阿曼达伯恩城市开放空间奖。

20多名德高望重的艺术家，包括费尔南德·利哥、马克·迪·塞维罗、凯斯·哈林、马丁·普利尔、吉姆·戴恩、托尼·史密斯和阿尔斯蒂德·马约尔等，为这个项目创造了作品。他们的作品提升了公园作为公共聚会空间的价值，成为人人向往的地方。"城市花园"还再次提升了人们对理查德塞拉不朽的雕塑"Twain（一堆）"的关注度，这座雕塑就在新雕塑公园旁边。

"城市花园"帮助复兴了圣路易斯长久以来沉寂的社区，方便到达，更具有审美情趣。"它的精神和特质是令人放松和友好的，"卡罗尔说，"而且，虽然这个地区有惊人的美景，但它一点也不装腔作势。"

沉睡在市中心的土地

冯莉

难以想象圣路易斯市中心的这两块地会沉睡了这么多年，好在艺术家们重新发掘出了他们的价值。通过艺术家的设计、改造，沉寂多年的土地焕发出新的生命力。公共艺术在这里发挥着至关重要的作用。项目成功之处在于对当地自然生态的模拟并与当代雕塑完美结合，令处在市中心的这块区域展现出自然与艺术之美，给人仿佛触摸木材般温暖的感受。和周围高楼林立的人工环境形成鲜明对比。正如基金会管理人员所说："它的精神和特质是令人放松和友好的，而且，虽然这个地区有惊人的美景，但它一点也不装腔作势。"这是对这一公共艺术项目最好的评价。社区重塑是当代公共艺术的主题之一，好的公共艺术不仅能改变社区环境，更能重塑文化生态，将人们从纷乱、浮躁的城市节奏中解救出来，享受温暖、平静。城市花园即是典型代表。

Artist: Nelson Byrd Woltz

Forms and Materials: Sculpture,Landscape Art

Location: St. Louis, Missouri Sat Vacnt, USA

Date: 2009

Budget Source/Commissioner: Gateway Foundation

Commissioners: Gateway Foundation

Researcher: Laine Bergeson

For 15 years, a two-block swath of land in downtown St. Louis, Missouri sat vacant. The area near Gateway Mall was just idle lawn space, lacking visual interest and human activity.

That all changed in 2009 with the opening of Citygarden, a three-acre park that houses large-scale contemporary sculpture (by some of the world's most highly regarded artists) and site-specific landscaping that celebrates the region's ecology. Commissioned by the Gateway Foundation, this sculpture garden and public gathering space, which is free and open to the public 365 days a year, has helped revitalize the downtown area by attracting visitors of all ages and socioeconomic groups.

One of the standout features of the project is its approachability. Children and the young at heart are invited to touch, climb on, and run under (one of the sculptures is a series of 100 vertical jets of water) the artwork. Twenty species of shade trees dot the park and provide escape from the hot summer sun, and a 1,150-foot-long seat wall weaves through a garden of perennial and shrub plants. A glass café sits high up on one of the park' s bluffs. The project's aim, says Carol Cordani, an administrator at the Gateway Foundation, is to inspire visitors to appreciate public art and urban spaces, and to develop community. "In short, Citygarden appeals to every segment of society," says Cordani. "It is a profoundly democratic space."

The landscape architect Nelson Byrd Woltz designed the garden in a way that explores, mirrors, and celebrates the area's natural history and ecology. A series of rain gardens helps treat the site's storm water; three distinct ecological precincts within the park— River Bluffs, Flood Plains, and River Terrace—reflect their natural occurrence in and around St. Louis; and a 550-foot long wall of Missouri limestone evokes the nearby river bluffs. Regionally native plants also help connect the park and its visitors to the local geography.

Appreciation for the park has helped spur reinvestment in the surrounding neighborhoods. "The project has revived civic spirit, pride, and optimism," says Cordani. She adds, Citygarden is fomenting the development of other projects in the area, including renovation plans for the Arch and Riverfront areas. In 2011, Citygarden was recognized with the Urban Land Institute's Amanda Burden Urban Open Space Award as a catalyst for development.

Over 20 highly regarded artists, including Fernand Léger, Mark di Suvero, Keith Haring, Martin Puryear, Jim Dine, Tony Smith, and Aristide Maillol, created pieces for the project. Their work elevates the park from a public gathering space to a destination in its own right. Citygarden has also generated renewed interest in the monumental Richard Serra sculpture, Twain, which sits adjacent to the new sculpture garden.

Citygarden has helped transform a long-dormant part of St. Louis by being both accessible and aesthetically pleasing. "Its spirit and character are disarming and friendly," says Cordani. "And, despite the place's stunning beauty, entirely unpretentious."

摩尔街市场
Moore Street Market

艺术家：多名艺术家，由 ISCP 组织
地点：美国纽约布鲁克林
形式和材料：装置艺术
时间：2012 年
委托人：国际工作室和策展项目（ISCP）的参与性项目计划
推荐人：莱恩·柏格森

在纽约，艺术家、居民项目和社区组织之间的协作并非人们想象的那么常见。国际工作室和策展项目的特殊项目经理朱莉安娜·科佩指出：资金缺乏是一部分原因，另一部分原因则为规避风险，尤其在完成临时或短暂作品时会这些问题更加突出。

于是 艺术家米尼亚·古、弗朗西斯科·蒙托亚、苏育贤、洛特·凡·登与国际工作室和策展项目（ISCP）以及布鲁克林经济发展公司合作，直面上述挑战，为公众呈现出了"摩尔街市场"。2011 年和 2012 年这些参与性作品出现在露天市场上，帮助打破界限，为参观者及购买者营造出一种社区氛围。

分别来自德国和中国台湾的艺术家弗朗西斯科蒙托亚和苏育贤合作了作品"身体和灵魂"。这件作品试图通过一系列干预手段，满足市场上商贩的需求。两位艺术家将雨伞取下挡在玻璃屋顶上，为工人挡住耀眼的阳光。他们还在其他地方安装了手提风扇，以驱赶果蝇。"身体和灵魂"的第三个部分是一个神物铺子的广告视频，在这个视频中，业主描述了待出售的有形与无形商品。艺术家努力试图在艺术家、购买者和市场之间发展出一种默契。

在"大西洋—太平洋"中，米尼亚古建立了一个功能商店，专门出售"爱冒险的探险家的战利品和掠夺品"科佩说。讲述这些小说中探险家探险故事的航海日志和地图被放置在这些幻想商品的旁边。古希望这些从布鲁克林收集到的物品能帮助参观者重新诠释它们的价值。艺术家还希望收集展示过去殖民地贸易公司的图像，以说明商品和人一样，也具有自己的社交生活。

比利时艺术家洛特·凡·登在她摩尔街的装置艺术中充分利用了环境和氛围。"Potentialis"，荷兰语意为"潜在方法"正符合作品主题。装置的一部分采用磷光乙烯基轮廓，挂在市场员工通道口。这件艺术作品的另外一个元素就是磷光性丝网印刷，重复印制"或多或少"等词语，铺满整个墙上，除非点燃并从特定的角度观看，否则几乎看不到。"这件作品的含蓄性暗

示出，公共空间中的艺术作品不需要处于支配或压倒性的地位，"科佩说，
"关键是如何停留在那里，呈现已经存在的，并产生优雅的效果。"

"这些装置艺术品含蓄但深刻地影响了市场"，科佩说。帮助商家及游客
以全新的方式看待这一空间。或许，总体上对社区最显著的影响是，这个
项目为该地区吸引了更多的公共艺术项目并因此带来收益。ISCP 已经采用
"摩尔街市场"项目的成功模式，以在将来为艺术家和策展人主导的项目
提供支持。包括在新城溪沿着超级基金场地展示的一系列艺术作品，以及
与布鲁克林的住房权利组织 Los Sures 的长期合作。

氛围营造
冯莉

"摩尔街市场"中公共艺术作品在环境中的地位悄悄发生着改变。而这一
改变正暗示着公共艺术的日趋成熟，体现在艺术表现方式的多元化及公众
看待公共艺术的方式。早期公共艺术中主导、地标性质的艺术作品渐渐消失，
隐蔽、含蓄的作品，精巧的设计与恰如其分的氛围营造成为公共艺术新的
表现形式，这在"摩尔街市场"中得到充分的体现。正如比利时艺术家说的：
"这件作品的含蓄性暗示出，公共空间中的艺术作品不需要处于支配或压
倒性的地位，关键是如何停留在那里，呈现已经存在的，并产生优雅的效果。"
而公众，作为公共艺术作品的受众，对这一转变的接纳与认可，自然会带
来经济收益，从而产生正反馈，形成良性循环。"摩尔街市场"的成功正
在于艺术家的对艺术作品在公共空间地位的准确定位。

Artist: Multiple, organized by ISCP
Location: Brooklyn, NY, USA
Forms and Materials: Exhibition of installation art
Year Completed: 2012
Budget Source/Commissioner: International Studio & Curatorial
Program (ISCP)'s Participatory Projects initiative
Researcher: Laine Bergeson

In New York City, collaborations between artists, residency programs, and community organizations are not as common as one might expect. It's partly due to a lack of funding, notes Juliana Cope, special projects program manager at the International Studio & Curatorial Program, and partly an aversion to risk-taking. These concerns are amplified when collaborations produce temporary or ephemeral works.

Artists Minja Gu, Francisco Montoya Cázarez, Su Yu-Hsien, and Lotte Van den Audenaeren, together with the International Studio & Curatorial Program (ISCP) and the Brooklyn Economic Development Corporation, overcame these structural and financial challenges to present the Moore Street Market Projects. Installed at the open-air marketplace in 2011 and 2012, these participatory works helped collapse boundaries and foster a sense of community for visitors and shoppers.

Francisco Montoya Cázarez and Su Yu-Hsien, based in Germany and Taiwan, China respectively, presented Body and Soul. This work was a series of different interventions that helped address the needs of the marketplace vendors. In one location, the two artists placed the fabric of broken umbrellas over the glass roof to shield workers from the bright sunlight. In other areas, they installed portable fans to discourage fruit flies from hovering near the produce. A third component of Body and Soul involved a video installation featuring an advertisement for a botánica in which the owner describes both tangible and intangible goods for sale. The artists endeavored to develop a rapport between artists, shoppers, and the market itself.

In Atlantic-Pacific, Minja Gu created a market within a market. She set up a functional store within the market that sold "the plunder and booty of adventuresome explorers" says Cope. A logbook and map that told the adventures of the fictional explorers sat adjacent to these fantastical wares. Gu hoped the objects, collected from around Brooklyn, would help visitors reinterpret their value. The artist also hoped to summon images of historic colonial trading companies and to show that commodities, much like people, have their own social lives.

Belgian artist Lotte Van den Audenaeren played with context and atmosphere in her Moore Street installation. Potentialis, which is Dutch for "potential modus," is an apt name for these suggestive, sometimes secretive, installations. One component of the installation was a phosphorescent vinyl cutout, hung in a space only accessible by the market's workers. Another element of the artwork, a phosphorescent silkscreen repetition of the words "more or less or," covered an entire wall, yet was nearly invisible unless lit and viewed from a certain angle. "The subtlety of the work shows how art in a public space need not dominate or overwhelm, says Cope. "But how it can linger on and make present what is already there, creating elegant effects."

These installations have impacted the Market in subtle but profound ways, says Cope, helping vendors and visitors see the space in a new way. Perhaps the most significant impact for the community at large was the project's success in generating interest in—and money for—more public art projects in the region. The ISCP has already used the success of the Moore Street Market Projects to leverage support for future artist and curator-led projects, including works along the Superfund site at Newtown Creek and a long-term collaboration with the housing rights organization, Los Sures, which is also based in Brooklyn.

卡马拉 蓝道玛
Camara Lambdoma

艺术家：阿里艾尔·古兹克
地点：墨西哥墨西哥城
形式和材料：管风琴，声音装置
时间：2010 年
委托人：墨西哥市自然历史博物馆和环境部
推荐人：里奥·谭

由墨西哥艺术家阿里艾尔·古兹克设计的"卡马拉 蓝道玛"（又称：蓝道玛室），出现在第 55 届威尼斯双年展上，是安装在查普特佩克公园中的一个永久发声装置。"蓝道玛室"由一套精心设计的复杂声波机械构成。看上去是由两套管子制成的管风琴，能发出和声。该作品从本质上来说是一个声音装置，一个可放置水下的水传感器用于收集反馈，气象塔中的管风琴基于蓝道玛矩阵发出声音，该矩阵是古哲学家毕达哥拉斯发明的一个数学网格，产生声音。Pythagoras 的网格绘制了整数比率和反分数。在音乐这一背景下，网格规定了比率，用与基音相关的音符来表示。

墨西哥市自然历史博物馆和环境部将"卡马拉 蓝道玛"作为融入民间社会团体协调工作的一部分，来重振墨西哥最大的、最受欢迎的公园——查普特佩克公园，以恢复和保护两个标志性的作品："特拉洛克喷泉"（特拉洛克是阿兹特克人的雨、富饶和水之神）以及迭戈·里维拉的壁画"水，生命的起源"。早在 1951 年，建筑师里卡多·里瓦斯完成了特拉洛克喷泉和这幅壁画的创作，为了纪念将水输送给墨西哥城的水利工程系统 Lerma（水系统缓冲器）。里维拉的壁画几十年来因长期被水侵蚀，其所在的建筑物已被关闭了 20 年。

作为一个公共项目，古兹克的"蓝道玛室"，相较于数目较多的永久性视觉和雕像公共纪念碑，在墨西哥仅有的几个永久音响设施装置之中脱颖而出。它的复杂性以及这一复杂性与场地特异性的完美和谐令人印象深刻。共振装置是该艺术家 20 多年研究的结晶。作为功能是输送重要物质——水的建筑群修复的一部分，它直接接入水源从而发出声音，在重新使用具有 2 500 年历史的数学矩阵的同时实现发声。古老的希腊矩阵来源也在建筑中找到了回声，其门廊和圆顶使人想起了古代文明的万神殿。

据这位艺术家所述，室内产生的催眠音乐"力图在倾听者内心营造一种反思的状态"，为在墨西哥城，甚至全球范围内唤起人们对水运动的意识。其目的在于保护"宫殿的原始图案：沉思和意识"。它具有相应的复杂程度，

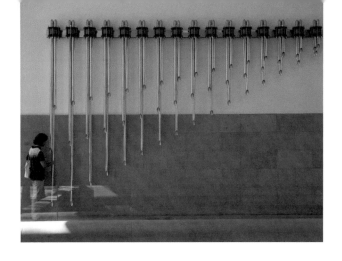

从审美学上敏感地向水表达了敬意，对迭戈·里维拉来说，272 平方米的壁画本身是一种技术和艺术创新，因为它会被永久淹没在水下，静静讲述着自己的故事。

据 OECD（经济合作与发展组织）报告，到 2030 年，受"严重水荒"影响的人数会超过 39 亿。我们所处的时代，非可持续的应用和管理造成水资源越来越匮乏，对重要资源无节制的商品化日趋严重。坐落于游人众多的公共空间中的"卡马拉 蓝道玛"，其对现存的文化记忆的干预是难能可贵的。官方统计报告，迄今为止，共有 72 400 人参观了"蓝道玛室"。通过音乐，它利用独特的共振方法，提醒了游客水对地球上生命的重要性。该项目赢得了"第四届墨西哥国际新媒体艺术节"（2011）的头奖。

水的低吟
冯莉

管风琴历史悠久，最早的水压式管风琴出现在公元前 250 年的古希腊，它以阿基米德原理，将水汽通入多根音管内，从而可以使键盘放平，发出优美的乐音，因为体积极小（一管一音，音域为一个半八度），所以便于携带。4 世纪出现了桌子型的伴唱琴，以风箱式装置替代水压式，增强了音量。5 世纪末，管风琴扩充为房间的体积，并引入教堂。它是最能激发人类对音乐产生敬畏之情的乐器，也是最具宗教色彩的乐器。

"蓝道玛室"借助管风琴，通过古老数学矩阵的变幻，传达出了水的声音。音乐、喷泉、水中壁画完美结合，还有墨西哥城与水的特殊感情，这一切表达出人们对水的敬畏和感恩之情。相信每一个参观者，在室内聆听管风琴音乐时都会有所触动，音乐像是水在低声倾诉，告诉人们水的圣洁与珍贵，提醒人们珍惜地球水资源。

公共艺术通过营造这一特殊的公共空间，传统与现代，艺术与科技，充分表达了当下墨西哥城的公众态度与城市气质。

Artist: Ariel Guzik
Location: Mexico City, Mexico
Forms and Materials: Pipe organ, Sound Installation
Year Completed: 2010
Budget Source/Commissioner: The Museum of Natural History and the Ministry of Environment of Mexico City
Researcher: Leon Tan

Cámara Lambdoma, or Lambdoma Chamber, by Mexico's recent representative artist to the 55th Venice Biennale, Ariel Guzik, is a permanent artwork installed in Cárcamo de Chapultepec. The Lambdoma Chamber consists of a complex set of painstakingly crafted sonic machinery. The most visible component is an organ made up of two sets of pipes, producing harmonies and sub-harmonies. The work is essentially a sound installation; feedback from a submersible water sensor and a meteorological tower produces sounds by the organ based on the Chamber's namesake, the Lambdoma Matrix, a mathematical grid developed by the ancient philosopher Pythagoras. Pythagoras' grid plots whole-number ratios and inverse fractions. In the context of music, the grid specifies ratios, represented by notes, relating to a fundamental tone.

The Museum of Natural History and the Ministry of Environment of Mexico City commissioned Cámara Lambdoma as part of coordinated efforts involving civil society groups, to revitalize Mexico City's largest and most popular park, Bosque de Chapultepec, and to restore and conserve two iconic works: the Tlaloc Fountain (Tlaloc is the Aztec God of rain, fertility and water) and Diego Rivera's fresco Agua, el origin de la vida (Water, origin of life). The Tlaloc Fountain and the fresco were originally commissioned in 1951 by the architect Ricardo Rivas to accompany the hydraulic engineering system Cárcamo del Sistema Lerma, (Lerma System Water Buffer), which delivers water to Mexico City. Rivera's mural was eroded by the passage of water over decades, and the building that housed it had been closed to the public for 20 years.

As a public project, Guzik's chamber stands out as one of only a few permanent sound installations in Mexico, relative to the large number of permanent visual and sculptural public monuments. It is remarkable for both its complexity and the harmonization of that complexity with the specificities of the site. The resonance machinery is the culmination of more than two decades of research by the artist. As part of the restoration of an architectural complex whose function was the passage of a vital substance—water— it directly taps into sources of water to produce its sounds, and it does so while resurrecting a 2,500-year-old mathematical matrix. The ancient Greek source of the matrix also finds an echo in the architecture, whose portico and dome recall the Pantheon of ancient civilization.

According to the artist, the hypnotic music produced in the chamber "seeks to create in listeners a state of introspection," to evoke awareness of the movement of water through Mexico City, and indeed, the earth. Its purpose is to preserve "the original motifs of [the] enclosure: contemplation and awareness." It is an appropriately complex and aesthetically sensitive homage to water, and to Diego Rivera, whose 272 square-meter mural was itself a technical and artistic innovation, since it was to remain permanently submerged and viewed through the substance it celebrated and whose story it told.

According to an OECD report, by 2030, the number of people affected by "severe water stress" will exceed 3.9 billion. In an era of increasing water scarcity due to unsustainable use and management, and the unchecked commodification of this vital resource, Cámara Lambdoma, sited in a popularly frequented public space, is commendable for its intervention in living cultural memory. Official statistics report that 72,400 have visited the chamber to date. Through music, it reminds its audiences of the importance of water to life on earth via uniquely resonant means. The project won first prize in the International Festival of New Media Arts Transitio-MX 04 (2011).

音乐秋千
Balan coires (21 Swings)

艺术家：Daily / Tous les Jours 设计团队
地点：加拿大魁北克
形式和材料：多媒体技术，秋千
时间：2011 年
委托人：The Quartier des Spectacles Partnership
推荐人：卡梅伦·卡蒂埃

"21 音乐秋千"由设计团队 Daily（负责人为梅利莎·蒙盖和穆娜·安德鲁奥）设计，并被蒙特利尔 Quartier des Spectacles 指定参加 2011 年的 Promenade des Artistes（艺术广场）。珍妮·曼斯与圣于尔班大街之间的迈松纳夫街道上的狭长地带过去是一条通向地铁的道路。它位于歌剧院后边与魁北克大学蒙特利尔分校理科大楼之间，经历过大规模的基础建设后才建造了该艺术广场，由步行道与小亭框架组成区域，成为商贩的食品摊和工匠们的展示区间。

该设计团队希望能创造一件这样的作品，它能够连接艺术与科学，促进合作以及吸引人们回到这个因长期建设而无法进入的地方。

所有日常的公共艺术项目都要能体现其娱乐性。起初该团队试图使用秋千作为交互式组件，但是，在现场参观后他们决定使用小亭框架作为起点，以免与现场已有组件冲突。之后的设计是一套另建的 21 条交互式秋千，当观众来回摆动时每条秋千会发出不一样的音符。秋千安装到小亭框架上，每年春天安装它们需耗时八周，是该地区春季和夏季的一件大事。

魁北克大学蒙特利尔分校的动物行为学家卢兰-阿兰·吉尔罗迪欧指出，秋千需要配合与协作，他也对该作品的创作提供了帮助。当地的作曲家拉德温·加齐·穆内被委任创作声色部分。最初，九条秋千弹奏钢琴的音符，六条秋千弹奏电颤琴的音符，剩余六条秋千弹奏吉他的音符。过了几年，竖琴的音符也添加进来了。

当观众随意摆动"音乐秋千"，可以弹奏出粗腔横调的音符；当秋千一致摆动时，它们就可以弹奏出和谐的旋律来。

"音乐秋千"确实很受欢迎。最初，它只是个一次性的临时项目，由于公众的需要，才成为每年都举行的公共活动。随着时间的推移，设计团队一

点点精进该工程；增强秋千缆索，改善秋千座位的照明组件，拓宽乐谱的音符，并于 2012 年加入了互动投影系统——"21 障碍"。

"21 障碍"是一个大型数字弹球游戏，投影在面向"21 音乐秋千"建筑的墙上。游戏由摆秋千的人与路上行人一起开启。行人使用手机就可以发射一枚弹球，"音乐秋千"的运动指引着障碍的移动，而弹球必须穿越这些障碍。在弹球和障碍进行交互时运动时，就形成了视觉效果。

"21 音乐秋千"是 Printemps numérique 艺术广场 2014 年的一部分，后者由蒙特利尔 Conférence régionale des élus (CRÉ) 发起，它与 50 多家机构合作呈现出蒙特利尔当地数字艺术的生气勃勃与创造力。

交互——公共艺术的生命力
冯莉

"音乐秋千"采用多媒体技术捕捉秋千的运动，以音符的形式再现，成功地将运动转化为音乐，这一大胆尝试使传统游戏——秋千幻化出了新的生命力。赢得了大众的一致好评。而设计团队精益求精，在第一阶段成功的基础之上，进一步开发出数字弹幕游戏，使互动的参与者从荡秋千的人扩展到步行街上的路人，使影响范围由点及线直至面。这一公共艺术案例正符合当前公共艺术的趋势 —— 交互性和参与性。从早期只是作品的展示，到当下强调人的参与和互动；从早期的仅仅影响看到作品的人，到辐射整个街道、社区的效应。不仅如此，"音乐秋千"更以其新颖的形式吸引了从孩子到老人的广大人群，使得效应更具普遍性。 动物行为学家卢克 - 阿兰·吉尔罗迪欧指出了秋千这一运动的关键，即配合与协作，这正是交互的一种体现。 某种程度上，交互也是公共艺术生命力的关键。

Artist: Daily / Tous les Jours

Location: Quartier des spectacles, Montréal, Quebec, Canada

Forms and Materials: Multimedia, Swings

Year Completed: 2011 - Present

Budget Source/Commissioner: The Quartier des Spectacles Partnership

Researcher: Cameron Cartiere, Laine Bergeson

21 Balan coires (21 Swings) was designed by Daily / Tous les Jours (principals Melissa Mongiat and Mouna Andraos) and commissioned by the city of Montréal via les Quartier des Spectacles to launch the Promenade des Artistes in 2011. The strip of land on Maisonneuve Street between Jeanne-Mance and Saint-Urbain Streets was previously a thoroughfare to the metro. Situated between the back of the opera house (Place des Arts) and the science building of the University of Quebec at Montréal (UQAM), the area underwent extensive construction to develop the Promenade des Artistes, a pedestrian strip lined with kiosk frames that would allow for easy transformation by vendors and craftspeople into food stalls and exhibition spaces.

The design team wanted to create a work that would bridge the worlds of arts and sciences, inspire collaboration, and invite people back to an area that had long been inaccessible due to the lengthy construction.

Playfulness is embedded in all of Daily's public art projects. Originally the team explored using seesaws as the interactive component, but, following a site visit, they decided to use the kiosk frames as the starting point so as to not compete with existing site components. The resulting design is an extended set of 21 interactive swings, each playing a unique note as the user moves

back and forth. The swings mount onto the kiosk frames and are installed for eight weeks each spring, serving as a spectacular launch for the spring and summer season of outdoor activities in the quarter.

The work evolved with the assistance of animal behaviorist Luc-Alain Giraldeau from UQAM, who noted that the swings would invoke cooperation as well as collaboration. Local composer, Radwan Ghazi Moumneh, was commissioned to develop the sound element. Originally, nine swings played piano notes, six played vibraphone, and six played guitar. In subsequent years, a harp has been added.

When participants swing at random the notes can be cacophonous, but when they swing in unison, they can create a melody together.

The work has proven to be extremely popular. Originally commissioned as a one-time, temporary project, it was brought back by popular demand as an annual public event. The design team has refined the work over time; improving the swing cables, enhancing the lighting component in the swing seat, expanding the notes in the musical score, and in 2012 adding the interactive projection 21 Obstacles.

21 Obstacles is a giant digital pinball game projected on the side of the building facing 21 Swings. The game is activated both by participants on the swings and passersby. Using their mobile phones, passersby can launch a ball into the game and the movement of the swings directs the actions of the obstacles the balls must try and pass through. When the ball and the obstacle interact, visual effects are triggered.

21 Balançoires (21 Swings) is part of Printemps numérique 2014, an initiative of Montreal's Conférence régionale des élus (CRÉ), in partnership with more than 50 organizations that showcase the effervescence and creativity of Montreal's digital art community.

巴斯特空间
Spacebuster

艺术家：Raumlabor 团队
地点：美国纽约曼哈顿岛和布鲁克林
形式和材料：充气结构
时间：2011 年
委托人：临街屋艺术与建筑事务所与纽约歌德学院
推荐人：卡梅伦·卡蒂埃

"巴斯特空间"采用可移动的充气结构设计，目的是"将各类公共空间转化为社区聚会的点"。临街屋艺术与建筑事务所指定由位于柏林的跨学科团队 Raumlabor 创作的巴斯特空间，是从货车的后面飘出的大泡沫。它的充气结构最简单和容易移动。此类结构中很常见的是，半透明充气的形状和样式自动适应周围环境，在货车停靠的地方形成其他亭楼状的环境。Spacebuster2009 在纽约之行为期 10 天，占用了该市不少地方，从户外直到高架索公园下边的狭窄地带。

Raumlabor 建筑团队认为，所有人都可以建造他们自己的"巴斯特空间"，因为所用的材料相对简单和容易获得：风扇、塑料薄膜 (Visqueen)、双面胶，使泡沫成形并托住的重块。它是一种均分结构，能根据当下全球经济面临的挑战提出合理的观念。该项目目的是回收再利用公共空间，把人们团结起来再创立新的团队意识。Spacebuster2009 在曼哈顿和布鲁克林举办了总共为期 10 天的讲座、展会和电影放映。

"巴斯特空间"于 2011 年回到纽约，成为在该市举行的"点子节"的组成部分。其中公众参与的活动之一就是吸引他们制作和组装为该活动专门准备的木椅。这些木椅后来用于膨胀空间举行的演讲和表演。

"巴斯特空间"是乌托邦式的游牧民建筑实例之一，这种结构源于 20 世纪 70 年代的充气建筑，如日本大阪世博会上的富士集团馆（1970 年）与蚂蚁农场充气建筑（1971 年），后者还衍生了一本类似设计的入门著作《充气建筑手册》。对于蚂蚁农场，该充气结构展现了"当人将它置于手上，感受它并塑出新形状时，所在的环境可以变化成什么样。"

"巴斯特空间"的第一版叫作"厨房纪念公园"，最初安装在柏林，用作社区厨房，以增强居民的参与度。附近的人获邀到公共场所为彼此做饭吃。2006 年，"厨房纪念公园"穿越三个国家（英国、德国、波兰），到达杜伊斯堡、吉森、汉堡、利物浦、米尔海姆和华沙的公众视野中。

作为建筑实践的总体原则，Raumlabor 设计团队感兴趣的是人们如何与建筑和空间互动的。他们设计了几个充气移动式设备（移动催化剂），包括 Rosy（芭蕾舞女演员罗西）for Portavillion(2010 年，英国)，它出现在了伦敦的 15 个地方。罗西为伦敦建筑节的舞会、表演和讨论会提供了场地空间。另一个移动催化剂是大爆炸 (2009 年，韩国)，它的每个移动式设备里有充气囊。

2012 年 8 月，Raumlabor 将"巴斯特空间"带到了底特律，它是底特律设计节中的小组讨论"建筑类文化转移"的内容，探索了这座受困惑城市的再生战略。目前 Raumlabor 的日程表中并没有有关"巴斯特空间"的其他活动计划安排，但活动计划及其他充气元素的发展，是当下建筑团队 Raumlabor 柏林的社区参与和社会实践的研究方法。

乌托邦式的游民建筑

冯莉

艺术家们对公共艺术空间的探索从来没有停止过，"巴斯特空间"这一移动的自动适应充气结构被用来回收再利用公共空间，从最初的社区厨房开始，发展为公众活动聚集地，进而参与到城市再生战略中。因其可移动的特性，自适应环境的特点被称为乌托邦式的游民建筑。其便捷的特性，尤其适合在拥挤的城市中划分空间，创造相对独立的封闭环境。

球形结构带给人安全感，半透明的材质又使得人们与外界环境并非完全隔绝。外界光影的变化，为里面的人们带来梦幻般的感受。从外部观察，柔性材质和周围坚硬的建筑更形成鲜明对比。更重要的是，该作品能在短时间内充气完成，可以快速搭建与拆卸，仿佛人们随身携带的移动空间，大大拓展了公共艺术空间的概念。

Artists: Raumlabor
Location: Manhattan and Brooklyn, New York, USA
Forms and Materials: Inflatable architecture
Date: 2011
Budget Source/Commissioner: Storefront for Art and Architecture in
collaboration with Goethe Institute New York
Researcher: Cameron Cartiere

Spacebuster is a mobile, inflatable structure designed to "transform public spaces of all kinds into points for community gathering." Commissioned by Storefront for Art and Architecture and created by Berlin-based interdisciplinary group Raumlabor, Spacebuster is essentially a giant bubble that expands from the back of a van. It is inflatable architecture in its simplest and most mobile form. As is common with such structures, the shape and form of the translucent inflatable adjusts to its surroundings, creating a different pavilion-like condition wherever it is parked. Over its ten day run in New York City in 2009, Spacebuster occupied a range of urban conditions—from an open field to a narrow lot beneath the High Line.

The Raumlabor architecture team believes that anyone can build their own Spacebuster, as the materials used are relatively simple and readily available: a fan, plastic sheeting (Visqueen), double-sided tape, and weights to shape the bubble and hold it in place. It is a type of egalitarian architecture that offers up an appropriate concept in light of the ongoing economic challenges of our current global economy. The goal of the project was to reclaim public space, bring people together, and create a new sense of community. The 2009 version of Spacebuster hosted lectures, exhibitions, and screenings for a total of ten events throughout Manhattan and Brooklyn.

Spacebuster returned to New York in 2011 as part of the Festival of Ideas for the New City. One of the socially engaged activities offered in Spacebuster invited the public to construct and assemble wooden chairs designed specifically for the event. The chairs were

subsequently used for the lectures and performances held in the inflated space.

Spacebuster was an example of quasi-utopian, nomadic architecture that sees its roots in the 1970s inflatables such as the Fuji Group Pavilion at the Expo in Osaka, Japan (1970) and Ant Farm inflatables (1971) which also produced a guide to similar designs titled, Inflatocookbook, For Ant Farm, the inflatable was a vision of "what environment can mean when a person takes it in his own hands, feeling it and molding new forms."

The first iteration of Spacebuster, known as Kitchen Monument (Küchenmonument), was initially installed in Berlin and was used as a community kitchen as a means of creating community engagement. Neighbors were invited out into public spaces to cook for one another. In 2006, Kitchen Monument traveled across three countries (England, Germany, Poland) and engaged with the public in Duisburg, Giessen, Hamburg, Liverpool, Mülheim, and Warsaw.

As an overarching principle of the architecture practice, Raumlabor is interested in how people can interact with architecture and space. They have design several inflatable mobile units ("mobile activators") including Rosy (the ballerina) for Portavillion (2010, United Kingdom) that traveled to 15 locations in the London. Rosy provided space for dance events, performances, and discussions at the London Festival for Architecture. Another mobile activator was Big Bang (2009, South Korea), which featured two inflatable spaces in one mobile unit.

In August 2012, Raumlabor took Spacebuster to Detroit as part of a panel discussion called The ArchiCULTURAL Shift as part of the Detroit Design Festival, exploring regeneration strategy for the beleaguered city. There are no other events planned for Spacebuster on the current Raumlabor schedule, but the development of mobile activators and other inflatable elements is an ongoing methodology in the community engaged and social practice of the architecture team of Raumlabor Berlin.

冲突餐厅
Conflict Kitchen

艺术家：乔恩·罗宾和道恩·威利斯基
地点：美国宾夕法尼亚州匹兹堡
形式和材料：餐厅，菜单，店面设计
时间：2010 年
委托人：Kickstarter/ 艺术家的自发项目
推荐人：卡梅伦·卡蒂埃

"冲突餐厅"位于匹兹堡卡耐基博物馆附近公园中的凉亭内，既是一家餐厅，又是一个只提供与美国存在冲突的国家风味美食的社会参与式公共意识项目。该项目，由乔恩·罗宾和道恩·威利斯基创建，每几个月会根据当前任何地域的政治事件，包括古巴、伊朗和委内瑞拉等，对菜单和店面进行变换。冲突餐厅当前的主题是阿富汗。

"冲突餐厅"每一个主题的更换，由寻求扩大公众对焦点国家中文化、政治以及利益攸关问题的参与度的活动和讨论来补充。首次试验是强调伊朗这个国家的波斯食物，其目的在于"在大街上"创建一个公众能聚集在一起，讨论政治问题的空间。

除了提供食物外，"冲突餐厅"还通过多种方法创建了社交论坛，提供的食物是手持街头食品。食物的包装纸中摘录了对这一地区人民的采访以及其他信息，从而扩大用餐者对食物来源以及在网络电视的政治报道过程中总是会失去个性的国家的理解。

社交的另一种方式是冲突餐厅周三的就餐体验。在这次体验中，人们能加入到与"阿凡达"的讨论中，和"阿凡达"就餐——"阿凡达"是现场连接到当前冲突国家中的人的一名员工。扮演阿凡达的该名员工，为相隔半个地球之远的人与坐在阿凡达面前的人之间的交流架起了一座桥梁。

还会有一些 Skype（网络电话）晚宴，比如说，人们可以聚集在匹兹堡和伊朗，享用集体餐，而这种集会只有通过技术才能隔断。

"冲突餐厅"这一作品会在东利伯蒂原店面和 Schenley 广场现有亭子的基础上继续扩大。该项目也是可移动的，举办了诸如午餐时间（与匹兹堡大学荣誉大学协作）等活动；在这些活动中，学生可参与讨论，享用处于危机之中的焦点国家的食物，如乌克兰和埃及。

　　"冲突餐厅"还发起了一个称为"总统演说"的项目。该项目的第一期是"古巴演讲"，邀请了 40 多名古巴人和古巴裔美国人写他们希望奥巴马总统发表的讲话，然后聘用一名扮演奥巴马的演员来发表讲话。

2014 年"冲突餐厅"的主题会包括巴勒斯坦和以色列风味美食。

超越食物

冯莉

恐怕没有人会把食物和政治话题联系到一起，而冲突餐厅却打破常规，将餐厅的主题设定为当今日益严峻的冲突问题，并以此展开了一系列的社会参与尝试。于是餐厅、食物、社交尝试演化为一系列公共艺术活动，引发人们关注、讨论当下的社会热点问题，并尝试沟通以消除误解，消解隔阂。这一活动的意义早已超越食物、餐厅，也远非创造公共空间那么简单，它是公共艺术的大胆尝试，更让人们对如何解决冲突问题作深刻的思考。如何化解冲突？是以暴制暴，剑拔弩张还是角色互换，尝试沟通？是回避问题，歧视他国还是求同存异，化干戈为玉帛？冲突餐厅作了很好的尝试。

好的创意，就像一粒火花，一旦出现就引发一系列的讨论，自然会演变成一系列成功的公共艺术活动。真正好的公共艺术，不在于规模大小，资金投入，而在于它的变革性，冲突餐厅的成功也许正在于此。

Artists: Jon Rubin and Dawn Weleski
Location: Pittsburgh , PA , US
Forms and Materials: Restaurant, Menu and the Door design
Daet: 2010
Budget Source/Commissioner: Kickstarter/ artists' self-initiated project
Researcher: Cameron Cartiere

Located in a kiosk within the park surrounding the Carnegie Museums of Pittsburgh, Conflict Kitchen is both a restaurant and a socially engaged public art project that only serves cuisine from countries with which the USA is in conflict. The project, created by Jon Rubin and Dawn Weleski, rotates identities every few months in relation to current geopolitical events and has included, Cuba, Iran, and Venezuela. The current version of Conflict Kitchen is Afghanistan.

Each Conflict Kitchen iteration is supplemented by events, performances, and discussions that seek to expand the engagement the public has with the culture, politics, and issues at stake within the focus country. The first experiment was with Persian food highlighting Iran and the aim was to create a space "on the street" where the public could gather and debate political questions and be introduced to things they were not familiar with.

In addition to serving food, there are several means by which Conflict Kitchen creates a forum for social engagement. Much of the food served through Conflict Kitchen is hand-held street food. The wrapper that surrounds the food item contains quotes from interviews with people from the region and other information to expand the diners' understanding of not only where the food comes from, but also some insight about a country that is often depersonalized in the political reporting process of network television.

Another means of social engagement is through Conflict Kitchen's Wednesday dining experience, where people can join in a discussion and meal with a "human avatar," a staff member who is connected via live feed to an individual from the current conflict country. As the avatar, the staff member serves as a conduit for the individual half a world away to engage with the individual sitting in front of the avatar.

There are also group Skype meals where, for example, a group can gather in Pittsburgh and in Iran can gather for a collective meal that is only separated by technology.

The work of Conflict Kitchen has continued to expand beyond of the original storefront in East Liberty and the current kiosk in Schenley Plaza. The project is also mobile, hosting events such as The Lunch Hour (in conjunction with Pitt Honors College) where students can come for discussions and food focused on countries in crisis such as Ukraine and Egypt.

Conflict Kitchen has also developed a project called The President's Speech. The first version was "The Cuban Speech" for which Conflict Kitchen asked over 40 Cubans and Cuban Americans to write part of a speech that they would like President Barack Obama to deliver. An Obama impersonator was then hired to deliver the speech.

Future iterations of Conflict Kitchen in 2014 will include Palestinian and Israeli cuisine.

云之森林
Cloud Arbor

艺术家：奈德·卡恩
地点：美国宾夕法尼亚州匹兹堡布尔社区公园
形式和材料：不锈钢雕塑
时间：2012 年
委托人：Randall 兰德尔慈善基金会
推荐人：卡莉·安·克里斯滕森

"云之森林"位于宾夕法尼亚州匹兹堡儿童博物馆的前方，是一个在内部形成一片不锈钢杆森林的，直径为 20 英尺的雾球。这一雕塑全年运作，从通过高压阀将水变成雾的 30 英尺高的不锈钢杆处创建出云朵，云朵每隔几分钟就会出现、消失。该项目是艺术家奈德·卡恩与园林建筑师安德里亚·科克伦以及匹兹堡儿童博物馆合作设计的。

奈德·卡恩的作品，就像"云之森林"，将自然力量融入到了城市以及作品的文脉中。该作品的互动性和精度显示了对自然体系的科学、生物和工程方面的深入理解。根据现场的气候条件，雾可能会突然降临在游客的身上，或悬停在公园上方。孩子和大人一样，能与该雕塑互动，可以在炎热的夏日身着泳衣，或是在其午餐期间享受清新的喷雾。

奈德·卡恩与园林建筑师安德里亚·科克伦密切合作开发了该场地定制设施，将该设施充分融入到公园以及公园形象的基础中。在设计过程中，公园附近的社区主张，公园应作为公民空间来重建，而不仅仅是儿童博物馆的公共扩展。在雕塑和公园设计过程中，项目合伙人深受社区反馈的启发，开发了一个所有年龄阶段的人都喜欢、都可访问的地方。

该雕塑为公共艺术和设计如何携手共同创建巨大的公共空间提供了指南。据安德里亚所说，杠杆放置的间隔是五英尺，以鼓励游客穿过该雕塑，从而营造出一种渗透到公园其他区域的感觉。此外，该雕塑沿着该公园的一条最大交叉小路铺设，与儿童博物馆和该场地南边的办公大楼连成一线。基架系统使水能通过铺路石排出，而不需要人工池塘；还在路基下建立了一个空隙来放置电缆以及连接到雕塑中的排水装置。艺术家和设计师与照明设计师合作，在这次表层带内定制创建了一个灯架，将固定装置布局集成到杠杆和铺路石的网格中，创建了一个不仅在全年，而且在一天的不同时间段也能为人们带来享受的公共空间。

为了使改造的公园及其艺术品能实现长期的可持续性，儿童博物馆和匹兹堡市共同承担了维护、修复和保护公园的责任。儿童博物馆还创建了一个基金会来帮助支付"云之森林"的维修费。由于这一设施的水会作为科学

和艺术奇景来使用，喷嘴和喷泉的持续维护对该设施的长期成功至关重要。
人们将该基金会和公园的重新设计视为博物馆赐予这座城市的礼物。

该雕塑是公共艺术如何锚定公民公园空间的一个良好范例。"云之森林"是
Buhl(布尔)社区公园—— 一个聚集匹兹堡北面空间的充满活力的社区——
的焦点。2012 年，儿童博物馆带领大家将这一被忽视的、饱受摧残的和未
利用的城市公园改造成了附近居民可使用的、充满生气的绿色空间。

社区精灵
冯莉

公园改造是当下公共艺术的常见主题。很多几十年前兴建的绿地公园已经
满足不了当下的需要，改造迫在眉睫。公众的需求应该是设计首要满足的。
"云之森林"正是在改造前期做了大量的社区调研工作，收集社区反馈之
后准确定位了公园改造目标—— 一个公民空间而非儿童博物馆的扩建工
程。 在此基础之上，汇集了建筑师、设计师、雕塑家、自然科学家的集体
创造，终于使公园恢复活力，重新成为社区焦点。博物馆和基金会对"云
之森林"的细心呵护，更让那朵有灵性的云朵成为了社区的精灵和守护者。
公园的改造过程，从收集社区反馈到后期维护，增强了社区民众的凝聚力，
使公共艺术真正成为凝聚人心，人人参与的艺术活动。

Artist: Ned Kahn

Location: Allegheny Square, Buhl Community Park, Pittsburgh, PA , USA

Forms and Materials: Stainless steel sculpture

Daet: 2012

Budget Source/Commissioner: Charity Randall Foundation

Researcher: Carrie Ann Christensen

In front of the Children's Museum of Pittsburgh, Pennsylvania, Cloud Arbor is a 20-foot diameter sphere of fog that forms inside a forest of stainless steel poles. The sculpture functions year round and creates a cloud that appears and vanishes every few minutes from 30-foot tall poles that convert water into fog through high-pressure valves. The project was a collaboration between artist Ned Kahn with landscape architect Andrea Cochran and the Pittsburgh Children's Museum.

Kahn's work, like Cloud Arbor, integrates the forces of nature in urban and formal exhibition contexts. The interactive nature and precision of the piece display a deep understanding of the scientific, biological, and engineering of natural systems. Depending on the climatic conditions of the site, the fog may descend upon visitors or hover above the park in a sphere. Children and adults alike interact with the sculpture whether in a bathing suit on a hot summer day, or as a refreshing mist on their lunch break.

Kahn worked closely with landscape architect Andrea Cochran to develop the site-specific installation, fully integrated into the park, and a foundation of the park's identity. During the design process, the community around the park came forward to advocate that the park be redeveloped as a civic space, not just a public extension of the Children's Museum. The project partners were deeply inspired by community input during the sculpture and park design process and developed a place that is delightful and accessible to all ages.

The sculpture is a beacon for how public art and design can work hand-in-hand to create great public space. According to Andrea Cochran, the poles are spaced at four-foot intervals to encourage visitors to move through the sculpture and to create a sense of permeability with the rest of the park. In addition, the sculpture is sited along one of the park's major, cross-axial pathways, aligning it with the entrances to both the Children's Museum and the office building on the south side of the site. A pedestal system allows water to drain through the pavers without ponding and creates a void below grade to house the electrical conduit and water lines that run to the sculpture. The artist and designer worked with lighting designers to create a custom in-grade light support within this subsurface zone, integrating the fixture layout into the grid of poles and pavers, creating a public space that is not only enjoyable throughout the year, but also at different times of day.

To provide for the long-term sustainability of the renovated park and its artwork, the Children's Museum and the City of Pittsburgh share the responsibilities of maintaining, repairing, and protecting the park. The Children's Museum has also created an endowment to help pay for the upkeep of Cloud Arbor. Due to its use of water as a scientific and artistic wonder, ongoing maintenance of the nozzles and the fountain is critical to the long-term success of the installation. This endowment and the park redesign are considered gifts to the city by the museum.

The sculpture is a great example of how public art can anchor civic park space. Cloud Arbor is the focal point of Buhl Community Park, a revitalized community gathering space for Pittsburgh's Northside. In 2012, the Children's Museum led an effort to transform the neglected city park that was largely blighted and unused into a vibrant green space for the neighborhood.

南湾再生
South Cove Regeneration

艺术家：迈克尔·辛格
地点：美国佛罗里达州西棕榈滩
形式和材料：环境再生项目
时间：2012 年
委托人：未知
推荐人：莱恩·柏格森

一个世纪的城市化使佛罗里达西棕榈滩的沃斯湖只剩下一座空壳。该区域曾经是一个富饶的、多样化的自然湿地，而现在成为汇聚有毒物质（来自不透水外层或其他新建工程快速堆积的废水、废物）的储存库。野生动物死亡或逃走。海岸线开始受到侵蚀。

2005 年，迈克尔·辛格工作室的迈克尔·辛格为该区域提出了一系列再生环境干预措施，作为西棕榈滩海滨公共项目的一部分。辛格研究了对该区域最有利的策略，通过音乐会的形式与棕榈滩郡环境资源管理部（PBCDERM）合作。在 2008 年和 2012 年间，这位艺术家和市政当局通过工程和建造，创建 3 个不同的岛屿、2 英亩的红树林和米草栖息地，3.5 英亩的海草栖息地，以及 9 英亩的牡蛎礁，推动了该项目。

这些环境干预措施提供了自然过滤——每个牡蛎每天能过滤 40—50 加仑的水，从而改善水质。该项目还包括一个 556 英尺的高架木板路，16 英尺×16 英尺的瞭望甲板以及供公众来访的教育凉亭。

"对我来说，'南湾再生'项目最引人注目的方面是，它对人们持有的关于公共艺术的能力和目的，以及艺术家扮演角色的看法，提出了质疑，"艺术总监迈克尔·辛格说道。当有人认为该项目的主要目标是天然水系的积极振兴时，"有些人会问，艺术在哪里？"

该项目的其中一个杰出之处正在于它拓展了公共艺术的定义。"该项目倡导社区互动和理解，体现了生态群和政府环境资源管理机构的使命。"迈克尔·辛格说，"它有利于修复西棕榈滩近岸内航道南湾，这一区域因滨水道路、污染以及重要自然栖息地和野生动物的丧失而饱受摧残。"

在该项目中，除环境再生外，值得注意的还有社区和政府参与：川普大楼（俯瞰南湾的一个豪华合作公寓大厦）的居民，曾试图通过提起将他们与大型社区隔离开这一昂贵的诉讼来终止该项目。他们的行动不仅失败，这种公

开性还使得社区支持该项目。诉讼引起了该地区对这一项目的支持，"因而爆发了大规模的社区活动，公众更加了解该项目，了解位于这个城区中的新公共场所，优化其栖息地的重要意义。"

该项目有助于扩大公众对公共艺术的力量、可能性和本质的理解。迈克尔·辛格得出结论："'南湾再生'项目证明了公共艺术参与多功能问题解决、展示关键环境再生、满足基础设施需要、促进社区意识的能力。"

公共艺术与环境再生

冯莉

"南湾再生"项目充分说明了公共艺术在当下的一种发展趋势：由早期的雕塑、壁画、环境艺术的形式发展为多元化的艺术，成为参与区域建设并解决现实问题的一种运作机制。相应的，艺术家的角色也因项目的实现而渐渐由设计者、建筑师转变为组织者、政府和民间机构的协调人。

现在很多城市都面临着环境问题，越来越多的公共艺术案例以环境保护、环境再生为主题。仅仅引发人们的关注已远远不够，艺术家通过沟通协调政府、项目、民众，以最大化满足项目需要，利用可能的科技手段，完成环境再生的艺术创作。

社会进步，城市发展，社会问题的凸现，艺术家的本能与责任，这一切促成了公共艺术的演化，更对艺术家提出了更高的要求。无疑，"南湾再生"项目为我们作了一个很好的示范。

Artist: Michael Singer

Location: West Palm Beach, FL 33401, USA

Forms and Materials: Environmental regeneration

Daet: 2012

Budget Source/Commissioner: None provided

Researcher: Laine Bergeson

A century of urbanization left the Lake Worth Lagoon in West Palm Beach, Florida, a shadow of its former self. Once a rich and diverse natural wetland, the area had become a repository for toxic runoff (from a rapid pile-up of impervious surfaces and other new construction). Wildlife died or fled. The shoreline began to erode.

In 2005, Michael Singer of Michael Singer Studios proposed a series of regenerative environmental interventions for the area as part of the West Palm Beach Waterfront Commons project. Singer researched the most beneficial environmental strategies for the area and then, working in concert with the Palm Beach County Department of Environmental Resource Management (PBC DERM), began restoration work. Between 2008 and 2012, the artist and the municipality advanced the project through engineering and constructing, creating three distinct islands, 2 acres of mangrove and spartina habitat, 3.5 acres of seagrass habitat, and .9 acres of oyster reef.

These environmental interventions provide natural filtration—each oyster filters 40-50 gallons of water a day—and improve water quality. The project also comprises a 556-foot elevated boardwalk, a 16-by-16 foot observation deck, and an educational kiosk for public access.

"For me, the most compelling aspects of the South Cove Regeneration project are questions it raises about commonly held assumptions regarding the capability and purpose of public art as

well as the role of an artist," says lead artist Michael Singer. "Some would ask 'Where's the art?' when one considers the primary task of this project is the positive reinvigoration of a natural water system."

Part of this project's excellence is precisely the fact that it broadens the definition of public art. "This project advances community interactions and understanding, as well as manifests the missions of ecological groups and a governmental environmental resource management agency," says Singer. "It helps to replenish the South Cove of the West Palm Beach Intracoastal Waterway, which has suffered from the construction of a waterfront roadway, pollution, and loss of critical natural habitat and wildlife."

In addition to environmental regeneration, this project is notable for its community and governmental participation: residents of Trump Tower, a luxury co-op building overlooking the South Cove site, tried to terminate the project by engaging in a costly lawsuit that isolated them from the larger community. Not only did their action fail, notes Singer, the publicity fomented community support for the project. The lawsuit brought regional attention to the project, and "resulted in a large community campaign to enlighten the public about the project and the valuable benefits of this multifunctional water cleansing, habitat enhancing, accessible new public place located in an unexpected urban area."

The project has helped to expand the public's understanding of the power, possibility, and very nature of public art. Concludes Singer: "The South Cove Regeneration project demonstrates public art's ability to engage in multi-functional problem solving, demonstrate crucial environmental regeneration, and meet infrastructure needs as well as providing community awareness."

艺术棚屋项目
Art Shanty Projects

艺术家：多个艺术家（主管：詹妮弗·彭宁顿）
地点：美国明尼苏达州白熊湖
形式和材料：冰屋
时间：2012年
委托人：未知
推荐人：莱恩·柏格森

"艺术棚屋项目"是由艺术家发起的临时社区项目，每年二月，出现在明尼苏达州结冰的湖面上。由艺术家设计的冰屋组成（与冰上钓鱼的风格类似），该项目旨在探索如何利用非常规的冰湖公共空间，以一种新颖且挑战常规的方法，扩展艺术观念。

项目目标是成为集艺术画廊（不受任何因素胁迫）、艺术居住地、互动和社区经验于一体的地带。它的独特之处在于，允许艺术家与观众互相沟通并相互作用。该项目已从2004年启动时的两个棚屋，发展到近几年的20处棚屋和20 000名游客。2月份的每个周末，"艺术棚屋"都对外开放。

在强调艺术过程和产品质量的同时，ASP鼓励艺术家拓展他们的艺术范围。每个棚屋必须自我管理并且尊重湖边环境以及附近的邻居和渔民。在创建过程中，艺术家拥有灵活性和自主性，允许充分实现自己的想法。对于艺术家和观众来说，观众参与度被认为是成功至关重要的经验。

艺术家积极引导观众参与，如编织手工艺品、唱卡拉OK或以更加传统的方式为居民排忧解难。

2012年，各种不同的棚屋突然出现在湖中。艺术家萨拉·霍尼韦尔、亚当斯、安吉拉·琼斯和莫·霍尼韦尔设计了"淘气棚屋"，观察孔里展示了小巧、顽皮的场景，此外还有淘气的命运、顽皮小游戏，"谁从饼干罐偷了饼干"。还有弹弓制作的车站与糖果香烟诊疗所。

"篮球棚屋"是由艺术家莎拉·贝克、贝丝·杰斯·赫西、埃蒙·麦克莱恩和山姆制作的，作为更衣室的温暖小屋，游客观看Body Troupe的"中场表演"，Teen Wolf的案件重演和有游行乐队的赛前动员会。此外，还有由不同的艺术家展现的鼓舞人心的讲话、游戏计划、暧昧玩笑、团队照片和奖杯。

劳伦·鲍曼、梅根·威克、莫莉·巴尔科姆·罗利、阿历克斯·纽比、、阿比盖尔·莫里斯和阿瑞卡·罗创建了"棚屋堡垒",我们以孩子的视角重构虚构、亲密的氛围。游客们被鼓励使用一组滑轮来建立自己的堡垒、枕头和绳索。棚屋也包括了午睡一小时,参与者可以蜷缩在自己建的城堡中休息。在点心时间,游客们会玩食物游戏之后才吃掉它,而跟唱参与式故事时间也针对成年人和孩子开放。

20多个多样性和创造性兼具的棚屋,使人们在这深冬时节聚集在公共空间,享受艺术带来的乐趣。

冰面

冯莉

公共艺术案例遍及地上,地下、天空、海岸。这一次,艺术家们选择了明尼苏达州结冰的湖面——冰面,这一未被公共艺术涉及的处女地。尽管只能在每年深冬建造这些只可在二月周末开放的艺术冰屋,艺术家们仍旧乐此不疲,主要原因在于不受限制的创作自由和与观众零距离的交流。于是各种艺术棚屋陆续出现,像是冰面上相继开放的花朵,艺术形式多彩纷呈,吸引了更多的公众参与其中。20多个棚屋汇聚了两万多公众及游客,公共艺术的作用渐渐呈现。和被规划的地面、湿冷的水面相比,冰面上的艺术创作自由度更大,寒冷的气候、光滑的冰面带给艺术家和观众独特的创作灵感和体验,这种极端的环境反倒使得艺术家与观众的互动更加密切。该案例挑战传统的公共艺术空间,开拓了公共艺术的新形式。

Artist: Multiple Artists, Director: Jennifer Pennington
Location: White Bear Lake, Minnesota, USA
Forms and Materials: Ice house
Date: 2012
Budget Source/Commissioner: None provided
Researcher: Laine Bergeson

Art Shanty Projects is an artist-driven, temporary community that pops up on a frozen Minnesota lake every February. Comprised of artist-designed ice houses (similar in style to those used for ice fishing), the project is dedicated to exploring how the unregulated public space of the frozen lake can be used in new and challenging ways, expanding notions of what art can be.

The project aims to be an art gallery (without the intimidation factor), artistic residency, and interactive and community experience all at once. The installation is unique in that it allows artists to interact with their audience—and the audience to interact with the artists. The project started with just two shanties in 2004, and grew to 20 shanties and 20,000 visitors in recent years. It is open to the public every weekend during the month of February.

With an emphasis on both the artistic process and the quality of the product, the ASP encourages artists to push their own artistic boundaries. Each shanty must be self-governing and respect the environment of the lake and the community of neighbors and fisherman that already inhabit the space. Artists are given flexibility and autonomy in the creation process, allowing for the full realization of their ideas. Audience engagement is considered essential to the success of the experience, for both artist and audience.

Artists can engage audiences in an active way—with craft projects like knitting or singing Karaoke or solving puzzles—or in a more passive, traditional gallery-esque way.

In 2012, a variety of different shanties popped up on the lake. "Naughty Shanty" by artists Sarah Honeywell, Aneesa Adams, Marieka Heinlen, Angela Maki-Jones, and Mo Honeywell, featured peep-holes that reveal tiny, naughty scenes inside; naughty fortunes; and naughty little games, like "who stole the cookie from the cookie jar." There was also a slingshot-making station and a candy cigarette dispensary.

"Basketball Shanty" by artists Sarah Baker, Beth Chekola, Jess Hirsch, Eamonn McClain, and Sam Hoolihan, served as a locker room warming hut where visitors could explore locker room dynamics "halftime performances" by Body Troupe, re-enactments of Teen Wolf, and pep rallies with live marching bands. There were also motivational pep talks by various artists, game plans, towel snapping, team photos, and trophy displays.

Lauren Herzak-Bauman, Megan Wicker, Molly Balcom Raleigh, Alex Newby, Abigail Merlis, and Areca Roe created "Fort Shanty," a space to recreate the imaginary and intimate places we made as children. Visitors were encouraged to build their own forts using an array of pulleys, pillows, and ropes. The shanty also featured a nap hour, where you could curl up and rest in the fort you built; a snack time where visitors were encouraged to play with their food and then eat it, and a story hour with participatory sing-alongs for both adults and kids.

The twenty varied and creative shanties, brought people together in a public space to engage with art—and each other—in the deep midwinter.

中东和中亚
Middle East & Central Asia

中东和中亚

每当谈及中东和中亚,人们不免会联想到丰富的油气资源、凸显的地缘政治位置、复杂的宗教冲突等。在这片广袤的土地上,存在着极为不平衡的国家发展状态。这里既有因富有而闻名世界的阿联酋、卡塔尔,这里也有经济困难、局势动荡的阿富汗、伊拉克。因此该地区的公共艺术作品题材广泛、内涵丰厚,立足于当下,直击社会现实,多围绕关注民生、文化认同、民众教育、地方重塑等主题展开。

随着油气资源的开采、现代工业的兴起以及对外贸易的发展,中东及中亚国家已不同程度的进入工业现代化的时期,城市面貌日新月异,物质生活趋于便利,人们既享受着社会进步所带来的福利,同时也忍受着诸多社会问题和矛盾。艺术家们不囿于成见,试图通过多元的角度去探索艺术重塑生活的可能,揭示底层人民的遭遇,通过公共艺术探寻一种“折中”、“缓冲”的表现形式,以可持续发展的理念,诠释现代城市应具备的文化精神,缓解城市发展中的各种矛盾,追溯本源的人文、自然状态,实现人文情怀的回归以及人际关系的改善。同时,来自中东及中亚的艺术家们呼吁,在未来,公共艺术应更多的兼备社会责任,以“重塑”的概念引导公众重新认识自己的生活环境,服务于民众的多元化需求,探寻人类文明发展的最高阶段。(马熙逵)

帕克斯通的记忆
A Pakhtun Memory

艺术家："暂定集体"团队
地点：巴基斯坦卡拉奇市
形式和材料：影像
时间：2011 年
委托人："暂定集体"团队
推荐人：格雷戈里·道尔

2011 年 12 月，艺术家与"暂定集体"团队招募音乐家，在巴基斯坦卡拉奇市外来人口聚居的棚户区附近的居民广场"违法"表演。他们演奏了来自帕克斯通以及许多迁移到当地的棚户区居民的乡村广场歌曲。闪光瞬间的事件变成了自发、快乐的聚会，其中包括棚户区居民的即兴舞蹈表演。在当地警察部队成员最初试图打破非法集会后也给予了他们默许。

像世界大多数地方一样，巴基斯坦正经历从农村到城市的人口迁移。特别是普什图移民组成了卡拉奇最大的社区，他们在国内劳动力市场承担主要作用。这种迁移导致的文化中断体现在乡村生活的手机视频中，它们作为农村留守纪念品从一个人传递到下一个人。这样的视频，描绘了普什图村民民间歌舞表演，是艺术家拉希德·汗，默罕默德·萨蒂坤·汗和亚米纳尼·纳西尔·柴奥德丽作为"暂定集体"集体合作的灵感。这个公共介入项目就是这样创建出来的。

公共场所的音乐和舞蹈的介入，在视频中可以获得，它们对于这些聚集的人产生了深远影响。年轻的棚户区居民加入表演，为传统舞蹈作贡献。虽然许多人聚集在一起可以分享自己的经验。虽然这种自发的事件是事先规划的，但并未获得官方许可。所以该项目最初建立时，导致了艺术家、棚户区观众和当地政府之间的冲突。起初警方勒令人群停止。但参与者坚持聚会只是为了好玩。一个男人抱怨说，巴基斯坦已经成为"甚至禁止表达幸福"的国家。这个事件轮流上演，所以警察只能同意并允许表演继续。

这样的介入可能是短暂的，但他们的深刻影响却可以持续。在卡拉奇和平集会，甚至是一起欢笑或听音乐都是被禁止的。因此策划这样一场集会带有风险。简单来说，"暂定集体"只是转移动态，以支持普什图流离失所的农民工，他们这些普遍经济难民面临文化错位和剥夺公民政治权利的事情。然而，通过执行这种转变，集体打开一个更深、更持久转变的可能性。

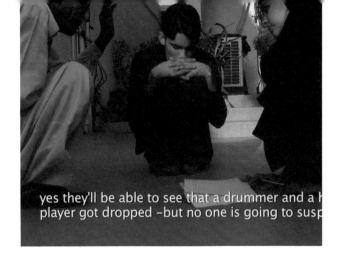

从审美和艺术的角度来看，项目代表一组迷人、重叠的文化规范和变化，导致众多质疑的假设。如智能手机经常作为"真正的"文化敌人遭到蔑视。然而在普什图移民的案例中，他们不仅能共享传统的民间舞蹈，还能保存并进行延续这些东西。相反，在保存歌曲的同时，"暂定集体"可以帮助传统文化应对人口转变的压力。

在莫霍克哈得逊地区展览、大学艺术博物馆、UAlbany 中，以此事件为原本的电影获得了评委奖，并被选为 2012 锡拉库兹国际电影节拍摄奖。

来自乡土的翩跹

马熙逵

基于乡村的传统文化，曾经兴盛于各个家庭和邻里之间，但随着国家城市化的步伐，这种传统文化在逐渐消亡。在巴基斯坦，文化传统就像寮屋居民的乡村歌曲，正在经受着威胁。飞速的城市化进程，意味着作为国家文化根基的乡村生活将迅速消失，同时伴随它消亡的还有传统和历史。巴基斯坦作为尚不发达国家为谋求发展，将城市化与现代化及经济增长等同起来，却忽视乡村所蕴含的国家文化基础。摧毁乡村及其内在的文化传统，同时也揭示了一个更深的偏见——对乡村的漠视与不屑，然而，生活在乡土间的农民中却隐居着国家文化传统的精英。新的环境下，基于乡村的传统文化不再集中并且不再传播给下一代，这些文化遗产一旦消失，将是国家走向富强道路上的一件极大憾事。寮屋居民的即兴乡村歌曲表演，是民间自发的"呐喊"，包裹着对于国家现代化发展的"爱"与"背叛"。

Artist: Tentative Collective
Location: Shirin Jinnah Colony/Sikanderabad, Karachi, Pakistan
Forms and Materials: Image
Date: December 2011
Commissioner: Tentative Collective
Researcher: Gregory Door

In December of 2011, artists associated with the Tentative Collective recruited musicians to perform—illegally—in a public square near a squatters' colony in Karachi City, Pakistan. They played a song that derived from Pakhtun, the rural home-place from which many of the local squatters had migrated. The flash event turned into a spontaneous, joyful gathering that included impromptu dance performances by squatters. Members of the local police force, after initially trying to break up the illegal gathering, gave the meeting their tacit approval.

Like most of the world, Pakistan is experiencing a population migration from rural to urban areas. In particular, Pakhtun migrants make up one of the largest communities in Karachi, where they serve in the domestic and physical labor force. The cultural disruption resulting from this migration is reflected in cell phone videos of village life that are passed from one person to the next as mementos of the countryside left behind. One such video, depicting Pakhtun villagers performing a folk song and dance, served as the inspiration for this public intervention by artists Rasheed Khan, Mohammad Saddique Khan, and Yaminay Nasir Chaudhri, working in collaboration as the Tentative Collective.

The music/dance intervention in public space, which is captured on video (http://vimeo.com/35740223), had a profound impact on those gathered. Young squatters came and joined the performance, contributing traditional dance, while numerous people gathered to share the experience. The spontaneous event was planned without official permission, and initially set up a conflict between the artists

what's going on here?
We're just playing music- nothing else is happening.

taken permission from anyone?

No.

o ask Sahib-   -Listen listen, just let us play one last song then we'll leave--just 5-10 minutes--
-don't ask Sahib

s you can't do this here,
reparing for President Zardari [to visit the hospital]
                          In this country even celebrating happiness has become difficult for

't you celebrate at home?
t a place to celebrate happiness
                    let us play one more please!   Come on man   Come on man
            just let us do one more   one more
                    we pray to allah for such a big revolution that-   one song please
olution can come in the next 3 minutes if it wants to!   one song one more please!
                                    one song   one more   Come on man
        I am also very tired of this country! Let the revolution come faster!
                                                    -Let us do the song m

and squatter/spectators and local authorities. Police at first tell the crowd to cease. Participants insist that the gathering is just for fun. One man complained that Pakistan had become a country "where even displays of happiness are forbidden." At this, the event takes a turn, with the police agreeing with the comment and allowing the performance to continue.

Interventions such as this one may be fleeting in duration, but their lasting impact can be profound. In the case of Karachi, where peaceable assemblies, even to share laughter or music, are prohibited, simply orchestrating such a gathering carries a risk. The Tentative Collective briefly shifted the power dynamics in favor of the displaced migrant workers of Pakhtun, who face the cultural dislocation and political disenfranchisement common among economic refugees. By enacting such a shift, however briefly, the collective opens the possibility of a deeper and more permanent shift.

From an aesthetic/artistic standpoint, the project represents a fascinating set of overlapping cultural norms and changes—and calls into question numerous assumptions. For instance, smart phones are frequently disdained as an enemy of "real" culture; yet in the case of Pakhtun migrants, they enable not only the sharing of traditional folk dances, but their preservation and perpetuation. In preserving the songs, the Tentative Collective, in turn, helps perpetuate traditional culture in the face of crushing population shifts.

The film of the event won the juror's prize award at the Mohawk Hudson Regional Exhibition, University Art Museum, UAlbany, 2012 and was selected for screening at the Syracuse International Film Festival of 2012.

在路边
On the Side of the Road

艺术家："暂定集体"团队
地点：以色列 / 巴勒斯坦，西岸和特拉维夫
形式和材料：影像
时间：2011 年至今
委托人：捐赠和个人资金
推荐人：格雷戈里·道尔

像许多当代艺术家一样，摄影师们自发组成名为"暂定集体"团队，正在探索替代几百年来一直控制展览的画廊系统。与那些严格注重审美的介入措施不同，"持续活性"团队关心社会变化，尤其是中东地区，因为很多摄影师精选于以色列和巴勒斯坦。

集体的目标是清晰的。这个群体的网上宣言是："作为摄影师，我们相信照片的力量，能够塑造公众态度，提高平常公共话语问题的认识。我们把它们视作反对一切形式压迫、种族歧视和基本权利和自由侵犯的一部分。"不出意料，"持续活性"发现了传统的画廊方法——在限制社会转型的目的下，"建筑"将观众尽可能置于艺术品的主体。因此，这个群体推出了"在路边"，这是一个正在进行的街头展览系列。

"2005 年我们首次发起这一行动时，街头摄影展览并不常见，""持续活性"成员希拉·格林鲍姆说，"现在在以色列和巴勒斯坦，这些已经众所周知。"在反对约旦河西岸城镇碧林以色列建造的"隔离墙"时，"在路边"的思想产生。摄影师发现主流媒体对他们记录的抗议活动的照片不感兴趣，所以他们就自己动手了，将其打印在 A4 信纸上，并将其钉在公共墙上。

大众反响很迅速，包括涂鸦、涂改以及积极的反馈。各种反响向群体证实，这些方法是值得追求的。

"从很早开始，我们就知道工作的每一个回馈都是成功的标志，"格林鲍姆说，"不管我们的工作是否被诋毁或赞扬，我们认为将图片展出是重要的场所政治化。现在公民政治领域参与度很低，我们看到，每一个行动即使是为了对付我们的观点，但作为斗争的一部分，也能不断大众化公共空间，扩展了言论自由的范围。"

项目开始的时候，社会媒体照片共享激增，"持续活性"的图像照片获得了声誉。因此这些图片的摄影师也经常出现在主流和替代媒体中。

事实上，"在路边"的想法已经被该地区的其他组织采用。"它成为了一个模型，现在被以色列和约旦河西岸其他当地活动家等不同的群体使用，开始移动展览，分享关于他们斗争的视觉和文本信息，"格林鲍姆说。其中的一些展览组成是从集体"租借"的图片，其他由模仿者则从头开始创建。

在吸引公共场所观众方面，他们保持着能力。"最让我们满意的是"格林鲍姆说，"左翼和右翼观众继续参与我们的照片并给予反馈。"

以色列国际摄影节期间，该群体被邀请参加项目部分展览的。

墙下的困惑

马熙迮

为了预防巴勒斯坦激进分子的袭击，以色列强行修建"隔离墙"加强警戒，将西耶犹太人居住区与东耶阿拉伯人居住区隔离开，以此来维持西耶犹太人的安全。这种行为遭到国际社会的普遍谴责，在某种程度上也激化了以色列与巴勒斯坦间的矛盾。"隔离墙"割裂巴勒斯坦的社会和家庭，导致严重的人道主义危机。很多时候人们为难民鸣不平，提抗议，但却没有解决他们最初为什么会成为难民的问题。

作品"在路边"试图通过艺术介入，引发人们对公共问题的探讨与认识，从侧面反映以色列在约旦河西岸建造"隔离墙"所引发的各种抗议声音，它呼吁人们关注到身边正在发生的一切压迫与歧视行为。一味地选择默许，只是面对乱象的暂时对峙法，解决的途径是主动地面对问题。

Artist: Tentative Collective
Location: West Bank and Tel Aviv, Israel / Palestine
Forms and Materials: Image
Date: December 2011
Commissioner: Donation, Personal Funds
Researcher: Gregory Door

Like many contemporary artists, the photographers who make up
the collective Activestills are exploring alternatives to the gallery
system that has dominated exhibition for hundreds of years. Unlike
those who focus on strictly aesthetic interventions, however, the
Activestills collective, as its name implies, are concerned with
social change—specifically in the Middle East, as many of the
photographers involved are of Israeli and Palestinian extraction.

The collective is explicit in its aims. The group's web description
serves as a manifesto: "As photographers, we believe in the power
of images to shape public attitudes and to raise awareness on issues
that are generally absent from public discourse. We view ourselves
as part of the struggle against all forms of oppression, racism, and
violations of the basic right to freedom." Not surprisingly, Activestills
found the traditional gallery approach—which "curates" an
audience as much as a body of artwork—limiting to their aim of
social transformation. As a result, the group launched On the Side
of the Road, an ongoing street exhibition series. Each intervention
consisted of large-scale printed photographs displayed on walls in
public spaces traversed by thousands.

"When we initiated this practice in 2005, street photo exhibitions
were not common in the region," says Shiraz Grinbaum, an
Activestills member. "Today, they are well known in Israel and
Palestine." The idea for On the Side of the Road came during
demonstrations against the Israeli-built "separation wall" in
the West Bank town of Bil'in. When the photographers found
mainstream media outlets were uninterested in their photographs
documenting the protests, they took matters into their own hands
by simply printing them out on A4 letter paper and tacking them on
public walls.

The reaction was immediate, and included graffiti and defacement, as well as positive feedback. The variety of responses convinced the group that the method was worth pursuing.

"Very early on, we understood that every reaction to the work is a mark of success," says Grinbaum. "Whether our work is vandalized or praised, we see this as an important politicization of the space in which the images are exhibited. With today's low level of civil involvement in the political sphere, we see every action, even when it comes to counter our point of view, as part of the struggle to continuously democratize public space and to extend the boundaries of freedom of speech."

Since the start of the project, social media photo sharing has exploded, and Activestills has developed a reputation for its imagery of the conflict. Its photographers are regularly featured in the mainstream and alternative press.

In spite of these developments, the On the Side of the Road exhibitions continue. In fact, the idea has been adopted by other groups in the region. "It's become a model that is now used by other local activists in Israel and the West Bank as different groups initiate mobile exhibitions to share visual and textual information regarding their struggles," says Grinbaum. Some of these exhibitions consist of images "on loan" from the collective; others are created from scratch by imitators.

And they retain their ability to attract audiences in public spaces. "To our great satisfaction," Grinbaum says, "both left-wing and right-wing audiences continue to engage with our photos and react to them."

The group was invited to exhibit part of the project during Israel's International Photography Festival.

Y 项目
Project Y

艺术家：拉维·阿加瓦尔
地点：印度新德里
形式和材料：景观
时间：2011 年 11 月 9 日—11 月 23 日
委托人：未知
推荐人：格雷戈里·道尔

"Y 项目"是汉堡易北河畔与新德里的亚穆纳河畔的公共艺术与外展计划的一部分。组织者说，位于印度新德里的"Y 项目"是该市为数不多的城市公共艺术项目之一。因此，该项目的发展目标就更加引人关注。"Y 项目"属于公众介入项目，负责人拉维·阿加瓦尔说："我们要处理好现场特性，历史、政治、艺术及未来的不确定性。因为它就坐落在新德里的亚穆纳河畔，涉及到当下公共话语、政策、法权与维权行动。"

亚穆纳河已遭污染，这在很大程度上仍未引起人口快速增长的新德里市的重视，该市目前拥有约 1 700 万人口。人口增长给亚穆纳河带来的压力是双重的：随着城市居民日益增多，河流流域受到的污染也日益加剧。再者，发展也给原来的湿地造成了威胁，其结果是，湿地没有了水，上面建起了人行道，更加破坏了河流奔流不息的景象，与此同时，也让那些过去居住在河岸上的低收入渔民和农民流离失所，无家可归。

"Y 项目"旨在通过多方途径包括讲习班和研讨会以及沿河摆放临时艺术展品增强人们对上述问题的意识。阿加瓦尔说，该项目的一个关键目标就是直接向场所营造方面靠拢——并不是为了环境建设，而是为了提高环保意识。该项目意把河畔改造为公共空间，吸引更多潜在公众，并注重"生态美"，尽管这条河流已被污染。

亚穆纳河岸持续两周时间的临时工程包含了各种物品与体验：濒危物种雕塑，传统音乐演奏以及由垃圾建造的浮船。该作品推广到地区学校和公共大众当中后，产生了相当的影响。"该项目吸引了成百上千来自各行各业的人们，"阿加瓦尔说，"不少人是第一次见到亚穆纳河，发现它竟然就在这个城市里，然后就感到很惊讶。有的人说现场'很漂亮'，设计的景观应当保留下来。"

该项目增强意识的目标带来了实实在在的效果，尤其是对于那些生活在这座城市但过去忽视亚穆纳河存在的人们。"该项目也对决策者产生了影响，"阿加瓦尔说。对该艺术展品现场拥有管辖权的新德里发展局停止了当地的大型施工建设，希望能够更好地做好沿岸的生态恢复与自然保护。

水蕴万物之悠
马熙逵

亚穆纳河发源于喜马拉雅山脉贾姆诺特里附近，是恒河最长的支流。然而，随着城经济发展，人口数量的几何增长，现代工业废水的乱排乱放以及城市垃圾污染等因素，造成亚穆纳河污染问题日益严重，并导致多条河流的生态环境遭到破坏。"Y计划"旨在通过一系列公共空间的改造计划，以艺术介入公众生活，揭示生态平衡的重要性，唤醒公众对于环境保护的意识。艺术家希望更多的人们能够懂得，对于水资源的保护，单以加强环境执法和提升水资源利用率是不够的，更重要的是公众的参与和认识。自然环境是人类栖息之地，也是人类生活的物质之源。保护好人类的生态资源、创造优美的生活、娱乐环境，不仅是为我们这一代人着想，而且关乎我们子孙后代的生存、生活条件。

Artist: Ravi Agarwal
Location: New Delhi, India
Forms and Materials: Landscape
Date: November 9-23, 2011
Commissioner: unknown
Researcher: Gregory Door

Project Y was one half of a public art and outreach program that took place on the banks of Hamburg's Elbe River and New Delhi's Yamuna River. Organizers say that Project Y, based in New Delhi, India, is one of the city's very few public art projects. The ambitions of the project are the more notable for this fact. Project Y was "a public intervention," says curator Ravi Agarwal, "dealing with site specificity, its history, politics, aesthetics, and possible futures. It was located within the ongoing public discourse, policy, legal, and activist initiatives in Delhi about the river."

The polluted Yamuna River is largely overlooked by the growing population of the city of New Delhi, which boasts a population now hovering around 17 million. The pressures of this population growth on the river are twofold: as the number of urban-dwellers grow, so does the amount of pollution in the river waters. At the same time, development pressures are encroaching on former wetlands that are being drained and covered with pavement, further robbing the river of its ability to thrive while simultaneously displacing the lower-income fishermen and farmers who formerly dwelled on the riverbanks.

Project Y was designed to raise awareness of these issues through a multifaceted approach that included workshops and seminars,

as well as a temporary art installation along the river. A key aim of the project, according to Agarwal, was directly connected to placemaking—not in the form of built environment, but in the form of raising awareness. "The project was meant to help situate the river front as a 'public' space with its many potential publics, which is 'ecologically beautiful' even though the river is polluted."

The temporary works that stood on the banks of the river for two weeks included a range of objects and experiences: sculptures depicting threatened species, performances by traditional musicians, and a floating vessel constructed from litter. Combined with outreach to area schools and the general public, the artworks had an impact. "The project attracted hundreds of people from all walks of life," says Agarwal. "Many visited the river for the first time, and were surprised that the river existed in the city at all. Several described the site as 'beautiful' and a landscape worth preserving."

The goal of raising awareness has resulted in tangible consequences, especially for people living in the city who had previously overlooked its river. "The project also had an impact on policy makers," says Agarwal. The Delhi Development Authority, which has jurisdiction over the site of the art installation, jettisoned large-scale construction in the area in favor of more natural restoration/preservation of the shoreline.

帕拉延
Palayan

艺术家：阿克沙伊拉吉·辛格·拉索尔
地点：印度拉贾斯坦邦帕尔塔普尔
形式和材料：景观
时间：2010 年 1 月 17 日
委托人：参德巴艺术工作室
推荐人：格雷戈里·道尔

"帕拉延"在印地语里对应的意思是"迁移"，它是印度裔艺术家阿克沙伊拉杰·辛格·拉索尔在驻地参德巴艺术工作室时创作的特定场域的交互式介入项目。拉索尔要求穆斯林居民与印度居民在穿越两边共有的街道之前，用白漆把双脚浸湿。由此产生的路就形成了一幅反映两边居民迁移情况的临时地图。

阿克沙伊拉杰·辛格·拉索尔住在印度拉贾斯坦邦班斯瓦拉县帕尔塔普尔镇。尽管帕尔塔普尔镇从严格意义上来说隶属于拉贾斯坦邦，它却位于古吉拉特邦与中央邦的边境处，两邦各有自己的语言和文化。特别值得注意的是，印地镇是很多穆斯林居民的居住地。

布哈瓦迪穆斯林居民区引起了拉索尔的格外注意。该地区是布哈逊尼派穆斯林的聚集地，居民主要是妇女和儿童，因为男子都到中东去找工作了。"布哈瓦迪的边境地带就是印度教居民与穆斯林居民互相碰到的地方，"拉索尔写道，"这里充满了竞争，两边的居民都认为这里属于他们自己所在的地区。"他选择了一条布哈穆斯林居民与印度教居民共用的街道来实施他的计划。

在此期间，有一天，拉索尔要求所有穿越街道的人用白漆把双脚浸湿。随着这一天的过去，"候鸟迁移"的模式出现了。

"对观众来说，这是很有趣的体验，因为情况会变得复杂起来，"拉索尔写道。"在早上，他们可以辨别出第一个人的脚印；到了中午，已经很难辨别出某个人的脚印；等到了晚上，看起来已经成了一条白色的路，就像森林里动物的踪迹一样，你根本很难准确分辨出是哪个动物先来过。"

自然而然地，阿克沙伊拉杰·辛格·拉索尔在实施计划的时候，和无数两边的居民交谈起来，他还集合了一批围观者参与进来。"该计划本身很具有参与性，促进了大量人与人之间的互动，是非常有趣的交流体验，"他说，"它改变了参与者对宗教、地区归属及土地所有等狭隘观念的看法。

这个计划通过让人们更直接地互相接触和解决争端，引起了社会对于不同宗教地区矛盾的关注。同时，留在街道上的赤足脚印凸显了人类的普遍性。它也涉及了环境艺术，强调了人类穿越空间的非线性基本方式。几天之后，来往穿梭的脚步拭去了先前的脚印，让人不禁觉得人生短暂。

正如他所说，拉索尔常常从事于此类短暂的或"不走寻常路的"项目。"艺术仍然可以与社会展开有意义的对话，"就是如此，拉索尔说道。

理解是沟通之源
马熙迨

生活在帕尔塔普尔镇的居民，同时存在两种宗教信仰，即印度教与伊斯兰教。因教义的差异，双方在历史上有多次冲突，抱怨和怀疑的情绪意识已深深埋根于印度教徒与穆斯林之中。艺术家拉索尔希望通过"帕拉延"的实施，化解因宗教分歧而潜藏在社区的邻里矛盾。该计划旨在通过艺术介入公共空间的形式，让不同信仰的人们参与到群体互动，在交流中改变人们原有对宗教、地区归属及土地所有等狭隘观念的看法。争端的解决需要双方的接触与谅解，心灵的沟通增进相互的理解与信任。"帕拉延"通过艺术与社会的对话，引发群体间的交流体验，引导人们以一种冷静的处世态度去化解矛盾。在人类的世界里，永远忌讳的是猜忌和不信任，人与人相互的理解是给予人性最大的关怀。

Artist: Akshay Raj Singh Rathore
Location: Partapur, Rajasthan, India
Forms and Materials: Landscape
Date: January 17, 2010
Commissioner: Sandahbh Art Residency
Researcher: Gregory Door

Palayan, named for the Hindi term for "migration," was a site-specific, interactive intervention by Indian-born artist Akshay Raj Singh Rathore during an artist residency through Sandahbh Artist Workshop. Rathore asked Muslim and Hindi residents of two adjacent neighborhoods to dip their feet in white paint before traversing a street common to both communities. The paths that resulted served as a temporary map of the migration patterns of the neighborhoods.

Rathore's residency took place in the town of Partapur, in the Banswara district of the state of Rajasthan, India. Although Partapur is technically in Rajasthan, it is situated at the border of the states of Gujarat and Madhya Pradesh, each with its own culture and language. In particular, the Hindu town is home to a significant minority of Muslims.

The Muslim neighborhood of Bohrawadi particularly captured the artist's attention. Home to a sect called the Sunni Bohra, the neighborhood was populated primarily by women and children as the men had all migrated to jobs in the Middle East. "The edges of Bohrawadi are where the Hindus and Muslims come face to face," Rathore writes, "they are the contested spaces, where the two communities lay their claim." Rathore chose one of these border streets shared by the Bohra and their Hindi neighbors to enact the project.

Over the course of one day, the artist asked each person traversing the street to allow his or her feet to be dipped in white paint. As the day progressed, "migratory" patterns emerged.

"It was an interesting experience for the audience, as the work kept on getting complex," writes the artist. "In the morning, they could identify the first man's trail. By noon, it had become difficult to identify the individual, and by the evening, it all started looking like white paths, like a game-trail in a forest, where you can't pin-point which animal made it first."

Naturally, Rathore spoke with numerous members of the community and collected a crowd of onlookers while accomplishing the project. "The participatory nature of the project enabled great interaction with the people and very interesting exchanges," he says, "it changed the audience's perspective on narrow ideas of religion, community belonging, and ownership of earth."

The piece drew attention to the differences among the religious neighbors by putting people into more direct contact and addressing their difference directly. At the same time, the bare footprints painted on the street emphasized human universality. The piece also spoke to environmental concerns, emphasizing the non-linear and organic ways that human beings traverse space. After a few days, the paint was obliterated by foot-traffic, suggesting human transience.

Rathore frequently works on ephemeral projects of this sort, or "outside of the white box," as he puts it. That's where "art can still provide meaningful dialogs to society," he says.

布的力量
The Power of Cloth

艺术家：罗可汗·阿派迪亚
地点：印度拉贾斯坦邦瓦嘎达
形式和材料：装置艺术
时间：2009 年
委托人：印度当代艺术基金会
推荐人：格雷戈里·道尔

在"布的力量"中，艺术家罗可汗·阿派迪亚征集印度拉贾斯坦邦瓦嘎达农村地区的居民来建立装置，用来探索和质疑该地区基于种姓制度的婚姻传统。阿派迪亚是瓦嘎达本地人，在当地手工艺人的帮助下，他建立了有关传统婚礼结构的艺术渲染。结构包括一个钢脚手架，挂上由当地捐赠的剩余布料，作品的形式是传统的婚礼帐篷或沙米娜（布天篷）。里面有两个巨大的椅子，在传统婚礼上是由新娘和新郎坐的。在结构中结合了来自本地婚礼的照片。

瓦嘎达农村地区是传统部落族地区，很多仍然实施嫁妆制度（自 1961 年起不合法），在该制度下，要求新娘家给新郎家礼物。一开始，这个制度是为了规避禁止女儿继承的问题，嫁妆制度受到了匹配，导致了针对一些女性的暴力行为，并物化所有女性。这个制度还被指责以契约约束有女儿的家庭，并提供诱因基于性别终止怀孕。由于非法嫁妆经常藏在或掩盖在布的礼品中，"布的力量"为正在进行嫁妆制度作恶，损害女性。

不是通过对抗的形式解决这些问题，阿派迪亚设计了一个参与式过程，在他的项目中邀请村民参加。"根据他们遵循的社交礼仪和传统，这个项目遵循了一系列当地公众的对话和讨论，"艺术家解释道。

在实际建造和安装过程中，阿派迪亚邀请了当地手工艺人的帮忙。一些艺人提供了金属加工或木器，一些艺人提供了用于帐篷壁的织物。这个协作方法是项目影响的关键，艺术家指出。首先，它提供了一个机会，"在纯创意层面"，通过把他们引进艺术创作过程，"为当地手工艺人发现新的机会"。作为回报，这些炼铁工人、裁缝和木匠告知项目的概念。

同样重要的是，项目的协作形式提供了机会来分享和参与作品的主题。尤其是，阿派迪亚征集了"一组社区妇女来帮助建立（作品）结构"，阿派迪亚说。在这个建造阶段，这些妇女"通过分享她们有关婚姻和相关问题的长对话，发展了友谊。"

邀请社区讨论的另外一个策略就是阿派迪亚对沙米娜布的设计。帐篷壁通常印有华丽的设计和漂亮的风景，但是在这个情况下，阿派迪亚印上了从当地居民结婚画册收集来的照片。结果，以人性化和富有审美情趣的方式强调了选择、传统以及地区的种姓差异。村民乐意发现他们的老照片被复制——并置使他们能够就本地传统进行进一步对话和反思。

通过提供多个切入点，切入点的作品获得各种不同的反应。"观众以不同的水平参与其中，"他说，"有些人通过坐在椅子上或者在沙米娜壁上发现他们的照片——或者唱歌跳舞，就像他们在婚礼上那样——在装置中发现自己。"

这件作品从瓦嘎达农村地区穿行到斋蒲尔市，从那里又到一个国际学校，最近在特拉维夫市展出。

婚姻是一抹神圣的轻纱

马熙逵

印度种姓制度将人进行等级划分，由于等级差异，人们在社会、宗教、法律中享受的权利均有所不平等。基于种姓制度而产生的内婚制，划清了不同种姓间的界线，又进一步剥夺了女性在婚姻中的自由与平等。由于瓦嘎达尚属于农村部族区域，仍实行不合理的婚嫁制度，男权的社会中，女人是绝对的私有财产，"男尊女卑"的现象屡见不鲜。"布的力量"是对这种损害女性权利行为的叩问，是对现时代下愚昧种姓划界的反思。艺术家希望"布的力量"能给人们带来启示，揭示非法婚嫁制度的肮脏和卑劣，同时团结当地女性，寻求在婚姻中的平等权利。现时代的哲学告诉我们，完美的婚姻需要男女双方的感情投入与悉心爱护，恩爱的夫妻必须建立在彼此间的尊重之上。女性的解放不仅是女人份内的事情，更需要全社会的共同努力。

Artist: Lochan Upadhyay
Location: Vagad, Rajasthan, India
Forms and Materials: Installation Art
Date: 2009
Commissioner: Foundation for Indian Contemporary Art
Researcher: Gregory Door

In The Power of Cloth, artist Lochan Upadhyay enlisted residents of rural Vagad, Rajasthan, India in creating an installation that probed and questioned the caste-based marriage traditions of the region. Upadhyay, a native of Vagad, built, with the help of local craftspeople, an artistic rendering of a traditional wedding structure. Constructed of a steel scaffold hung with cloths made from local, donated cast-offs, the piece is in the form of a traditional wedding tent, or shamiana. Inside are two oversized chairs, which in a traditional wedding would belong to bride and groom. Photographs from local weddings were incorporated into the structure.

The rural Vagad region is home to traditional tribesman, many of who observe a dowry system (illegal since 1961), under which the bride's family is expected to give gifts to the husband's. Originally intended as a way to get around the prohibition of passing inheritances to women, the dowry system has been criticized for contributing to violence against some women while objectifying all. It's also blamed for indenturing families with daughters and providing an incentive to terminate pregnancies based on gender. Because illegal dowries are often hidden or disguised under cloth gifts, one "power of the cloth" is to perpetrate the ongoing dowry system to the detriment of women.

Rather than address these issues in a confrontational manner, Upadhyay devised a participatory process to involve the villagers in his project. "The project followed a series of conversations and discussion amongst the local public in context of the social norms and customs that they live by," explains the artist.

In the practical construction of the installation, Upadhyay invited

local craftspeople into the process. Some provided metalwork or carpentry, while others gave the textiles used to create the walls. This collaborative approach was key to the project's impact, the artist says. For one thing, it provided the opportunity, "on a purely creative level, to unearth new opportunities for local craftspeople" by introducing them to the artistic process. In turn, the ironworkers, tailors, and carpenters informed the project's concept.

Equally importantly, the collaborative nature of the project gave the opportunity for sharing and engagement with the subject matter of the piece. In particular, Upadhyay enlisted "a group of community women to help build up the structure" of the piece, says Upadhyay. During this construction phase, the women "developed relationships of friendship and exchange through long conversations sharing their experiences with marriage and the issues related to it."

Another strategy to invite community discussion was Upadhyay's design of the shamiana's cloth. The tent walls are typically printed with ornate designs and beautiful scenes, but in this case, Upadhyay printed photographs collected from local resident's wedding albums. As a result, the choices, traditions, and caste differences of the area were highlighted in a humanizing and aesthetically pleasing manner. Villagers enjoyed finding their old photographs reproduced—and the juxtapositions allowed further conversation and reflection on the local traditions.

By providing these multiple entry-points, Upadhyay's piece allowed a variety of responses. "The audience participated at different levels," he says. "Some identified themselves in the installation by sitting on the chairs or finding their photographs on the side walls of shamiana—or singing and dancing as they do for marriage ceremonies."

The installation traveled from rural Vagad to the city of Jaipur, and from there to an international school, and more recently to Tel Aviv.

卡利·卡麦
Kali Kamai

艺术家：万佛朗·迪恩托
地点：印度西隆
形式和材料：装置艺术
时间：2009—2011 年
委托人：印度当代艺术基金会
推荐人：格雷戈里·道尔

摄影师兼艺术家万佛朗·迪恩托探讨了"卡利·卡麦"的社会空间——印度种族多样化西隆地区的公共出租车。迪恩托以一种实际的形式，创建一个交互式、移动展览，发挥"卡利·卡麦"的作用，艺术家专注于口语、音乐传统、跨文化交际和种族间紧张关系的处理。

在印度，西隆地区的文化多样性有一些原因。首先这是卡西族的故乡——一个深深植根于地区的土著部落。迪恩托认为这是世界上现存最古老的母系文化之一。殖民统治时期，印度的西隆地区以其自然资源地位不可撼动，成为一个卡西族以外包括英国、孟加拉、尼泊尔、比哈尔、马拉尼等人们居住的重要地方。

万佛朗·迪恩托说："联婚以及与外部的贸易和商业，一直塑造着我们的身份。自 20 世纪 70 年代以来，西隆已经经历了多次种族骚乱。"他继续补充，"这种种族骚乱不仅存在于处于共享状态的各种文化之间，而且存在于卡西族各派系间。在乡村地区，更传统的部族成员反对'外来人员'，而各种政治利益探索个人利益的这些分歧。"

卡西族代表为数不多的公共空间，迫使这些不同的群体亲近。"这些出租车在西隆穿梭工作，接送附近的人们，费用也相对较小"迪恩托解释道，"该车可以乘坐五名乘客（包括司机），但车里通常会塞满 8—9 人，他们全部挤在一起，不分年龄、性别、种族或阶级，是所有社会部门强烈模糊的完美缩影。"

对于在西隆出生和长大的艺术家，"卡利·卡麦"代表一个奇怪的社会空间，行为的正常标准受到挑战。"我发现，观察人们在这个空间的行为——肢体语言、口头交流是非常有趣的，"他解释说，"在这些空间，阶级界限一度模糊，无法接受外来人员的恐惧变得无关紧要，因为人们别无选择，只有互动。"

最初，迪恩托计划了一个音频项目，由"卡利·卡麦"记录的对话在其他位置重播组成。在探索项目的同时，这变成了一个移动交互式的"卡利·卡麦"。

为了做到这些，迪恩托获得一辆汽车并改造它为一个旅游展览以及功能出租车。艺术家在汽车的外观，画上了传统的卡西口述历史，而在汽车内部也有各种互动元素包括一个可以显示从现在到历史性的西隆场景的触屏电脑，一个混合卡西族谈话记录和20世纪70和80年代电台广播的音乐。这些不但不收费，司机还会给出"科维"，这是一种带着包叶的槟榔传统小吃，被普遍视为欢迎的方式。

艺术家认为，这个项目是西隆地区第一个公共艺术，也是一个独具创造性和有趣的创意场所营造方式。迪恩托介入解决现实生活中影响参与者日常生活的事情。同时，特定场地巧妙地利用资源，以熟悉的方式再利用，使参与者去质疑自己的假设。

处世的艺术
马熙迤

卡西族是印度的少数民族之一，在婚姻方面仍然保持着母系氏族组织社会。丈夫随妻而居，但时常回到自己母亲的氏族生活。他们实行母系继承制，家庭中的一切财产，均由幼女继承。印度殖民统治时期，迫于殖民者的划定，自然资源开采成为当地单一式经济形态的主要内容，此地混杂卡西族以外包括英国、孟加拉、尼泊尔、比哈尔、马拉尼等人群居住。自20世纪70年代以来，因信仰分歧，已引发西隆地区多次种族骚乱。

作品"卡利·卡麦"专注于民族和谐的回归，它通过穿梭于西隆的出租车空间，向当下公众展示传统卡西的口述历史，包括混合卡西族谈话记录和20世纪70—80年代电台广播的音乐。艺术家希望通过这一系列的艺术行为能影响到人们最日常的生活状态中，促进不同群体间的亲近和理解。

Artist: Wanphrang Diengdoh
Location: Shillong, India
Forms and Materials: Installation Art
Date: 2009-2011
Commissioner: Foundation for Indian Contemporary Art
Researcher: Gregory Door

This project, by photographer and artist Wanphrang Diengdoh, explored the social space of the Kali Kamai—the communal taxicabs of India's racially diverse Shillong region. Diengdoh created an interactive, mobile exhibit in the form of an actual, functioning Kali Kamai that focused on oral and musical traditions, cross-cultural communication, and racial tensions.

Within India, the Shillong region is culturally distinct for a number of reasons. It is home to the Khasi people, an indigenous tribal group with deep roots in the region and, according to Diengdoh, one of the oldest surviving matrilineal cultures in the world. During the colonial period, the Shillong region held vital importance for India' s natural resources and became a key point of settlement for non-Khasi, including the British, Bengali, Nepali, Bihari, and Marwari.

"Intermarriage, as well as trade and commerce with outside communities, has always shaped our identity," says Diengdoh. "And since the 1970s, Shillong has witnessed repeated bouts of racial turmoil." This turmoil, he continues, exists not only between the various cultures who share the state, but also within various factions of the Khasi. In rural areas, more traditional tribalists rail against "outsiders," while various political interests exploit these divisions for personal gain.

The Kali Kamai represent one of the very few public spaces that force these disparate groups into close—often extremely close—proximity. "These taxis ply their trade all over Shillong, ferrying people around for a relatively small fee," explains Diengdoh. "Made to seat five passengers (including the driver), they are usually stuffed

with eight to nine people, all squeezed in with little regard for age, gender, race or class—a perfect microcosm where any and all social divisions are forcibly blurred."

For the artist, who was born and raised in Shillong, the Kali Kamai presented a curious social space where normal standards of behavior became challenged. "I find it extremely interesting to see how people behave in these spaces—the body language, verbal exchanges, and interactions that take place," he explains. "In these spaces, class boundaries are blurred and the fear of the unwanted outsider becomes irrelevant as people have no choice but to interact."

Diengdoh initially planned an audio project consisting of recorded conversations taken from within Kali Kamai replayed at some other location. As he explored the project, he expanded it to become a mobile, interactive Kali Kamai.

In doing so, Diengdoh obtained an automobile and altered it to become a traveling exhibition—as well as a functioning taxicab. Artists painted traditional Khasi oral histories on the exterior of the car, while inside a variety of interactive elements were deployed including a touch-screen computer that displayed scenes from both present-day and historic Shillong, and a soundtrack that mixed music with recorded conversations from other Kali Kamai and found-sound from radio broadcasts of the 1970s and 1980s. Instead of collecting a fare, the driver gave out kwai, a traditional snack of betel nut with paan leaf that is universally seen as a gesture of welcome.

According to the artist, the project was the first public art installation in the Shillong region—and it was also a creative and interesting expression of creative placemaking. Diengdoh's intervention addressed real-life issues of substance that impact the daily lives of participants. At the same time, the piece makes ingenious use of a site-specific resource, repurposing the familiar in a way that causes participants to question their own assumptions.

阿富汗的街道
Streets of Afghanistan

艺术家：珊侬·高玭
地点：阿富汗伊斯塔立夫、喀布尔、潘杰希尔峡谷、巴博花园、喀布尔动物园
形式和材料：影像
时间：2011 年
委托人：未知
推荐人：格雷戈里·道尔

"阿富汗的街道"是珊侬·高玭发起的项目。珊侬·高玭是一位经常在阿富汗工作的杰出的活动家、自行车运动员以及女权倡导者。珊侬·高玭"山到山"组织的创始人，与美国和海外的摄影师合作，收集了一系列照片，将它们安装在大幅面轻便式织品上，安排在阿富汗各地进行街头展览。

"阿富汗的街"最初的构思是举办博物馆画廊展览，从而使美国更加了解阿富汗人的复杂需求，之后迅速演变成阿富汗人民的街头展览。

"在巴黎或伦敦举办一个画廊展览非常容易"，珊侬·高玭在被评为 2013 年十佳"本年度冒险家"后在国家地理的采访中解释道，"阿富汗人民没有杂志或画廊。他们看不到摄影师和记者拍摄的他们国家的无数照片。我希望阿富汗人看到他们自己的文化，他们自己的美丽。"

在快闪街头展览中，"阿富汗的街道"由"重点突出这个国家的美丽和心碎"的 32 张 10 英尺 ×17 英尺的照片构成。这些特约摄影师有托尼·迪·紫诺、贝斯·沃德、波拉·布朗斯坦以及纳吉布拉·穆瑟佛。展览遍及多个地点，包括阿富汗的城市和农村。这些照片，目前可通过书籍形式获得（一部分销售用来支持高玭的基金会），包括画像、街头摄影和自然景观。这些照片，在大幅面中复制出来，具有不可否认的真实性，以摄影传统工艺为基础。

从关于公共艺术的大规模讨论、社会实践、场所营造和社会行动这一背景来看珊侬·高玭是一个有趣的人。她最常把自己标记为女权活动分子和山地车手。她没有接受过正式的艺术培训，或至少在她的公共资料里没有这一方面的信息。

她所在组织与她的个性和头衔密切相关，正如她对科罗拉多的一家报纸所说，她关注了"以教育和艺术，给妇女、小孩、年轻人一个表达权利的平台"项目。她获得了巨大的成功，国际媒体在显著的位置刊登了她的信息，而且她拥有筹款的影响力。她负责筛选不可否认的好作品，很多作品都是

开创性的，如她帮助阿富汗妇女形成该国第一个女性单车活动联盟（宗教保守派对妇女的单车活动做出了制裁）。

同时，珊侬·高玭的作品在西方背景下，就东方视角提出了重要问题。她的作品特别是变恶为善的西方叙事：作为一个从强奸事件中恢复过来的年轻女子，她在过去的五年中一直致力于为阿富汗妇女提供更好的条件。在很多方面，她对西方式的女性权利赋予的执着，让人耳目一新。但是，如果她避开对文化的相对性进行细致的讨论，她也可能无法真正理解能帮助她更好地服务于她想帮助的人们的文化脉络。

虽然珊侬·高玭未将她自己标记为艺术家，但她确实和在公共艺术、街头干涉主义领域中担任策展人、联系人和推动人，和有着丰富的社会实践的艺术家一样，是"阿富汗的街道"进行巡回图片展的催化剂（在该图片展中，摄影师托尼·迪·紫诺被列为画册的共同作者）。

考虑到艺术家、社会活动家角色的流动性，珊侬·高玭作为一位非实践艺术家，选择获得公共艺术奖是非常有趣的。她的这一选择会对公共艺术和场所营造所扮演的角色、定义和身份提出更加深入的、更为复杂的问题。

蕴藏的美丽

马熙逵

阿富汗问题异常复杂，民族关系、部族关系、宗教关系以及与外部势力的关系，这一困境难以在短时间内解决。随着美军在阿富汗的撤离，发展经济、扩大开放、改善民生成为当下阿富汗的发展目标。作品"阿富汗的街道"题材随意，反映的现象是基于民生的真实，包括人物画像、街头随拍以及自然景致。艺术家珊侬·高玭希望通过街头的艺术展示，引发群体间的大规模探讨。通过真实的图像，反映阿富汗富有内涵的文化传统，让阿富汗民众感悟本国风光的无限美丽，树立民族自信。国家富强的道路，绝不会是一帆风顺的，它曲折崎岖，荆棘丛生，甚至会面对巨石断崖般的阻隔。此时，人们需要的正是以民族自信作为精神支柱，在探求国家富强的道路上披坚执锐，勇往直前。民族自信，是一种蕴藏的美丽。

Artist: Shannon Galpin

Location: Eight exhibitions in public spaces including: Istalif, Kabul, Panjshir Valley, Barbur Gardens, Kabul Zoo, Afghanistan

Forms and Materials: Image

Date: 2011

Commissioner: unknown

Researcher: Gregory Door

Streets of Afghanistan is a project launched by Shannon Galpin, a prominent activist, cyclist, and women's rights advocate who frequently works in Afghanistan. Working with photographers from the United States and abroad, Galpin, founder of the organization Mountain to Mountain, collected a series of photographs, mounted them on large-format, portable fabric, and arranged for street exhibitions of the works throughout Afghanistan.

Originally conceived as a museum gallery exhibit to raise awareness in the United States for the complex needs on the ground in Afghanistan, Streets of Afghanistan quickly morphed into a street exhibit for Afghanistan instead.

"It's easy to do a gallery show in Paris or London," Galpin explained in an interview with National Geographic, after being named one of the magazine's ten "Adventurers of the year" in 2013. "The Afghan people don't have magazines or galleries. They don't see the myriad of images that are taken of their country by photographers and journalists. I wanted Afghans to see their own culture, their own beauty."

Streets of Afghanistan consisted of 32, 10-by-17-foot images that "focused on highlighting the beauty and the heartbreak of the country" in a pop-up street exhibition. Among the photographers featured were Tony Di Zinno, Beth Wald, Paula Bronstein, and Najibullah Musafer. The exhibit traveled to a wide variety of locations, both urban and rural, in Afghanistan. The photos, which are currently available in book form (a percentage of sales supports Galpin's foundation), include portraits, street photography, and landscapes. Reproduced in large-format, the images have the undeniable authenticity and craft of the photographic tradition as a foundation.

In the context of the larger conversation about public art, social practice, placemaking, and social activism, Galpin is an interesting figure. She most commonly identifies herself as a women's rights activist and mountain-biker; she has no formal art training, or at least none that figures in her publicity materials.

Her organization, which is closely identified with her as a personality and figurehead, focuses on projects that "center around the idea of giving a voice to women, the children, youth through education, training and art," as she told a Colorado newspaper. She has met with remarkable success, and is prominently featured in international media and possesses fundraising clout. She is responsible for an undeniable number of good works, many of them groundbreaking. For example, Galpin is instrumental in helping women in Afghanistan form that country's first female cycling league. (Religious conservatives sanction women for riding bicycles).

At the same time, Galpin's work raises important questions about western perspectives in eastern contexts. Hers is a particularly western narrative of turning bad into good: Recovering from a rape as a young woman, she has devoted herself for the past five years to bettering women in Afghanistan. In many ways, her single-minded focus on Western-style empowerment of women is refreshing. But if she avoids getting bogged down in nuanced discussions of cultural relativity, she also risks missing cultural contexts that might help her better serve the people she seeks to help.

Even though Galpin doesn't identify herself as an artist, she has served as the catalyst for the Streets of Afghanistan traveling photo exhibit in much the same way that artists with a social practice serve as curator, connector, and enabler in the public art/street interventionist realm. (Photographer Tony Di Zinno is listed as a co-author of the coffee-table book based on the photo exhibit.)

Given the fluidity of the role of artist/social practitioner, Galpin, as a non-practicing-artist, would make an interesting choice to receive a public art award. Her selection would pose further and more complicated questions about the role, definitions, and identities surrounding public art and placemaking.

由内而外
Inside Out

艺术家：JR
地点：全球多个地方，包括厄瓜多尔、尼泊尔、巴勒斯坦和墨西哥
形式和材料：影像
时间：2011 年至今
委托人：艺术家发起
推荐人：里奥·谭

JR 是一个在巴黎、以色列和巴勒斯坦等备受国际关注大规模街头艺术作品的匿名艺术家。2011 年，他获得 TED 大奖，这使他继续推出基于全球社区参与、同一国际标准下的项目——"由内而外"。一直以来，艺术家各方合作，而"由内而外"也不例外。人们被邀请通过自拍展现他们的表情，分享他们的生活、故事、担忧和激情。用艺术家的话说，"这个项目的概念让每个人都有机会与世界分享他们的表情和他们的立场"。

"由内而外"作为一个国际项目，在自拍的肖像照片基础上，通过网站在线提交的形式收集更多照片。"由内而外"团队打印出数字图像并运送给参与者，他们将粘贴到自己选择的网站海报上。艺术家也会在公共场所、博物馆和画廊配备带有打印机的照相亭，这些亭子附近的人可以自发参与。最后，项目网站开始服务并继续服务，作为一个公共档案和所有提交作品的展览以及现场纪录片材料。

在众多"由内而外"国际例子中，有两个脱颖而出。特别值得提到的是："印度教育"和"真正的阿富汗庆祝活动"。第一张照片中，孩子举起小黑板指出他们的职业目标，如医生、警察、数学老师等。街上张贴的大型海报是为了提醒次大陆教育的重要性。鉴于印度强烈的父权文化和妇女在追求职业生涯面临的困难，图片中带有职业野心的女孩尤为重要。

"真正的阿富汗庆祝活动"发生在喀布尔古城，塑造了日常阿富汗人肖像，运用主流媒体和政治领袖传播对比这些阿富汗人。许多图像提醒街上和在线观众，尽管存在巨大的困难和政治动荡，但希望和欢笑依然存在。可以想象，这个项目帮助阿富汗的人性化的观念，允许将人们视作整体性，而不是恐怖分子、鸦片耕种者、贩子或需要国际援助的受害者。

如果论走访的数量，似乎"由内而外"是一个非常成功的公共艺术项目。艺术家说，自 2011 年推出后的两年来，来自 10 000 个城市超过 140 000 人参加了"由内而外"项目。艺术家一直出现在有影响力的平台如 TED 和

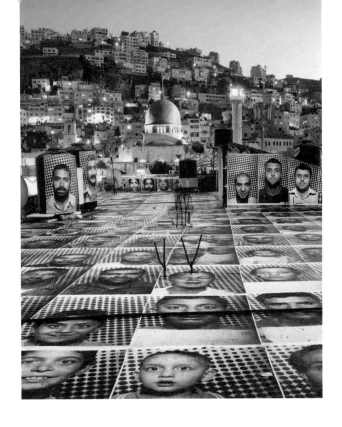

TEDx，而项目电影也于 2013 年在翠贝卡电影节首映。这个项目值得称赞，鼓励参与激活社区，同时为当地街头艺术行为提供一个高度自治。然而，它确实有争议性，在一定范围内，一些街头艺术行为在一些国家被视为违法。所以，一方面，它被视为犯罪推动者，另一方面，是公共空间回收再利用的促进者。

反思的历史

马熙逵

作品"由内而外"是一系列的艺术影像展示，题材包括文化、教育、民生等。艺术家希望通过这种形式向公众传递一种价值观念，即揭露现实的偏见，寻求民权抗争的合法性，同时流露出艺术家基于人道立场上对"底层"遭遇的同情。介于相互了解的不足，在西方社会的视阈下，尚未觉察第三世界国家的社会变革，很多印象蒙蔽于虚假的社会现象。"由内而外"不是简单的图像展示品，它所反映出的现实包含一种对全球历史观的呼吁，即作为世界公民的平等与自信。公民自信是一种较高层次的文化自觉，也是超越个体、地域和时空的心灵操守，面对现实不卑不亢，秉持自信的精神家园，更是一种民族意志力的比拼。作为世界整体的一个有机组成部分，第三世界的人民理应得到最广泛的尊重和认同。

Artist: JR

Location: Global, multiple locations, including Ecuador, Nepal, Palestine, and Mexico

Forms and Materials: Image

Date: 2011-present

Commissioner: unknown

Researcher: Leon Tan

JR is an anonymous artist who came to international attention for large-scale street art works in Paris, Israel, and Palestine. He received the TED Prize in 2011, enabling him to launch Inside Out, an international project engaging individuals from diverse communities worldwide in creating street art wheat-pastes in their localities. The artist has consistently worked with faces, and Inside Out is no different in this respect. People were invited to share their lives, stories, concerns, and passions through self-composed photographs showing their faces. In the words of the artist, "The concept of the project is to give everyone the opportunity to share their portrait and a statement of what they stand for, with the world."

As an international project, Inside Out functioned on the basis of portrait photographs submitted online through the Inside Out website. The digital images were printed by the Inside Out team and shipped back to participants, who pasted the posters in sites of their own choosing. The artist also installed photo-booths equipped with printers in public spaces, museums, and galleries, enabling people in the vicinity of these booths to participate spontaneously. Finally, the project website served, and continues to serve, as a public archive and exhibition of all the submitted works, together with on-site documentary material.

Of the many international examples of Inside Out, a couple stood out, which are worth specific mention: "Educating India" and "The Celebration of the Real Afghanistan." The first consisted of photographs of children holding up little blackboards noting their

professional ambitions, e.g. doctor, policeman, math teacher. The large format posters pasted on the street served as a reminder of the importance of education in the subcontinent. The images of girls with professional ambitions were particularly significant, given the fiercely patriarchal culture of India and the difficulties women face in pursuing professional careers.

"The Celebration of the Real Afghanistan" took place in the old city of Kabul, and featured portraits of everyday Afghanis in order to contrast these depictions of Afghanistan with the ones propagated by mainstream media and political leaders. Many of the images reminded street and online audiences that hope and laughter could exist in spite of tremendous hardships and political turmoil. Conceivably, the project helped to humanize perceptions of Afghanistan, allowing its people to be seen in their wholeness, and not just as terrorists, opium cultivators and traffickers or victims needing international aid handouts.

If numbers were anything to go by, it would seem that Inside Out was a highly successful public art project. According to the artist, over 140,000 people from 10,000 cities have participated in Inside Out in the two years following its launch in 2011. The artist has been featured on influential platforms such as TED and TEDx, while a film on the project premiered in 2013 at Tribeca Film Festival. The project is commendable for the way in which it activated communities by encouraging participation, while affording a high degree of autonomy to local street art actions. It does, however, have a contentious dimension, insofar as street art actions are considered illegal in some countries. On the one hand, it may be seen as an enabler of crime, and on the other, a facilitator of community reclamation of public spaces.

72 小时都市行动
72 Hour Urban Action

艺术家：凯雷姆·郝柏瑞特、吉利·卡耶夫斯基
地点：以色列巴特亚姆、德国斯图加特、意大利特尔尼
形式和材料：空间改造
时间：2010 年，2012 年
委托人：同当地合作伙伴协作委托，包括当地市议会
推荐人：里奥·谭

战略都市生活指的是越来越受欢迎的实践，即在都市环境中设计和实施简单的、相对便宜和迅速变化的生活方式，以展示长期转变的可能性。2010 年，在以色列，作为巴特亚姆景观都市主义双年展的一部分 "72 小时都市行动（72HUA）" 参与其中，并将自己描述为 "全球首个实时建筑竞赛，根据当地需求，十个国际团队用三天三夜的时间在公共空间设计和建造项目。" 在 2010 年到 2012 年期间，共有三次 "72HUA" 活动：首个活动在以色列巴特亚姆，第二个在德国斯图加特，第三个在意大利特尔尼。

首次竞赛发生在 2010 年 9 月。向十个国际团队发放项目简介，每个团队有三天三夜的时间来创建建筑干预，以改进特定公共空间的质量。一个被忽视的街道被选择用来作为场所营造的目标，团队根据当地的需求，如遮阴的座位区或者需要更新垃圾填埋区。这些团队不仅面对很紧的截止日期，还需要面对有限的预算和空间。另外，根据团队成员数量，他们需要生活和工作在现场，因此会很快融入当地社区的日常生活。根据组织者，这些困难的限制条件是草根项目的特点（相对于资金充足的从上而下的城市规划或再发展项目），并根据社区自我感知的需求，协助创新思考和想象。巴特亚姆都市行动的结果包括改造的公交站台、花园、为老年人建立的塔式建筑以及到达商业区的新入口。

斯图加特都市行动发生在 2012 年 7 月，也包括十个参赛的国际团队（120 多人），在三天三夜的时间内，在中心城区的沃根哈伦区域和诺德巴霍夫街的公共空间设计和建造建筑干预项目。特尔尼行动发生在 2012 年 9 月，这次包括五个团队，在相同的时间限制内改造一个住宅街区。此后，每年举行都市行动。

"72HUA" 是一个富有想象力的实用项目，在国际上很受欢迎。它受欢迎程度至少部分上与比赛的高度本地化场所营造干预在参赛者身上所产生的

赋权感相关，包括当地社区的成员。组织者正确地认识到时间、空间和财务的限制是促进集体创造重要因素。同时不幸的是，此类限制条件也可以抑制创意和阻碍集体行动。在"72 小时都市行动"中与众不同的一点可能就是对预算、时间和空间的限制采取了乐观的态度，并且简介明确识别了本地需求。该项目使得公共空间的集体实验成为可能，在各个社区的关键和创意干预中，使那些通常不会参与城市设计的人参与其中。这样一来，这成了草根场所营造的一个范例，当地社区可以针对不同的都市复兴挑战进行部署（在竞争框架范围内）。

创意与释放

马熙达

作品"72 小时都市行动"是艺术介入空间改造的项目，在 72 小时内完成对特定公共空间的改造，包括空间利用率、舒适度、生活品质等因素的考量，涉及巴特亚姆、斯图加特、特尔尼三个城市。项目的开展结合了当地的需求，以最低成本去探求空间改造的可能性，注重成员的协作与创新意识。客观的思考和对艺术的审美活动是人类追寻纯粹快乐的途径，繁忙的都市生活缺乏的正是这个层面上的内涵，打破人之价值的禁锢便是对知识、理念创新的欲求。在"72 小时都市行动"的创意干预计划中，以创新为理念，围绕社区和谐展开各种改造计划，包括功能改造和精神抚慰。同时与城市设计没有直接关系的社区成员也转化为计划的执行者，通过公众参与，互助交流，增进彼此理解，营造乐观和谐的社区氛围。

Location: Global, multiple locations, including Ecuador, Nepal, Palestine, and Mexico
Forms and Materials: Building, Installations
Date: Projects took place in 2010 and 2012 in the space of 72 hours.
Commissioner: Commissioned in collaboration with local partners including local municipal councils
Researcher: Leon Tan

Tactical urbanism refers to the increasingly popular practice of devising and implementing simple, relatively inexpensive, and quick changes in an urban environment in order to demonstrate possibilities for longer-term transformation. Founded in Israel in 2010 as part of the wider Bat-Yam Biennale of Landscape Urbanism, 72 Hour Urban Action (72HUA) belongs to this movement and describes itself as "the world's first real-time architecture competition, where ten international teams have three days and three nights to design and build projects in public space in response to local needs." Between 2010 and 2012, three 72HUA events took place: first in Bat-Yam and subsequently in Stuttgart, Germany, and Terni, Italy.

The inaugural competition took place in September 2010. Project briefs were issued to ten international teams, each of which had three days and three nights to create architectural interventions to improve the quality of specific public spaces. A neglected street was chosen as a placemaking target with briefs highlighting local needs for things such as sheltered seating areas and the renewal of a rubbish-filled plot of land. Teams not only faced tight deadlines, but also had to contend with limited budgets and space. Furthermore, they had to live and work on-site, depending on a local team member, and thus became rapidly immersed in the daily life of the local community. According to organizers, these difficult constraints are characteristic of grassroots initiatives (as opposed to well-funded top-down urban planning or redevelopment initiatives), and facilitate creative thinking and imagination closely aligned with

a community's self-perceived needs. Results of the Bat-Yam urban action include a transformed bus stop, a garden, a tower block for the elderly, and a new entrance to the business district.

The Stuttgart (Germany) urban action took place in July 2012, and once again consisted of ten competing international teams (of over 120 people) designing and constructing architectural interventions in public space in the Wagenhallen area and Nordbahnhof Street in the central city over the space of three days and three nights. The Terni (Italy) action took place in September 2012, this time involving five teams working within the same time constraints to transform a residential block. Annual urban actions have taken place each following year.

72HUA is an imaginative and pragmatic initiative that is proving to be internationally popular. Its popularity may be at least partially related to the sense of empowerment that the competition's highly localized placemaking interventions engendered in participants, including members of the local communities at stake. The organizers correctly identified extreme temporal, spatial, and financial constraints as facilitators of collective creativity. On the other hand, such constraints can also act as inhibitors of creativity and barriers to collective action. What made the difference in the case of 72HUA actions is perhaps the adoption of an optimistic attitude towards limits in budget, time, and space, and briefs clearly identifying local needs. The project made possible collective experiments in public space, engaging those who would not otherwise normally be involved in urban design in critical and creative interventions in a variety of communities. As such, it is an exemplary model of grassroots placemaking that may be deployed (inside or outside the framework of a competition) by local communities facing different urban revitalization challenges.

伊斯坦布尔的音乐故事

Yekpare

艺术家：德尼兹·卡德尔、Candaş Şişman
地点：土耳其伊斯坦布尔海达尔帕夏火车站
形式和材料：景观
时间：2010 年，两场展出，每场持续 3 天
委托人：伊斯坦布尔 2010 欧洲文化之都
推荐人：里奥·谭

"伊斯坦布尔的音乐故事"是由艺术家德尼兹·卡德尔与 Candaş Şişman 受伊斯坦布尔欧洲文化之都 2010 委任，为伊斯坦布尔海达尔帕夏火车站 设计的大型投影节目。毫无疑问，伊斯坦布尔有着古老的历史。在古代它 曾是希腊的一座城市，叫作拜占庭。当它成为罗马帝国的新首都后，君士 坦丁大帝将它重命名为君士坦丁堡（330 年）。直到 1453 年，它被划入 奥斯曼帝国。到了 1923 年，伊斯坦布尔并入新成立的土耳其共和国。"伊 斯坦布尔的音乐故事"通过宗教的符号讲述了这座城市古老的历史。

"伊斯坦布尔的音乐故事"位于伊斯坦布尔的地标海达尔帕夏火车站，占 地 2 500 平方米，这使人联想到该车站作为连接中东与西方通道所发挥的 作用。它是土耳其最繁忙的车站，事实上它也是东欧最繁忙的车站，也就 是说，伊斯坦布尔的音乐故事凭借其迷人的图像、配乐和声响，吸引了数 目众多的多元化的观众。从卡德柯伊海岸也可以观赏到该作品，因为投射 在海面上形成了影像，使人印象深刻。与不同观众和宗教相关的投影符号， 加上投影现场的繁忙景象，它在该市中心所在的位置及其独特的历史，构 成了伊斯坦布尔过去和现在的完美画卷。

"伊斯坦布尔的音乐故事"体现了场景营造的创造性，它的成功之处在于 它对伊斯坦布尔多元化历史和社会大众的敏感性以及作品本身上乘的制作 水准。能够将"伊斯坦布尔的音乐故事"的尺寸准确地投射出来是一项技 术上的壮举，这是这件作品能取得成功的一个重要因素，要知道在新媒体 史上有不少质量低下的城市投影节目。2010 年"伊斯坦布尔的音乐故事" 投射了两次，每次为期三天，它向观众展现了现场诗意化的奇观，吸引许 多经过的人停下脚步来，陷入了沉思和思考。该作品用艺术而不是铺天盖 地的商业广告暂时中断了人们习惯性的交通和商业生活，改变了车站及其 周围环境的体验。这种广告媒体艺术的移位是值得称道的，尤其是在我们 这样一个媒体与广告刺激消费已经不幸地占满日常生活的时代。

该作品的成功与伊斯坦布尔作为欧洲文化之都的地位是相得益彰的，同时也考虑到土耳其正怀有成为欧盟成员国的雄心。"伊斯坦布尔的音乐故事"突显了与不同宗教即基督教、伊斯兰教以及古老异教有关的符号，融合了土耳其悠久的历史，强调了其与欧洲和东方的联系，准确地表达出这个国家当下复杂的政治和经济诉求，不满足于简单的二元性，土耳其是融合欧洲或者中东共同属性的新兴国家。

东与西的见证

马熙逵

在伊斯坦布尔，欧洲板块与亚洲板块的结合已远超地理上的意义，东方文明与西方文明在这里的交汇融合让伊斯坦布尔独具魅力，彰显神秘。作为丝绸之路的终点，伊斯坦布尔见证着中西交流的繁荣与鼎盛，为人类文明的发展史写下了浓厚的一笔。

作品"伊斯坦布尔的音乐故事"是一场视觉与听觉的盛宴，炫目的灯光与悠扬的音乐交织，让人们感受着欧亚文明交汇的脉搏，体味着两种文明融合的底蕴。遥望东方，仿佛还能清晰听到远古时期丝绸之路茶马古道上传来的一串串驼铃声。作为欧洲文化之都系列文化活动的承办城市之一，"伊斯坦布尔的音乐故事"也满怀土耳其成为欧盟成员国的勃勃雄心，彰显着其作为中西交融中心的独特地位以及复杂宗教背景下的国家凝聚力。

Artist: Deniz Kader and Candaş Şişman
Location: Haydarpaşa Train Station, Istanbul, Turkey
Forms and Materials: Landscape
Date: Two showings, lasting three days each, in 2010
Commissioner: Istanbul 2010 European Capital of Culture Agency
Researcher: Leon Tan

Yekpare (Monolithic) was a large-scale projection by artists Deniz Kader and Candaş Şişman on the Haydarpaşa Train Station in Istanbul, commissioned by the Istanbul European Capital of Culture 2010 Agency. Istanbul is, of course, an ancient city. It was once a Greek city called the Byzantium. When the Emperor Constantine made it the new capital of the Roman Empire, he named it Constantinople (330 AD). It was only in 1453 that it became a part of the Ottoman Empire, and then in 1923, Istanbul became part of the new Republic of Turkey. Yekpare was a narrative work recounting this ancient history through its Pagan, Christian, and Islamic symbols.

Sited at the iconic Haydarpaşa Train Station, the 2,500 square meter projection evoked the station's role as a conduit between the Middle East and the West. The station is Turkey' s busiest, and indeed, one of the busiest in Eastern Europe, meaning that Yekpare reached an extremely large and diverse public audience with its captivating imagery, music, and sound. It could also be seen from the Kadıköy Coast, as the projection produced haunting reflections on the surface of the water. The relevance of the projected symbols to different audiences and faiths, combined with the bustle of the projection site and its position in the heart of the city and its history, made for a successful dramatization of Istanbul's past and present.

As a placemaking initiative, Yekpare's excellence consists of its sensitivity to the diverse history of Istanbul and its publics, as well as the high quality of its execution. Accurately mapping a projection of the size of Yekpare to the train station constituted a technical feat, and must be considered an important factor in the work's success, especially as there have been many poorly mapped and low quality urban projections in the recent history of new media. Performed twice in 2010, for three days each time, Yekpare provided to its audiences with poetic encounters with the site, causing some observers to slow down or pause in contemplation and reflection. The project transformed public experiences of the station and its surrounds by interrupting habitual routines of commuting and business, not with commercial advertising, which tends to monopolize technologies of urban projection and spectacle, but with art. This displacement of advertising by media art is commendable, particularly in an era when media and advertising fuelled consumption has become an unfortunate feature and preoccupation of daily life.

The alignment of the project with Istanbul's status as European Capital of Culture is worth mention, especially given Turkey's ambitions to become a member nation of the European Union. By highlighting symbols relating to the different religions, Christian and Islamic, as well as older Pagan faiths, interwoven in Turkey's long history, and by emphasizing the city's links with both Europe and the East, Yekpare accurately reflects the complexity of the Republic's contemporary political-economic claims, refusing simple binaries, that Turkey is either European or Middle Eastern, for example, in favor of the idea that it could be both.

可处理房子
Disposable House

艺术家：萨姆德拉·卡娇·赛琪亚
地点：印度阿萨姆邦古瓦哈蒂
形式和材料：装置艺术
时间：2012 年 2 月 20 日
委托人：印度当代艺术基金会
推荐人：里奥·谭

　　"可处理房子"是由印度当代艺术基金会于 2010 年委任艺术家萨姆德拉·卡娇·赛琪亚创建的一个移动公共艺术项目，在 2012 年 2 月 20 日竣工。该项目的开发耗时多年，它将个人和文化记忆中对家影像的公开探索作为其理论的基础。"可处理房子"出现在横穿印度阿萨姆邦古瓦哈蒂市的长 22 公里的宗教仪式队伍中，它们让不同种姓和阶级的沿街观众见到了与实物一样大小的房子，总共有五间，全部是由当地的材料如木料、泥土和干草建筑而成。

　　记录材料的调查表明，该项目作为公共景观是非常成功的，它用惊喜抓住了沿街观众，吸引他们上前，或提出问题，或展开对话。赛琪亚本人则自封为先知或大师解答疑难。仪式队伍开放后，房子周围围满了人，于是整个行程充满了"苏菲"与"黛哈塔瓦"的歌声，"道塔拉"（一种带弦民间乐器，类似于曼陀林曼陀林）与"考尔"（当地的变种鼓）的音乐，给这景象赋予了精神层面的含义。阿萨姆邦成立的第一个电制片厂"乔迪·琪绰宛"建造的这类房子，表明仪式队伍的精神寓意比任何其他公共行为都要深。然而，介于公众对次大陆上很多承袭或自封的专家、圣徒都报以广泛的支持，上述精神寓意或许只是该项目受到热烈好评的部分原因。

　　这五间房子本身都具有象征性的意义。每间都代表了一个特定的主题。所以它们是"苏菲"（音乐）、城市、位移、社会规范与"堪霍瓦"（演员）。在布拉马普特拉河沿岸的仪式队伍最后，现场安装了这些房子，一首歌曲献给当地的流浪家庭，这是涉及到家概念的作品的最终一幕。仪式队伍使观众去思考不同类型的房子和家的概念以及对实物家的看法，尽管很难去客观地评定他们的参与程度，可以想象的是，至少这些概念在部分公众看来并未丢失，至少在那些无家可归的人收到房子或目睹这慷慨之举的旁人看来确实如此。

该项目其成功之处在于通过表演呈现出虚构的体验，然后提出关于现实
生活环境的问题，如有家可住与无家可归，占有与失去，财富与贫穷。
显然它改变或打破了毫无戒备的观众的感知习惯，最起码它能促进路人
之间进行一次和蔼可亲的对话。就场景营造而言，它在一个饱受宗教间
冲突之苦的国家带来了公共空间的欢乐，这是非常值得称道的。此外，"可
处理房子"也确实给无家可归的人创造了生活场所，从而使该艺术活动
有益于社会。

"家"是一份责任

马熙逵

作品"可处理的房子"是以"家"为中心展开的一系列文化探索，并以公
共景致的形式映入观众的视线，由于新奇、大胆的创意造型，引来沿街观
众的围观，人们在此展开问题和对话。艺术家强调人是富有感情的高级动
物，人内心最为牵挂的便是家园。"家"的一切永远铭记于心中，因为，
父母把我们生养在这个地方。由于作品整个"行程"充满当地特色的歌声，
整个艺术活动的进行被赋予了精神层面的含义，即"家"是永恒讨论的主题。
同时，艺术家从多个主题出发，去探寻"家"与城市、社会的关系，由此
引发观众参与到对"家"这一概念的探讨中。"可处理的房子"通过表演
呈现出虚构的体验，以"家"为反思对象，揭示社会资源分配不公的现状
以及民众饱受宗教冲突、种姓压迫的惨痛经历。

Artist: Samudra Kajal Saikia
Location: Guwahati, Assam, India
Forms and Materials: Installtions art
Date: February 20, 2012
Commissioner: Artist initiated, supported by the
Foundation for Indian Contemporary Art
Researcher: Leon Tan

Apna hi Ghar Samjho (Mobilizing the House or Disposable House) is a mobile public art project by Samudra Kajal Saikia, supported by a public art grant from the Foundation for Indian Contemporary Art in 2010, and realized on February 20, 2012. The project was developed over several years, and took as its conceptual premise the public exploration of images of home in individual and cultural memory. Based on a 22-kilometer ritualistic procession across the city of Guwahati, in the Indian state of Assam, Disposable House literally confronted street audiences across caste and class divides with life-sized physical homes, five in total, made from local materials such as timber, mud, and hay.

A survey of documentary materials suggests that the project was highly successful as a public spectacle, captivating street audiences with an element of surprise, and evoking both questions and conversations. The artist himself took on a role as a kind of self-styled prophet or guru. The procession was ritually opened, groups circumambulated the houses, and the entire journey was accompanied by Sufi and Dehatatwa songs with dotara (a stringed folk instrument similar to the mandolin) and khol (a local variety of drum) music, all of which conferred on the spectacle a sense of spiritual import. That the houses were constructed in Jyoti Chitravan, the first film studio established in Assam, suggests that the spiritual overtones of the procession were more theatrical dramatization than anything else, a public "act," so to speak. This dramatization,

however, may be partly responsible for the project's successful reception, given the widespread public support for any number of hereditary or self-appointed gurus or saints in the subcontinent.

The five houses themselves were adorned with symbolic references. Each represented a specific theme. Thus there was a Sufi house, an urban house, a house of displacement, a house of social norms, and the Kankhowa's (actor's) house. At the conclusion of the procession on the banks of the Brahmaputra River, the houses were installed on site, and a number donated to homeless families in the area, a fitting final act for a work that dealt with the notion of home. While it is difficult to objectively ascertain the extent to which the procession engaged audiences in reflection on the notion of different kinds of houses and homes, and indeed, on the idea of the body itself as home, it is conceivable that these notions were not lost on at least some of the public, not least, the homeless who received the gift of housing, and those who witnessed this generosity.

As a project, its excellence lies in the manner in which it engaged the fictive dimension of experience through performance, in order to pose questions concerning real life conditions, those of home and homelessness, possession and dispossession, property and poverty. It clearly transformed or disrupted the perceptual habits of unsuspecting audiences, and at the very least, stimulated amiable conversations between passersby in public. In terms of placemaking, the facilitation of social conviviality in the public spaces of a nation that is regularly traumatized by interreligious differences is highly commendable. Furthermore, Disposable House quite literally made living places for the homeless, rendering the aesthetic event socially useful.

和我说话
Talk to Me

艺术家：翟斯珉·帕萨娅
地点：印度班加罗尔、新德里、加尔各答和其他城市
形式和材料：行为互动
时间：2005 年至今，临时干预措施，时间不同（约 4 小时）
委托人：艺术家发起
推荐人：里奥·谭

2012 年 12 月，23 岁的女学生在印度的首都新德里一辆汽车中被轮奸，最终死在新加坡一家医院。该事件激起了国内外民愤，袭击的具体细节被传播到世界各地的网络媒体。最后，印度政府出台了新的反强奸法律，印度法庭判处强奸者死罪。虽然事件异乎寻常的残酷，但它绝不是一个孤立事件。更准确地来说，这可以描述为整个次大陆普遍和强烈的父权社会状况。其他现状包括许多印度人委婉地称之为"戏弄夏娃"，指的是公众男人对妇女的性骚扰和猥亵。"空白噪声"是班加罗尔的一个项目，2003 年作为一种对广泛的骚扰、调戏和强奸女性的艺术和政治回应。翟斯珉·帕萨娅发起了这个学生毕业项目，多年来势头强劲，成为全国社区艺术运动。

"空白噪声"包括并继续涉及印度公共对话问题上的性暴力。它也指出了一些城市独立的介入措施，依靠的这些志愿者被称为"行动英雄"。艺术家说，"行动英雄"指的是"越来越多的社区公民，包括将时间和精力捐赠到集体事业中的男性和女性"。一个明显的介入就是"和我说话"（2012），这个项目由斯利史提艺术和技术学院学生帕萨娅花了一个多月在班加罗尔设计。在"和我说话"项目中，参与者（耶拉汉卡行动英雄）会在耶拉汉卡附近发现觉得有性威胁的地段：没有公共照明的 25 公里街道或"调戏夏娃"经常发生在晚上的商店中。

行动英雄设置桌椅，标示"行动指南"，在"安全通道"邀请公众参与讨论"性暴力"等长达 4 个小时的公众接触。谈话中会有欢乐的精神、茶和萨莫萨三角饺提供。对话结束后，参与者会得到一朵花作为礼物。通过参与谈话，"行动英雄"有机会重新定义他们的脆弱性感知和威胁。此外，行动也暂时开拓了公益路段上的车辆以及与当地人、陌生人、跨阶级的壁垒、种姓、宗教、性别和语言之间对话，从而为社区凝聚力和情感投入播下了种子。"和我说话"一直在其他城市包括新德里、加尔各答举办。

　　"空白噪声"的干预措施，包括但不限于"和我说话"，由于其在公众的参与度，在解决根深蒂固的社会问题上非常值得称道。此外，这个项目可以高度访问，"行动英雄"的形象能很好识别，"行动词汇"可以很快理解（例如坐下来喝茶，吃萨莫萨三角饺聊天）。"空白噪声"可能会被认为是在一个模式敏感和响应印度的复杂状况，一个女权主义介入，一种始于美国20世纪70年代初的女权主义艺术的持续传统。TED和"阿绍卡"奖学金以及广泛的艺术家说明，"空白噪声"在线和印刷媒体必要性，同时证明了集体的成功和社会持续的相关性。

守护夏娃

马熙逵

印度的性侵现象之严重，已经成为了一种社会现象。在印度日常英语中甚至有个专门的说法"戏弄夏娃"（Eve-teasing），就是特指各种性侵——从口头性骚扰到性袭击。作品"和我说话"是对印度妇女遭受性侵问题的叩问，它引发公众对性暴力的探讨，呼吁人们参与到保护妇女安全的行动中。在传统种姓制度下，印度家庭往往视女性为家庭的负担，妇女的境遇极差，这是诱发女性受侵害的社会背景。对于强奸问题的严重，印度法治无力严惩犯罪分子，致使恶性事件屡禁不止。同时，由于受害者担心社会污名，印度绝大多数的性侵案件得不到报道，但随着女权保障思维的觉醒，越来越多的人开始关注这一现象。作品"和我说话"通过对话的形式，重新链接种姓、宗教、性别和语言之间的对话，为作为志愿者的"行动英雄"标示方向，一同触发和响应印度目前敏感和复杂的社会状况。

Artist: Jasmeen Patheja
Location: Bangalore, New Delhi, Kolkata, and other cities in India
Date: temporary interventions with varying duration
Forms and Materials: Interactve Behavior
 (approximately 4 hours), 2005-present
Commissioner: Artist initiated
Researcher: Leon Tan

In December 2012, a 23-year-old female student was gang-raped in a moving bus in India's capital city of New Delhi. She eventually died from her injuries in a hospital in Singapore. The event sparked outrage both nationally and internationally, as details of the attack were relayed across the media networks of the world. The Indian government introduced new anti-rape laws as a result, and an Indian court sentenced the attackers to death. While the event was singularly brutal, it was by no means an isolated incident. It is more accurately characterized as a symptom of pervasive and fiercely patriarchal social conditions to be found throughout the subcontinent. Other symptoms include what many in India euphemistically call "eve-teasing," referring to the public sexual harassment and molestation of women by men. Blank Noise is a project that emerged in Bangalore in 2003 as an artistic and political response to the widespread harassment, molestation and rape of women. It was initiated by Jasmeen Patheja as a student graduation project, and over the years gained momentum to become a nationwide community arts movement.

Blank Noise involved and continues to involve public conversations on issues of sexual violence in India. It has also realized numerous discrete interventions in several cities, relying on volunteers known as 'action heroes.' According to the artist, 'action heroes' are 'a growing community of citizens, male and female' who donate time and energy to the collective's cause. One notable intervention was Talk to Me (2012), a project facilitated by Patheja with Srishti School of Art, Design and Technology students over one month in Bangalore. In Talk to Me, participants (Yelahanka Action Heroes)

identified a site in the Yelahanka neighborhood in which they felt sexually threatened: a quarter-kilometer stretch of street without public lighting or shops in which eve-teasing often took place at night.

The action heroes set up tables and chairs signposted with "action guidelines" inviting the public to engage in conversation about "anything except sexual violence" on "Safest Lane" as a 4-hour-long public encounter. In a convivial spirit, tea and samosas were provided for the conversations. Participants were all gifted a flower at the end of each conversation. By engaging in the conversation, the action heroes were given an opportunity to reframe their perceptions of vulnerability and threat in relation to the site. Furthermore, the action also functioned to temporarily reclaim the stretch of road for public good, brokering encounters between locals, strangers, across the barriers of class, caste, religion, gender, and language, thereby sowing the seeds for community cohesion and emotional investment in the site. Talk to Me has subsequently been staged in other cities including New Delhi and Kolkata.

Blank Noise's interventions, including but not limited to Talk to Me, are highly commendable for their engagement of the public in tackling a deeply entrenched social problem. The projects are highly accessible, the image of the action hero being one that lends itself well to positive identification, and the vocabulary of action being immediately comprehensible (sitting down for tea and samosas to chat, for example). Blank Noise may be considered a feminist intervention, continuing a feminist art tradition that began in the early 1970s in the US, in a mode that is sensitive and responsive to the complex conditions of India. TED and Ashoka Fellowships, as well as extensive coverage of the artist and Blank Noise online and in print media, attest to the collective's success and continuing relevance to society.

尤瑟夫壁画
Yousif Mural

艺术家：乔治·罗格里格斯-格拉达
地点：巴林麦纳麦城
形式和材料：壁画
时间：2012 年
委托人：奥宛 338 公共艺术节
推荐人：里奥·谭

"尤瑟夫壁画"是由美籍古巴艺术家乔治·罗格里格斯-格拉达创作的壁画。主要用于巴林举行的奥宛 338 公共艺术节 (2012 年)。据测量，它有 12 米到 15 米长，工作规模巨大，画有"尤瑟夫"肖像。画中模特为传统的巴林渔夫，现在几百渔夫中，只有几个在钓鱼并维护该地区居民和大海之间的原始关系。几个世纪以来，巴林人民依靠航海和贸易作为生计。肖像壁画属于艺术家的"特性"系列，旨在从商业广告画面、政客和名人开拓公共空间以及质疑约束公共领域问题的表达。此外，还以不朽的街头艺术匿名人的肖像为特写。因此，它是一种国际现状的一部分，被称为"文化干扰"。

壁画位于巴林最大的城市麦纳麦，通过过去渔夫的肖像，唤起该地区过去的情况，通过巴林从石油中心转换为金融中心的自我改造，作为快速侵蚀传统的象征。巴林的渔村灭绝，因为风景区让位于摩天大楼和带有国王、政治精英与商业产品形象的广告图片。

艺术家说，在巴林，唯一大规模图像是国王和其他权力象征。插入匿名人图像到公共空间，这些角色更属于巴林的过去而非未来。在某种意义上，艺术家帮助改造公共空间，尤其是记忆的边缘、巴林后石油时代的经济。考虑到巴林参与"阿拉伯之春"的背景，本质上是一系列持续的抗议，人民以暴力和非暴力的形式反对中东地区的政治统治，这幅画很可能被视为一个值得称赞、介入城市经济的标志，所以很多画像构成政治（君主）的权力表达方式。这是一个项目的卓越地方。

事实上，项目另一个值得称道的方面，是艺术家的街头艺术实践，他们使用不破坏环境的材料。因此，所有的"特性"工作，包括"尤瑟夫壁画"，由于风和水的侵蚀，最终随着时间褪色消失。对比典型的喷涂料进行的涂鸦，这方面的工作可能不仅有毒而且具有相对永久性。对于场所营造而言，"尤瑟夫壁画"使得麦纳麦公共领域的一部分更宜居住。至少从一个存在主义的层面上来说，他们通过剥夺政治和商业网站的图像以及相关力量，用面对的未知代替。以一个"普通人"的形象盯着从巴林的进退，"尤瑟夫壁画"引起了巴林人们对未来日常地位的质疑。

酝酿中的革新

马熙逵

作品"尤瑟夫壁画"以渔夫肖像作为表现对象,这与权力表达方式下的君主式画像形成对比。艺术家希望通过"渔夫"为线索,展开对于巴林王国传统文化"灭绝"的调查,唤起民众对于传统文化的重视。随着世界对石油需求量的进一步提升,巴林王国也因丰富的石油储备跻身富强国家之列,然而不幸的是,经济高速发展的同时也伴随着基于"渔村"的传统文化快速消亡。

作品"尤瑟夫壁画"在某种意义上是对今日巴林王国历史变革的一系列持续性抗议,它呼吁以"阿拉伯之春"为代表的政治新秩序应当受到推崇,还权于民、关注民生,把注意力集中在审判和批评自己内部问题上,推动阿拉伯社会的日趋成熟。同时关注经济与文化发展的平衡,约束盲目的商业拓展,构建基于传统并自信开放的民主世界。

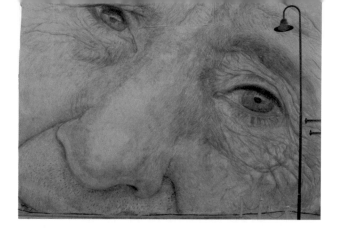

Artist: Jorge Rodríguez-Gerarda
Location: Manama City, Bahrain
Forms and Materials: Mural (Painting)
Date: 2012
Commissioner: Alwan 338 Public Art Festival
Researcher: Leon Tan

Yousif was a mural created by Cuban-American artist Jorge Rodríguez-Gerada for the inaugural edition of the Alwan 338 Public Art Festival (2012) in Bahrain. Measuring 12-meters by 15-meters, the work was monumental in scale, and featured a portrait of Yousif, a traditional Bahraini fisherman, one of only a few hundred still fishing and maintaining the region's ancient relations between the inhabitants and the sea. Bahrain has for centuries relied on seafaring and trade for the livelihood of its peoples. The portrait mural belongs to the artist's Identity series, a series intended to reclaim public space from images of commercial advertisement, politicians and celebrities, as well as to question constraints on expression in the public sphere by featuring monumental street art portraits of anonymous people. As such, it is part of an international phenomenon known as 'culture jamming.'

Sited in Bahrain's largest city, Manama, the mural evoked the region's past through the portrait of the fisherman, a symbol of the rapid erosion of tradition and the past by Bahrain's self-transformation through oil wealth into a financial hub. Bahrain's fishing villages are dying out, as the landscape gives way to skyscrapers and advertisements carrying images of the King and the political elite, together with those of commercial products. According to the artist, 'in Bahrain the only large scale images of heroic proportions are those of the king and others in power.'

By inserting the image of a relatively anonymous person into the public sphere—a character belonging more to Bahrain's past than to its future—the artist did in a sense help to reclaim public space, particularly for the memory of the marginalized, the nobodies in Bahrain's post-oil economy. Considered in the context of Bahrain's participation in the Arab Spring, essentially a series of ongoing protests, violent and non-violent, by 'the people' against political rulers in the Middle East, the portrait may well be seen as a laudable intervention in the city's economy of signs, so many of which constitute expressions of political (monarchical) power. This is one of the points of excellence of the project.

Another commendable aspect of the project, and indeed, of the artist's street art practice, is the use of materials that do not impact negatively on the environment. Thus, all the Identity works, including Yousif, fade and ultimately disappear over time, as a result of degradation/erosion by wind and water. This aspect of the work may be contrasted to graffiti executed in typical spray paints, which are not only toxic, but also relatively permanent. In terms of placemaking, Yousif made a part of Manama's public sphere more habitable, at least on an existential level, by depriving the site of political and commercial imagery and associated interpellative forces, providing in their stead the face of an unknown man, an 'everyman,' staring as it were, from Bahrain's receding past. In this way, Yousif also provoked questions about the position of everyday people in Bahrain's future.

气候大象
The Climate Elephant

艺术家：丹尼尔·当瑟
地点：印度德里，新德里，鲁奇·豪尔，瓦桑特·坤，瑞安国际学校
形式和材料：互动行为
时间：2010 年 11 月 28 日
委托人：Alwan 338 公共艺术节
推荐人：杰奎琳·怀特

在坎昆举办的 2010 年联合国气候变化大会前夜，来自印度新德里瑞安国际学校约三千名学生和老师参与到了"气候大象"的项目中。当他们的身体成为空中大象描绘的"画笔时"，就真的像俗语所说的，成为"房间里的大象"。通过鼓励观众重新想象他们和天空之间的关系，空中艺术变成一个特别恰当的媒介，使人们关注这个全球人民试图忽视的"大象"——人类活动所导致的全球变暖。丹尼尔·当瑟是一名来自美国的自我宣称的"空中艺术积极分子"，他在全球各个社区创造了 150 个"空中艺术"作品，"气候大象"是他到目前为止最重要的作品。

为了教育观众百万分之三百五十是地球上能安全维系生命的最高碳含量，当瑟在空中使用捐赠的起重机所拍摄的大象照片包含 350 这个数字。虽然当瑟之前在其他空中艺术作品中也包含了 350 这个数字，"气候大象"是首个也是唯一一个丹尼尔·当瑟和气候变化教育组织 350.org 正式合作的成果。"气候大象"是一个全球性项目，包括与通过 350.org 的协作所进行了十几个空中艺术项目，使人们在国际会谈前夕关注气候变化。讽刺的是，虽然创造大象图形时，有卫星正好从太空中飞过，能拍到照片，但是由于新德里的空气污染太严重，没有生成照片。

虽然丹尼尔·当瑟在学校举行的活动通常是三天，但是他只有一天的时间来创造"气候大象"。使用网格系统把大象简单地绘画放大，这也成为一个重要的教授孩子的机会。但是，丹尼尔·当瑟没有创造空中艺术图像所需的一个最基本的工具——卷尺。因此，他和帮助他的印度人（包括一些不会说英语的）不得不临场发挥，以 10 英尺的增量在绳子上做标志。使用粉笔画象牙，大象所站地方升起的水面是用蓝色布料描绘的。巧合的是，学生的校服是"大象"灰色。

一旦学生成形，他们还穿着不同颜色的衬衫，以象征大象变化的颜色，这样就可以被看到，不会被忽视。根据当瑟项目中的常规程序，项目参与者

需保持一分钟的安静。然后分别用英语和北印度语说："打开你的天空景象！"这是丹尼尔·当瑟天空艺术实践的六个教学之一，其他还包括意图、协助、互连、感激和无常。

丹尼尔·当瑟在印度创作"气候大象"尤其适当，因为大象最近被命名为印度的国家遗产动物。

"咖尼使"的乐章
马熙达

印度被世人称之为"象之国"，因为印度人供奉"大象神"。在印度，"大象神"的知名度和受欢迎的程度相当于中国的观世音菩萨，他是印度三大神之一的破坏神希瓦（Shiva）和雪山女神帕瓦蒂（Parvati）的儿子，印度人称之为"咖尼使"（Ganesh）。在印度，人们进行任何活动前均先礼拜"大象神"，因为它是创生和破除困难之神，也协助信众接近其他的神祇。

作品"气候大象"希望借助"大象神"的形态为媒介，鼓励公众重新界定人与自然的关系，呼吁人们关注全球气候变暖的现实。随着工业化的快速发展，人类大量使用煤、石油等化石燃料，在创造巨大物质财富的同时，也制造了大量污染物和温室气体，导致气候变化问题越来越突出。对于气候变化问题的严峻性，艺术家希望寄托于对年轻一代的启迪，通过教育让人们意识到环境恶化将带来的巨大生存挑战。

Artist: Daniel Dancer
Location: Ryan International School, Ruchi Vihar, Vasant Kunj,
New Delhi, Delhi, India
Forms and Materials: Interactive Behavior
Date: 350.org
Commissioner: Alwan 338 Public Art Festival
Researcher: Jacqueline White

On the eve of the 2010 United Nations Climate Change Conference in Cancun, approximately 3,000 school children and their teachers from the Ryan International School in New Delhi, India, literally became the proverbial "elephant in the room" when their bodies formed the "brushstrokes" for an aerial depiction of an elephant. Aerial art, which encourages viewers to re-imagine their relationship with the sky, is a particularly apt medium in which to bring attention to this particular "elephant that global citizens are attempting to ignore—the warming of the earth's atmosphere caused by human actions. Daniel Dancer, a self-professed "aerial art activist" based in the United States who has created 150 "sky art" formations in communities across the globe, claims The Climate Elephant as his most significant work to date.

To educate viewers that 350 parts per million is the highest carbon reading that will safely sustain life as we know it on earth, the elephant image, which Dancer photographed from above from a donated crane, includes the number 350. Although Dancer had previously included 350 in other sky art pieces, The Climate Elephant was the first and only time Dancer formally collaborated with the climate change education organization 350.org. The Climate Elephant was part of a worldwide installation of more than a dozen sky art projects coordinated by 350.org to bring attention to climate change on the eve of the international talks. Ironically, although creation of the elephant image was timed to coincide with

the passage of a satellite that would photograph the artwork from space, the air pollution in New Delhi was so dense that no such image resulted.

Although Dancer's school residencies are typically three-day affairs, Dancer only had one day to create The Climate Elephant. A very simple drawing of an elephant was enlarged to scale using a grid system, which also became an opportunity to teach this mathematical process to the children. However, Dancer did not have one of the most basic tools he needs to create a sky art image— a tape measure. Instead, he and the Indians helping him, including some who didn't speak English, had to improvise by marking a rope in ten-foot increments. The tusks were drawn with chalk, and the rising water that the elephant was standing in was depicted with blue fabric. Fortuitously, the student's school uniforms were "elephant" gray.

Once the students were in formation, they also donned multi-colored shirts, to symbolize the elephant changing colors so that it could be seen and no longer ignored. As is customary with Dancer's projects, the art participants observed one minute of silence. They also said, "Get your skysight on!" in both English and Hindi, which is one of the six teachings Dancer has identified as part of his Art For the Sky practice, which also include Intention, Collaboration, Interconnection, Gratitude, and Impermanence.

That Dancer created The Climate Elephant in India is particularly apt because the elephant had recently been named Indian's National Heritage Animal. It is worshipped in the form of Ganesha, "the remover of all obstacles."

边境
Border

艺术家：穆拉特·皋柯
地点：土耳其马尔丁（在土耳其和叙利亚边界上）
形式和材料：行为，影像
时间：2010 年
委托人：艺术家资助
推荐人：杰西卡·菲亚拉

边境界定了界限和模糊的领土，是具有身心和情感意义的无形真理。边界，隔开了允许入境与禁止入境人士，标志着冲突和争议土地的历史。但是，边境也有一定的荒谬性——这些任意的、可穿过的划界被作为不可避免的、永恒的界限来保护。边境是人类、思想和实践彼此交叉，建立一个独特地区——边境之地——的混合区域，而非完全分离的界线。

穆拉特·皋柯的 2010 年表演摄影"边境"，突显了边境作为严格划界的重要性这一不同方面，它记录了土耳其和叙利亚之间的边境——土耳其马尔丁的艺术行动。这个"边境"的围栏已经遭到破坏：一部分被撤除来放置吊床——在阳光的沐浴下，人们能侧卧在吊床静心放松。

由于这一地点的潜在危险，"边境"的公开表演是草率的，而该作品主要以照片形式存在和传播。"边境"需要现场表演，但通常观众能通过照片或视频文件体验的表演艺术类似，超越了表演摄影本身，作为一个公共艺术项目有着重要的作用。该作品的意义，虽然从很大程度上来说，很多未能亲临现场或亲自经历过，但它与地点和表演有着重要的联系。

这一人物的身体处于边界上。尽管这个人物是躺着的，但简朴的围栏提示着人们，他的存在和行动是未经许可的。因此他的冷静弥足珍贵，而这种冷静可能在任何时候被干扰。但是，穆拉特·皋柯的表演轻松地回应了边境的这一严格性。正如策展人西蒙·里斯所说，该作品"在其与周围环境发生关系时，幽默地反思了全球新闻报道的盲点"。边境与政府高层纠纷交织在一起，但是地球上的生命和政治领袖的夸夸其谈之间也有一个脱节，即照片对比牢固的围栏和现有结构的即兴重排时所突显的差距。

策展人切塞纳·欧泽门，评论了躺着的人和分割完整领土的边界的物理动态，强调了"皋柯质疑了不统一的概念，随着吊床摇动，我们的人会在边境的一边和另一边之间摆动"。这一表演中有一个幽默，但是正如欧泽门强调的，也是"一个默许的特权"。尽管边境是混合区域，但机构、能力和访问并

不是均享的。"边境"邀请观众质疑：是谁在任意地耍弄边境？当与边境相关的决策被过于小心、无情或简单地处理时，谁会受到损害？

这个人物摇摆于这一边和另一边之间，让人想起了定义的多变性，尤其是与土耳其的南部边界相关的定义。土耳其，离欧洲最远的国家，或中东的起点，通过某种或其他方式定义了自己，随着其参照的地理、历史、文化或政治的不同而变化。即使边境仍然存在，它们的意义也会随着周边诠释或背景的转变而迅速变化。在 2010 年拍摄这张照片时，有着一定的风险。现在，还在进行的叙利亚内战已经改变了边境的 500 英里延伸带的本质，而这一延伸带现在被一百多万个难民越过。穆拉特·皋柯的照片也有了全新的意义，随着人物的摆动，突显了稳定和冲突之间的鲜明对比。这一行动和照片与边境所在地有着复杂的联系，但是照片也鼓励着人们不断反思这种划界的出现和处理，而这种意义只有在该作品创建后才得以增强。

边境的默许

马熙逮

在政治学和地理学上，边境一般具有特殊的重要性，在战争期间边境会成为紧张地带，甚至成为战线的一部分。艺术家穆拉特·皋柯在土耳其和叙利亚之间发现一处特殊的边境，在这里，边境的围栏已经遭到破坏，一部分栅栏被撤除来放置吊床，该装置允许人们在阳光的沐浴下在吊床静心放松。

作品"边境"像是人在边境间的摇摆，此时摇摆的人获得了默许的特权，不受边境的概念的影响，突显了稳定和冲突之间的鲜明对比。一般情况下，边境地区多数设有出入限制，甚至被列为禁区，边境是国家关系的直接反映。艺术家希望通过作品"边境"引发公众的集体质疑，即在惨烈的边境冲突下，无辜的民众是否会沦为政府高层角力的牺牲品，决策者在制定边境政策时，是否应该更多考虑民众的境遇。

Artist: Murat Gök

Location: Mardin, Turkey (on Turkey / Syria border)

Forms and Materials: Behavior, Image

Date: 2012

Commissioner: Artist-funded

Researcher: Jessica Fiala

Borders are defining lines and nebulous territories, invisible absolutes that hold physical, psychological, and emotional weight. Separating in and out, they divide those welcomed from those barred from entry and mark histories of conflict and disputed lands. Yet there is also an absurdity to borders—these arbitrary and porous demarcations guarded as inevitable and perpetual. Rather than lines of complete separation, borders are regions of mixing, where people, ideas, and practices cross between and establish a unique zone—a borderlands.

These varied aspects of borders as strict divisions and regions of play are brought to the fore in Murat Gök's 2010 performance photograph Border, documenting an art action in Mardin, Turkey, on the border between Turkey and Syria. The severe border fence has been disrupted; one section removed to make room for a hammock in which a calm reclining figure relaxes in the sun.

Due to the potential danger of the location, the public performance of Border was brief, and the work exists and circulates primarily as a photograph. Akin to performance art that centrally entails live performance but is often experienced by audiences through photograph or video documentation, Border goes beyond being a staged photograph and has an important role as a public art project. Although largely not experienced on-site or in-person, the significance of the work is importantly tied to both place and performance.

The subject's body is on the line. Although the figure is reposed, the austere fencing serves as a reminder that his presence and action are unsanctioned. His calm is therefore precarious, menaced by potential disruption at any moment. Yet Gök's performance offers a lighthearted response to the severity of borders. As curator Simon

Rees noted, the work is "a humorous reflection on the blind spots of the global news cycle as it relates to peripheries." Borders are intertwined with high-level government disputes, but there is also a disconnect between life on the ground and the showboating of political leaders—a gap highlighted by the photograph's contrast between rigid fencing and the impromptu rearrangement of existing structures.

Commenting on the physical dynamics of both the reclining figure and borders cutting through unbroken territory, curator Şener Özmen underscored that "Gök questions the concept of disunity... as the hammock rocks, our man will swing between one and the other side of the border." There is a humor in this staging, but as Özmen underscores, also "a privilege being winked at." Although borders are sites of mixing, agency, capability, and access are not equally shared. Border invites viewers to question who gets to play with borders and also who suffers when decisions about borders are handled too carelessly, callously, or easily.

The sway of the figure between one side and another also amusingly calls to mind the changeable nature of definitions, particularly relevant for the southern border of Turkey. The farthest reach of Europe or the beginning of the Middle East, Turkey is itself defined one way or another, varying with different references to geography, history, culture, or politics. Even when borders remain, their implications change quickly with the shift of surrounding interpretations or contexts. In 2010 when the photograph was taken, there was a degree of danger. Today, the on-going Syrian civil war has transformed the nature of this 500-mile stretch of border, crossed now by over 1 million refugees. Gök's photograph, too, takes on a new meaning, highlighting the stark contrast of stability and conflict across which the figure swings. The action and photograph are intricately linked to the place of the border, but the photograph also encourages continued reflections upon the establishment and handling of such divisions, a significance that has only increased since the work's creation.

公园
The Park

艺术家：斯瑞娅塔·罗伊
地点：印度新德里，达克史普利
形式和材料：装置，壁画
时间：2008—2009 年
委托人：由艺术家发起，与安柯教育机构协作
推荐人：里奥·谭

"公园"是斯瑞娅塔·罗伊在 2008—2009 年期间负责的，位于新德里南部达克史普利工人阶级社区中的一个公共艺术项目。罗伊的这一项目，得到了印度当代艺术基金会（FICA）公共艺术拨款的支持，旨在振兴和改造达克史普利地区一个疏于管理的公园。这一公园过去曾是当地居民的公共休闲空间。但是，它现在变成了垃圾场和公共厕所，晚上经常有酒鬼和吸毒者在这里游荡。当罗伊开始该项目时，她面对的是怀有敌意，持怀疑态度的当地人，很多当地人害怕她会将这里的情况报告给警方，或害怕她是土地夺取组织的一员。她曾一度甚至被威胁，要求离开。

"公园"和许多传统的社区艺术项目一样，其目标的实现依赖于对当地对抗情绪的精心利用以及对该区域联盟的调解。在发起一个反对随地扔垃圾的横幅运动前，斯瑞娅塔·罗伊花了几个月的时间在 J-Block 及其邻近区域举办研讨会，与当地人建立关系，和协作者合作。在这个横幅运动中，邻里孩子和年轻人也参与了公园清理这一集体行动。她还将一大片墙涂成蓝色，向社区提供了画布使他们分享他们的想法。一名 13 岁的参与者说，"这堵墙邀请了所有人说出自己的想法。我们描写这个公园，画下女人们在榕树下休息，孩子们荡秋千、滑滑梯的场景。"

斯瑞娅塔·罗伊基于与社区之间的对话以及与市政当局的合作（比如说，获得不同干预机构的批准）有效地推动了公园的集体创作。该项目的协作本质意味着，艺术家需作出重大妥协。例如，在有些地方她希望种植更多绿色植物，但是当地人想架上铁丝网；有些地方她倾向于布置成公共空间，但女人们却希望将这个空间隔离从而"远离男人，获得安全感"。在罗伊看来，这些妥协是"创建一个真正属于人民的艺术品"所必需的。

"公园"作为模糊了艺术和行动主义的界限的公共作品，是值得称赞的。该项目，从开始到结束大约花了两年的时间，例证了社会和空间改建是一

个有组织的基层过程这一事实。它的一个杰出之处在于，该项目通过集体合作方式改变了公园的利用模式。另一点是艺术家在克制自己的欲望，从而满足当地的集体期望和需求中显示出的敏感性。斯瑞娅塔·罗伊理解了个人和集体创造、表达之间的区别。"公园"向我们证明了：艺术在重塑公共空间和其相关的社会习惯中发挥了一个关键作用；艺术家，而非政府机关或私人利益，通常被认为是解决紧迫的社会问题的领袖和催化剂。诸如此类的项目表明，在考虑城市复兴和社区发展相关项目时，艺术家需得到更多的重视。

艺术无疆

马熙逵

随着城市化的进程，人们享受着丰腴社会带来的各项便捷，同时面临诸多社会问题和矛盾。作品"公园"是艺术介入公共空间改造的形式，它结合当地民众的需求，将疏于管理的废弃公园改造成集艺术教育、生活娱乐等功能于一体的文化交流平台。人们在这里敞开心扉，建立和谐的人际关系，共同协作。

"公园"作为一种公共艺术形式，在某种程度上成为都市矛盾的"缓冲"形式，同时以可持续发展的理念，传达城市文化精神、消解人与人之间的隔阂。艺术家追求作品与环境相适宜的艺术性，追求文化精神与美的构建，同时强调社会性的体现。以"创建一个真正属于人民的艺术品"为目标，引导公众重新认识自己的生活环境，服务于公民多元化的需求，具有提高共鸣生活质量、增进公民的文化艺术福利功能。

Artist: Sreejata Roy
Location: Dakshinpuri, New Delhi, India
Forms and Materials: Installations, Mural(Painting)
Date: 2008-2009
Commissioner: Artist initiated in partnership with Ankur Society for Alternatives in Education
Researcher: Leon Tan

The Park was a public art project by Sreejata Roy in the working class neighborhood of Dakshinpuri in South Delhi over 2008-2009. Supported by a public art grant from the Foundation for Indian Contemporary Art (FICA), Roy's project aimed to revitalize and transform a neglected park in J-Block in Dakshinpuri. The park in question used to be a common recreational space for local residents. It had, however, become a rubbish dump and latrine, often frequented by alcoholics and drug abusers at night. When Roy began the project, she confronted hostile and suspicious locals, many of whom feared that she would report them to the police or that she was part of an organized land grab. At one point, she was even threatened and asked to leave.

Like many projects in the community arts tradition, The Park relied on the skillful navigation of local antagonisms, and the brokering of alliances in the area, to achieve its objective. Roy spent several months conducting workshops in J-Block and its immediate vicinity, building relationships with locals, and organizing with collaborators, before launching a banner campaign against public littering. This was accompanied by collective action involving neighborhood children and youth in the cleaning up of the park. She also painted a large wall blue, furnishing the community with a canvas to share their thoughts. A 13-year-old participant said, "The wall was an invitation to everyone to speak their minds. We wrote about the park, drew scenes of women resting under the banyan tree, children playing on swings and sliders."

Roy effectively facilitated a collective redesign of the park based on conversations with the community and engagement with the municipal authorities (for example, obtaining permissions for various interventions). The collaborative nature of the project meant that the artist had to compromise significantly. For instance, where she wanted more greenery, locals wanted barbed wire. Where she would have preferred an open space, women instead wanted a segregated space to "feel safe, away from the men." In Roy's opinion, these compromises were necessary "to create an artwork that truly belongs to the people."

The Park was commendable as a public work that blurred the boundaries between art and activism. Taking approximately two years from start to end, it exemplified an organic, grassroots process of social and spatial transformation. One point of excellence was the way in which the project collectively altered patterns of usage in the park. Another was the sensitivity the artist displayed in subjugating her own desires for the park to a collective set of local expectations and needs; Roy understood the difference between personal and collective creation/expression. The Park provides us with evidence that art can play a crucial role in reshaping public spaces and their associated social habits, and that artists, rather than government authorities or private interests, often prove to be the thought leaders and catalysts for solving pressing social problems. Projects like this suggest that artists should be given more importance in considerations relating to urban revitalization and community development.

第二届国际公共艺术奖评选会纪实

2014 年 9 月 14 至 15 日，第二届国际公共艺术奖评选会在上海大学举行。

"国际公共艺术奖"由中国《公共艺术》杂志和美国《公共艺术评论》杂志共同创办，首届颁奖仪式和国际公共艺术论坛于 2013 年 4 月在上海大学举行。第二届国际公共艺术奖评选会经过专家评述、案例讨论和公共投票，代表全世界 7 大区的 7 位评委从全球 19 位研究员推荐的 125 个公共艺术案例中评选出 32 个入围作品、7 个优秀奖和一个大奖。

本届国际公共艺术奖的评委有：大洋洲区评委 Rhana Devenport，新西兰戈维布鲁斯特艺术馆馆长，作家、策展人；东亚区评委汪大伟，上海大学美术学院院长、教授；欧亚大陆区评委 Ute Meta Bauer，新加坡当代艺术中心创始董事，南阳科技大学教授；非洲区评委 Jay Pather，南非开普敦大学教授，策展人、导演；南美洲区评委 Bill Kelley, Jr. 美国独立策展人；北美洲区评委 Chelsea Haines 美国独立策展人、作家；贝鲁特到孟加拉区评委：PoojaSood，印度 Khoj 国际艺术家联盟创始成员兼董事，策展人。

参加本届评奖的案例完成于 2007 年 1 月—2012 年 12 月之间。评选标准有三方面的要求：首先，必须是由艺术家为主导的、卓有成效的地方重塑、社区建设或社会实践艺术；其次，体现出最佳实践、创新设计和高水平的执行效果；第三，能够对所在地的长远或潜在发展产生积极影响。

从来自世界不同地区的案例中可以看到，"地方重塑"对社会现实的作用不同于注重物理材质的传统艺术，不仅以完全不同的思维方式和工作系统关照社会和人生，而且更积极地体现为一种艺术主张，这种主张不是建立在抽象的政治目标或道德说教之上，而是依托于关注人文的公共理想。由于所承载的社会功能和文化精神，决定了其美学趣味不可能是个人化的，而是从整体上介入公众生活，以一定的社会物质环境和社会精神环境为基础，传达公共社会理念和主导思想。

第二届国际公共艺术奖颁奖仪式和公共艺术论坛将于 2015 年6 月在新西兰奥克兰举行，由奥克兰大学美术学院、上海大学美术学院、山东工艺美术学院共同主办。第二届国际公共艺术奖在结束评选之后，由上海大学美术学院主办了"什么是好的公共艺术？"评委对话会，7 位评委嘉宾亲临现场，通过解读本届优秀案例来阐述他们对上述概念的理解，并回答听众提问。上海大学师生两百余人参加了对话会，对话会全程同声传译。

评委们围绕何为好的公共艺术展开充分阐述了各自的见解，大家认为，好的公共艺术首先要考虑到公众。公共艺术立项应该反映公众需求，项目公示需要真实采纳民意而不能被架空；另一方面，要有利于提高公众整体素质和审美水平。一个拥有真正公共艺术人口的社会，才能追求公共艺术的"好"；其次，艺术家的创作是一个独立的体验，而公共艺术面向公众，由此可能产生矛盾。并不只是大型的项目才好，很多小项目，往往能像火花一样引起一系列争论；再者，公共艺术理论变化很快。要关注并研究公共艺术未来的发展。公共艺术的作用主要体现在教育、激发和组织。重要的是艺术家如何来组织？是否能建立一个合适的组织，保证互动，良好的对话对于好的公共艺术非常重要。

作为本次活动主办方的上海大学美术学院院长、中国《公共艺术》杂志主编汪大伟教授还从学科建设的角度，谈到为什么上海大学美术学院把公共艺术列入重点学科。他说，这与学院和上海所处的地位有关，上海经济在发展过程中有各种问题出现，我们有责任面对这些问题，回答这些问题，甚至要去解决这些问题。当然，我们不可能用理工科的方法去解决，而是用艺术的方式和语言去介入社会的发展，解决这些问题。这些好的案例都是我们的教科书，学生能够从中得到很大的启迪。也希望学生能够在这方面付诸行动，这就是我们学校应有的态度。汪大伟院长表示，希望国际公共艺术奖在今后能吸引更多的专业人士共同参与，为中国城乡公共艺术的长远发展出谋划策。

后记

AFTERWORD